To Sheldon Winner

With friendship and
appreciation for your
participation in the
Distinguished Lecture
Series at the
Cole Eye Institute —

Cleveland
June 27, 2002

NEW INSIGHTS INTO RETINAL DEGENERATIVE DISEASES

NEW INSIGHTS INTO RETINAL DEGENERATIVE DISEASES

Edited by

Robert E. Anderson
University of Oklahoma Health Sciences Center
Dean A. McGee Eye Institute
Oklahoma City, Oklahoma

Matthew M. LaVail
Beckman Vision Center
University of California, San Francisco
San Francisco, California

and

Joe G. Hollyfield
Cole Eye Institute
Cleveland Clinic Foundation
Cleveland, Ohio

Kluwer Academic / Plenum Publishers
New York, Boston, Dordrecht, London, Moscow

Library of Congress Cataloging-in-Publication Data

New insights into retinal degenerative diseases/edited by Robert E. Anderson, Matthew
M. LaVail, Joe G. Hollyfield.
    p. ; cm.
    ISBN 0-306-46679-1
      1. Retinal degeneration—Congresses. I. Anderson, Robert E. (Robert Eugene) II.
LaVail, Matthew M. III. Hollyfield, Joe G. IV. International Symposium on Retinal
Degeneration (9th: 2000: Durango, Colo.)
      [DNLM: 1. Retinal Degeneration—genetics—Congresses. 2. Retinal
Degeneration—physiopathology—Congresses. 3. Retinal
Degeneration—therapy—Congresses. WW 270 N5317 2001]
    RE661.D3 N49 2001
    617.7′35—dc21

                    2001038936

Proceedings of the IXth International Symposium on Retinal Degeneration, held October 9–14, 2000 in
Durango, Colorado

ISBN 0-306-46679-1

©2001 Kluwer Academic / Plenum Publishers, New York
233 Spring Street, New York, N.Y. 10013

http://www.wkap.nl/

10 9 8 7 6 5 4 3 2 1

A C.I.P. record for this book is available from the Library of Congress

Richard Newton Lolley

May 25, 1933–April 3, 2000

Our friend Dick Lolley devoted his life to studies of inherited retinal degenerations. His pioneering work in the 1970s identified the biochemical defect in the *rd* mouse retina and showed us that indeed, it was possible to find the needle in the haystack.

PREFACE

Since 1984, we have organized satellite symposia on retinal degenerations that are held in conjunction with the biennial International Congress of Eye Research. The timing and location of our Retinal Degeneration Symposia have allowed scientists and clinicians from around the world to convene and present their exciting new findings. The symposia have been arranged to allow ample time for discussions and one-on-one interactions in a relaxed atmosphere, where international friendships and collaborations could be established.

The IXth International Symposium on Retinal Degeneration was held on October 9-14, 2000 in Durango, Colorado and was attended by over 100 scientists from six continents. This book contains many of their presentations. Several events of note occurred at this meeting. First, thanks to the generous support of the Foundation Fighting Blindness, we were able to sponsor the travel of 11 young scientists from six countries. Most of them have contributed chapters to this volume. The response to the travel program was so overwhelming that we will make it regular feature of our meeting. This will allow other bright, young investigators to be introduced to the world experts who study retinal degenerations. Second, about 40% of the scientists who attended this meeting were there for the first time. We believe that this indicates a growing interest in retinal degeneration research and ensures that new talent will be attracted to this important area of investigation.

The symposium received support from several organizations. We are particularly pleased to thank the Foundation Fighting Blindness, Hunter Valley, MD for its continuing support of this and the previous biennial symposia, without which we could not have held these important meetings. The travel program funded by the FFB was a wonderful addition to the meeting. We also thank David W. Parke II, MD, Chairman of the Ophthalmology Department at the University of Oklahoma Health Sciences Center and CEO of the Dean A. McGee Eye Institute, for his generous support of this meeting.

We are grateful to Holley Hunter and the staff of the Sheraton Tamarron for their assistance and advice in planning this meeting.

Lastly, we want to acknowledge the outstanding organizational skills and talents of Holly Whiteside, Administrative Manager of Dr. Anderson's laboratory at the OUHSC. This was our first meeting since 1988 that was organized without a local host. Holly was responsible for making all of the arrangements for the meeting from start to finish, including editing all of the chapters of this book. Without her dedication and hard work, this meeting would not have been possible.

We also thank Plenum Press for publishing this volume.

Robert E. Anderson
Matthew M. LaVail
Joe G. Hollyfield

CONTENTS

Part II. Treatment

Part III. Mechanisms of Retinal Degeneration

Part IV. Animal/Tissue Culture Studies

Part V. Population Studies

CLINICAL SPECTRUM IN AUTOSOMAL DOMINANT STARGARDT'S MACULAR DYSTROPHY WITH A MUTATION IN *ELOVL4* GENE

Yang Li[1], Linda A. Lam[1], Zhengya Yu[1], Zhenglin Yang[1], Paul Bither[2] and Kang Zhang[1]

1. INTRODUCTION

In 1909, Stargardt reported a recessively inherited progressive atrophic macular dystrophy associated with perifoveal flecks[1]. This condition, now termed Stargardt's macular dystrophy, presents insidiously within the first two decades of life with a gradual loss of central vision[1,2,3,4,5]. In the early course of the disease, fundus abnormalities on ophthalmoscopy may be minimal despite a marked reduction in visual acuity. Visual acuities often deteriorate to range between 20/200 to 20/400[6]. It is the most common form of hereditary macular dystrophy, and accounts for 7% of all retinal dystrophies[7]. The estimated incidence of Stargardt's macular dystrophy is 1 in 10,000[7,8]. Classic ophthlamoloscopic findings associated with Stargardt's macular dystrophy are bilateral atrophic macular lesions and yellow-white flecks at the level of the retinal pigment epithelium at the posterior pole[8]. Since that time, numerous reports have elaborated upon the genetics and clinical variations of Stargardt's macular dystrophy. Franceschetti later coined the term "fundus flavimaculatus" to describe a fundus morphology characterized by the presence of white-yellow retinal flecks, which were scattered throughout the posterior pole and may extend out to the midperiphery[9]. In fundus flavimaculatus, midperipheral flecks are more prominent than the central macular flecks. Both Stargardt's macular dystrophy and fundus flavimaculatus have been found to exhibit much clinical variation in phenotypic expression and rates of progression. Despite attempts to delineate a clear distinction between the two conditions, the clinical similarities, the occurrence of both within single pedigrees, and the results of linkage analysis all support the belief that Stargardt's macular dystrophy and fundus flavimaculatus both result from allelic mutations in the same gene[2,3,4].

[1] Cole Eye Institute, the Cleveland Clinic Foundation, Cleveland, OH
[2] Midwest Eye Institute, Indianapolis, IN

New Insights Into Retinal Degenerative Diseases
Edited by Anderson *et al.*, Kluwer Academic/Plenum Publishers, 2001

On fluorescein angiography, patients with Stargardt's macular dystrophy most often reveal a dark, or silent choroid with a variable central region of hypofluorescence that correlates with the severity of the disease[11]. It is believed that choroidal fluorescence is blocked by lipofuscin in RPE in Stargardt's macular dystrophy. Also hyperfluorescent spots in the posterior pole are observed on angiography, whose locations do not always correspond precisely to those seen on ophthalmoscopy[12,13,14]. At onset of visual loss, the results of ERG or EOG are often normal. However, as the disease progresses, ERG abnormalities may be observed ranging from mildly diminished photopic "b" wave amplitudes to markedly decreased photopic and scotopic "b" waves. Abnormal EOG ratios have also been described in patients with advanced Stargardt's and fundus flavimaculatus disease[11].

Most analyses have linked Stargardt's macular dystrophy to an autosomal recessive mode of inheritance. A gene associated with recessive Stargardt's macular dystrophy[7]. and fundus flavimaculatus[15] has been mapped to the short arm of chromosome 1, at locus 1p21, called STGD1. This gene, which has recently been cloned, has been identified as a photoreceptor-specific gene called *ABCA4*[16]. *ABCA4* encodes an energy-dependent transmembrane protein that is exclusively localized at the disc membrane of rod and cone outer segments[17]. This protein is a member of the ATP-binding cassette (ABC) transporter superfamily, and was previously described as rim protein (RmP)[18]. Mutations in the ABCR gene have also been found in other inherited eye diseases including cone-rod dystrophies[19], autosomal recessive retinitis pigmentosa[20], and age-related macular degeneration[21,22,23]. All studies of ABCR in ocular disease report a broad mutation spectrum and high allelic heterogeneity.

Numerous reports have also found Stargardt's macular dystrophy to be transmitted as an autosomal dominant trait[24,25,26,27]. Members with Stargardt's macular dystrophy who demonstrate a dominant inheritance pattern exhibit a progressive loss of central vision in the second and third decades of life[24,25]. The EOG and the ERG are relatively normal, with well-preserved b-wave amplitudes even late in life[25]. Studies of large North American pedigrees have mapped loci for autosomal dominant macular dystrophy to chromosome 6q14 (STGD3)[25,26] and 4p (STGD4)[28]. Both STGD3 and autosomal dominant macular dystrophy (adMD) have been mapped to 6q14, suggesting that same gene may be responsible for both diseases, which already bear close clinical resemblance[25,26,29]. Using a positional cloning approach, Zhang et al.[30] have identified a disease gene called *ELOVL4*, which is responsible for STGD3 and adMD. They identified a single five base-pair deletion in the coding region of *ELOVL4*, which generates a frame-shift mutation in all affected members of four independent STGD3 families and one adMD family. *ELOVL4* demonstrated cone and rod photoreceptor specific expression in the eye and encoded a putative transmembrane protein with similarities to the *ELO* family of proteins involved in elongation of very long chain fatty acids. In this chapter, we describe the clinical spectrum of STGD3 patients with a mutation in *ELOVL4* gene and propose a hypothesis for observed phenotypic variations.

2. METHODS

One family, spanning four generations, with 29 members affected with Stargardt's macular dystrophy is described in this study. Affected members demonstrated a spectrum of phenotypes, including the presence of yellow-white flecks at the level of the RPE with or without atrophic or bull's eye macular lesions. Pedigree analysis demonstrated an autosomal dominant form of inheritance. All persons included in this study underwent genotype analysis using short tandem repeat polymorphic markers and the disease trait was linked to the STGD3 locus on chromosome 6q14.

Patient demographics such as sex, age, age of onset, best corrected visual acuity, and fundus appearances were recorded. Noted from each fundus exam was the distribution of yellow-white flecks, the presence or absence of macular and periphery atrophy, and the presence or absence of bull's eye maculopathy. Those patients with disease classified as limited to the posterior pole demonstrated flecks or atrophy confined between the vascular arcades. Color fundus photographs were obtained from 19 patients. Fluorescein angiograms were obtained in two patients.

3. RESULTS

The age of onset, sex, and best-corrected visual acuity are summarized in Table 1. The earliest age of onset was 5, whereas the oldest age of onset was 46. The average age of onset was 16.9 years. 6 of the 19 patients were female (31.6%). A large variation existed among the visual acuities of patients in this family. Visual acuities ranged from 20/25 to 20/400. Generally, younger patients had better visual acuities than older patients. Also those with longer duration of the disease and those who had an earlier age of onset demonstrated worse visual acuity. Similar visual acuities were found usually in both eyes of each patient.

We also found relative symmetry in the fundus appearance between the eyes of each patient. However, there was much variation in the clinical fundus appearance among patients in this pedigree. Four main fundus appearances were noted in this pedigree: bull's eye maculopathy with and without flecks (Fig. 1), macular atrophy with flecks confined to the posterior pole and macular atrophy with flecks extending beyond the posterior pole (Fig. 2). It was noted that patients with a bull's eye lesion without flecks demonstrated the best visual acuity of the four groups. There was no correlation of age of onset or gender with a specific fundus appearance.

Fluorescein angiograms of two patients demonstrated a dark choroid (Fig. 3). Generally, the angiograms exhibited central hypofluorescence corresponding to the region of RPE and choriocapillaris atrophy, surrounding hyperfluorescent flecks, and characteristic dark choroid in the periphery.

Table 1. Clinical spectrum of STGD3 patients

Pt. ID	Sex/Age	Onset	VA		Fundus
			OD	OS	
913	M/15	11	20/60	20/50	Bull's Eye without Flecks
621	M/15	15	20/50	20/50	Bull's Eye without Flecks
243	F/25	13	20/120	20/120	Bull's Eye without Flecks
121	M/27	21	20/50	20/60	Bull's Eye without Flecks
232	F/23	22	20/240	20/240	Bull's Eye without Flecks
242	M/27	22	20/100	20/160	Bull's Eye without Flecks
23	M/51	22	20/200	20/200	Bull's Eye with Flecks
911	M/20	12	20/120	20/120	Bull's Eye with Flecks
2201	M/31	5	20/200	20/240	Bull's Eye with Flecks
112	M/31	19	20/60	20/60	GA with Limited Flecks
12	M49	35	20/280	20/280	GA with Limited Flecks
6	M/63	19	20/120	20/120	GA with Limited Flecks
9	F/58	13	20/120	20/160	GA with Limited Flecks
213	M/32	7	20/240	20/240	GA with Limited Flecks
93	M/37	18	20/120	20/120	GA with Limited Flecks
11	F/56	46	20/200	20/200	GA with Ext. Flecks
21	M/55	13	20/200	20/240	GA with Ext. Flecks
43	F/57	13	20/240	20/240	GA with Ext. Flecks
91	F/39	14	20/160	20/160	GA with Ext. Flecks

4. DISCUSSION

Few studies have investigated the phenotypic variation found among family members with autosomal dominant Stargardt's macular dystrophy. In this study, we found that 4 different clinical subtypes existed: bull's eye maculopathy with and without flecks, macular atrophy with flecks confined to the posterior pole and macular atrophy with flecks extending beyond the posterior pole. Despite the fact that all family members have the same mutation in the *ELOVL4*[30], a wide range of phenotypic expression was present. This variability existed even among family members with similar age at examination and age of onset. For instance, two family members who each had age of onset in early childhood (ages 5 and 7) and were middle-aged (31 and 32, respectively) at the time of examination exhibited dramatically disparate funduscopic phenotypes (patients 2201and 213 from Table 1). The first family member demonstrated a bull's eye type of maculopathy with minimal yellow-white flecks in the macula, whereas the second family member showed geographic atrophy with limited flecks in the macula. None of the four clinical subtypes found in this study was correlated with a specific age, sex, or visual acuity range.

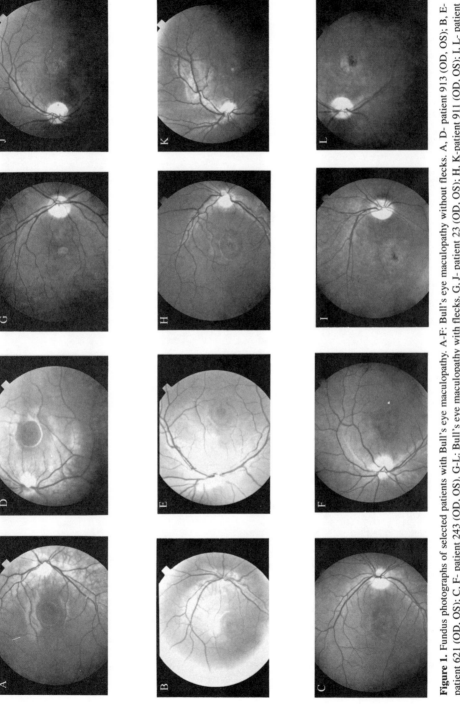

Figure 1. Fundus photographs of selected patients with Bull's eye maculopathy. A-F: Bull's eye maculopathy without flecks. A, D- patient 913 (OD, OS); B, E-patient 621 (OD, OS); C, F- patient 243 (OD, OS). G-L: Bull's eye maculopathy with flecks. G, J- patient 23 (OD, OS); H, K-patient 911 (OD, OS); I, L- patient 2201 (OD, OS). Patient numbers correspond to those in Table 1.

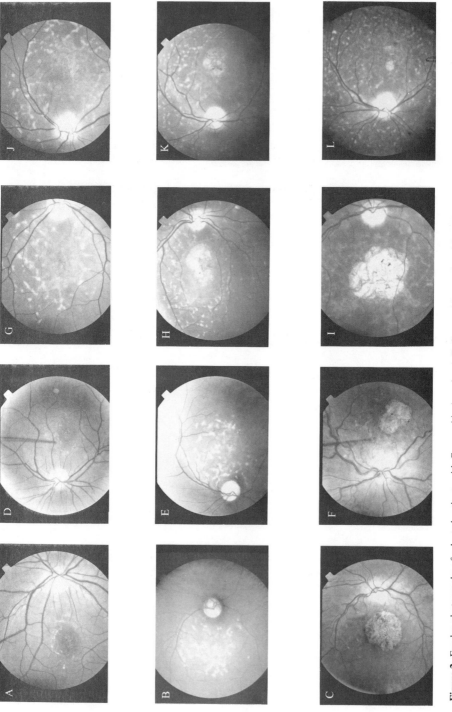

Figure 2. Fundus photographs of selected patients with Geographic Atrophy. A-F: Geographic atrophy without flecks. A, D- patient 112 (OD, OS); B, E- patient 12 (OD, OS); C, F- patient 6 (OD, OS). G-L: Geographic atrophy with flecks. G, J- patient 11 (OD, OS); H, K-patient 21 (OD, OS); I, L- patient 43 (OD, OS). Patient numbers correspond to those in Table 1.

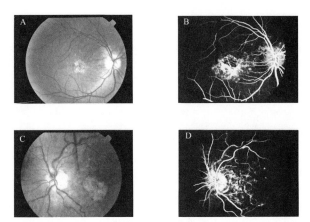

Figure 3.* Fundus photographs and flourescein angiogram of two patients (A & B, 93; C & D, 6) with STGD3.
A, C: Fundus photos depicting macular atrophy with limited flecks.
B, D: Fluorescein angiograms demonstrating central window defect and peripheral dark choroids.

Several investigators have documented wide intrafamilial phenotypic variation in members with recessively inherited Stargardt's macular dystrophy[31,32]. In addition to clinical appearance, there also exists great disparity in the age of onset and severity of the disease[3,9,31,32],. Only a few reports have examined the phenotypic variation found in the autosomal dominant form of the disease. Edwards et al.[26] studied two families with autosomal dominant Stargardt's dystrophy and subsequently found that both families joined genealogically to a common founder. In this pedigree, they found a phenotypic spectrum encompassing a broad range including: a pattern dystrophy-like appearance, starburst pattern, and diffuse geographic atrophy with and without flecks. By genetic linkage analysis, the disease gene was localized to the STGD3 locus[26].

Moreover, in our study cohort, we found that two members with STGD3 demonstrate a dark choroid on fluorescein angiography. Both patients demonstrated geographic atrophy with limited flecks on fundus exam. Recent studies[25,26,30] have demonstrated an absence of the dark choroid appearance on fluorescein angiogram in patients with STGD3 or adMD. This disparity may be explained by the age of the patients when they underwent fluorescein angiography. We noted that the patients in our study had relatively advanced stages of disease; therefore, it was possible that they had a longer time period for the accumulation of lipofuscin.

Historically, clinicians have often inferred the pathological basis of a disease from its phenotype. However, in recent years it has become evident that it is not possible to correlate a retinal dystrophy based on its ophthalmoscopic phenotype alone. Retinal flecks have been correlated with numerous disorders. Moreover, it is possible that patients with one phenotype may have another distinct phenotype at an earlier disease stage. Our classification into four phenotypes was not intended to denote a strict segregation of the different groups but rather a means to denote the wide morphological variability that was found in our cohort. A number of hereditary and toxic diseases could present with a fundus appearance similar to that found in Stargardt's macular dystrophy, including dominant progressive foveal dystrophies, progressive dominant cone dystrophy, pattern dystrophy, and toxic retinopathies. The wide phenotypic variation among family members indicates that a classification scheme based on morphological features alone would be insufficient in the diagnosis of retinal dystrophies such as in Stargardt's macular dystrophy.

One possible aspect accounting for the different phenotypes demonstrated in this study could be environmental factors, such as smoking and amount of sunlight exposure. Alternatively, it is probable that other modifier genes may influence the age of onset and severity of the disease in patients with an *ELOVL4* mutation.

Recently, it has been demonstrated that a single five base-pair deletion within the protein-coding region of a novel retinal photoreceptor-specific gene, termed *ELOVL4*, on chromosome 6q14 has been found in family members affected with STGD3 and adMD[29]. Sequence analysis of ELOVL4 gene revealed homology to a group of yeast and human proteins that may function in biosynthesis of long chain polyunsaturated fatty acids[33]. *ELOVL4* is the first gene involved in the biosynthesis of long chain fatty acids implicated in any form of macular degeneration. In light of the wide phenotypic variability, genetic analysis is a crucial clinical tool in establishing the diagnostic, prognostic, and management implications of a retinal dystrophy.

REFERENCES

1. K. Stargardt, Uber familiare, progressive degeneration in der maculagegend des auges, *Graefes Arch. Ophthalmol.* **71**, 534-550 (1909).
2. O.B. Hadden and J.D.M. Gass, Fundus flavimaculatus and Stargardt's disease, *Am. J. Ophthal.* **82**, 527-539 (1976).
3. A. E. Krill and A. F. Deutman, The various categories of juvenile maculadegeneration, *Trans. Am. Ophthalmol. Soc.* **70**, 220-245 (1972).
4. K. G. Noble and R. E. Carr, Stargardt's disease and fundus flavimaculatus, *Arch. Ophthalmol.* **97**, 1281-1285 (1979).
5. K. Zhang, T. H. Nguyen, A. Crandall, et al., Genetic and molecular studies of macular dystrophies: recent developments, *Surv Ophthalmol.* **40**, 51-61 (1995).
6. K. Zhang, H. Yeon, M. Han, et al, Molecular genetics of macular dystrophies, *Br. J. Ophthal.* **80**, 1018-1022 (1996).
7. J. Kaplan, S. Gerber, D. Larget-Piet, et al, A gene for Stargardt's disease (fundus flavimaculatus) maps to the short arm of chromosome 1, *Nat. Genet.* **5**, 308-311 (1993).
8. K.L. Anderson, L. Baird, R.A. Lewis, et al, A YAC contig encompassing the recessive Stargardt disease gene (STGD) on chromosome 1p, *Am. J. Hum. Genet.* **57**, 1351-63 (1995).
9. T. M. Aaberg, Stargardt's disease and fundus flavimaulatus: Evaluation of morphologic progression and intrafamilial co-existence, *Tr. Am Ophth Soc.* **84**,453-87 (1986).
10. A. Franceschetti, Fundus flavimaculatus, *Arch Ophthalmol.* **25**, 505-550 (1965).
11. D. C. Garibaldi and K. Zhang, Molecular genetics of macular degenerations, *Int. Ophthal. Clinics* **39**, 117-142 (1999).
12. J. D. M. Gass, Stereoscopic atlas of macular diseases: diagnosis and treatment. 4th ed. (Mosby, St. Louis, 1997), p. 326-333.
13. G. Fish, R. Grey, K. S. Sehmi, et al., The dark choroid in posterior retinal dystrophy, *Br. J. Ophthal.* **65**, 359-363 (1981).
14. A. E. Ullis, A. T. Moore, A. C. Bird, The dark choroids in posterior retinal dystrophies. Ophthalmology **97**, 1423-1427 (1987).
15. S. Gerber, J. M. Rozet, and T. J. R. van de Pol, A gene for late-onset fundus flavimaculatus with macular dystrophy maps to chromosome 1p13, *Am J. Hum. Gen.* **56**, 396-399 (1995).
16. R. Allikmets, N.F. Shroyer, N. Singh, et al, A photoreceptor cell-specific ATP-binding transporter gene (ABCR) is mutated in recessive Stargardt macular dystrophy, *Nat. Genet.* **15**, 236-246 (1997).

17. H. Sun and J. Nathans, Stargardt's ABCR is localized to the disc membrane of retinal rod outer segments. *Nat. Genet.* **17**, 15-16 (1997).
18. S. M. Azarian and G. H. Travis, The photoreceptor rim protein is an ABC transporter encoded by the gene for recessive Stargardt's disease, *Hum. Genet.* **102**, 699-705 (1998).
19. F. P. M. Cremers, D. J. R. van de Pol, M. van Driel, et al, Autosomal recessive retinitis pigmentosa and cone-rod dystrophy caused by splice site mutations in the Stargardt's disease gene ABCR. *Hum. Mol. Genet.* **7**, 355-362 (1998).
20. A. Martinez-Mir, E. Paloma, R. Allikmets, et al, Retinitis pigmentosa caused by a homozygous mutation in the Stargardt disease gene ABCR, *Nat Genet.* **18**, 11-12 (1998).
21. R. Allikmets, N. F. Shroyer, N. Singh, et al, Mutation of the Stargardt disease gene (ABCR) in age-related macular degeneration, *Science* **277**, 1805-1807 (1997).
22. Allikmets, R et al. Further evidence for an association of the ABCR alleles with age-related macular degeneration. *Am J Human Genetics*, **67,** 487-491 (2000).
23. N.F. Shroyer, R. A. Lewis, R. Allikmets et al, The rod photoreceptor ATP-binding cassette transporter gene, ABCR, and retinal disease: from monogenic to multifactoral, *Vis. Research* **39**, 2537-2544 (1999).
24. G. W. Cibis, M. Morey, and D. J. Harris. Dominantly inherited macular dystrophy with flecks (Stargardt), *Arch Ophthalmol* **98**, 1785-1789 (1980).
25. E. M. Stone, B. E. Nichols, A. E. Kimura, et al, Clinical features of a Stargardt-like dominant progressive macular dystrophy with genetic linkage to chromosome 6q, *Arch Ophthalmol.* **112**, 765-772 (1994).
26. A. O. Edwards, A. Miedziak, T. Vrabec, et al., Autosomal dominant Stargardt-like macular dystrophy: I. Clinical characterization, longitudinal follow-up, and evidence for a common ancestry in families linked to chromosome 6q14, *Am J. Ophthalmol.* **127**, 426-435 (1999).
27. P.Bither, L.Berns, Eominant inheritance of Stargardt's disease, Journal of the American Optometric Association 59,112-117 (1988).
28. M. Kniazeva, M. F. Chiang, B. Morgan, et al. A new locus for autosomal dominant Stargardt-like disease maps to chromosome 4, *Am J. Hum. Genet.* **64**, 1394-1399 (1999).
29. I. B. Griesinger, P. A. Sieving, and R. Ayyagari, Autosomal dominant macular atrophy at 6q14 excludes CORD7 and MCDR1/PBCRA loci, *Invest. Ophthal and Vis. Science* **41**, 248-255 (2000).
30. K. Zhang, M. Kniazeva, M. Han, et al, A 5-bp deletion in ELOVL4 is associated with two related forms of autosomal dominant macular dystrophy, *Nat. Genet.* **27**, 89-93 (2001).
31. J. D. Armstrong, D. Meyer, S. Xu, et al, Long-term follow-up of Stargardt's disease and fundus flavimaculatus, *Ophthalmology* **105**, 448-457 (1998).
32. N. Lois, G. E. Holder, F. W. Fitzke, et al, Intrafamilial variation of phenotype in Stargardt macular dystrophy- fundus flavimaculatus, *Invest Ophthal Vis Sci* **40**, 2668-2275 (1999).
33. D. C. Garibaldi, Z. Yang, Y. Li, et al., The role of fatty acids in the pathogenesis of retinal degeneration, In press: X International symposium on retinal degeneration (2001).

X-LINKED RETINITIS PIGMENTOSA: CURRENT STATUS

Debra K. Breuer,[1] Maurizio Affer,[2] Sten Andreasson,[3] David G. Birch,[4] Gerald A. Fishman,[5] John R. Heckenlively,[6] Suja Hiriyanna,[7] Dennis R. Hoffman,[4] Samuel G. Jacobson,[8] Alan J. Mears,[7] Maria A. Musarella,[9] Elena Redolfi,[2] Paul A. Sieving,[7] Alan F. Wright,[10] Beverly M. Yashar,[7] Ileana Zucchi,[2] Anand Swaroop[1,7,*]

1. INTRODUCTION

Retinitis pigmentosa (RP) is a clinically and genetically heterogeneous group of retinal degenerative diseases, characterized by nightblindness, progressive restriction of the visual field and pigmentary retinopathy.[1] At least 28 different genetic loci have been mapped for autosomal dominant, autosomal recessive, and X-linked forms of RP. [http://www.sph.uth.tmc.edu/Retnet/home.htm] The X-linked RP (XLRP) subtype is the most severe, with an early age of onset and more rapid progression, accounting for 10 to

[1] Dept. of Human Genetics, University of Michigan, Ann Arbor, MI
[2] C.N.R.-I.T.B.A., Milano, Italy
[3] Dept. of Ophthalmology, University Hospital, Lund, Sweden
[4] Retina Foundation of the Southwest, Dallas, TX
[5] University of Illinois at Chicago, Chicago, IL
[6] Jules Stein Eye Institute/UCLA, Los Angeles, CA
[7] Dept. of Ophthalmology and Visual Sciences, University of Michigan, Ann Arbor, MI
[8] Dept. of Ophthalmology, Scheie Eye Institute, University of Pennsylvania, Philadelphia, PA
[9] Dept. of Ophthalmology, Downstate SUNY Brooklyn Medical Science Center, Brooklyn, NY
[10] MRC Human Genetics Unit, Western General Hospital, Edinburgh, United Kingdom
* To whom correspondence should be addressed. All authors (except the first and the last) are listed in alphabetical order.

20% of RP families.[2,3]   XLRP is also genetically heterogeneous with at least 5 mapped loci: *RP2*, *RP3*, *RP6*, *RP23* and *RP24*, as schematically depicted in Figure 1. By linkage analysis, *RP2* is predicted to account for 10-20% of XLRP and *RP3* for 70-90%,[4-6] depending on the population. Genes for these two major loci have now been cloned. Our laboratory has been involved in the mutational screening and functional analysis of the two identified XLRP genes (*RPGR* and *RP2*), as well as the positional cloning of two other XLRP loci (*RP6* and *RP24*). This report summarizes these efforts as well as the current standing of XLRP research.

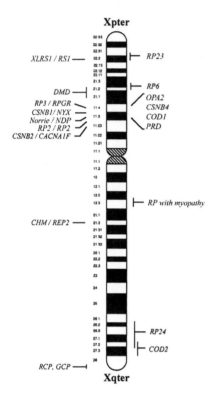

Figure 1. Location of genetic loci for retinal degenerative diseases on the X-chromosome

2. PATIENT RESOURCE

As the Resource Center for XLRP, supported by the Foundation Fighting Blindness, we have established a large database of well-characterized patients. We currently have 197 pedigrees catalogued as XLRP that have been clinically diagnosed by our collaborators, and an additional 79 families have been recruited from the patient registry of The Foundation Fighting Blindness.

2.1. RP2

RP2 was the first XLRP locus mapped, but the gene was identified only recently.[7] Cloning efforts were hampered by the difficulty in clinically and genetically distinguishing RP2 families from those of RP3, as well as the lack of any large chromosomal deletions in RP2 patients. To date, 18 mutations have been identified (summarized in Figure 2).[8-13] Depending on the study, the frequency of RP2 mutations has been shown to be 10-18% . In our initial published analysis we found 5 mutations in 51 families, however subsequently we have identified only 1 mutation in an additional 50 XLRP pedigrees. RP2 was thought to be nonexistent in North America, so this low mutation frequency is not unexpected.

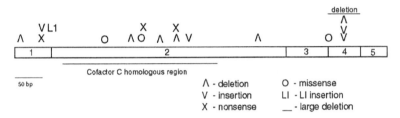

Figure 2. Summary of published RP2 mutations

The protein sequence of RP2 showed homology to Cofactor C, which is involved in the final folding step of ß-tubulin. In a recent study, the RP2 protein was localized predominantly to the plasma membrane in cultured cells and appears to be dually acylated.[14] This suggests that RP2 is not involved in tubulin biogenesis but may be associated with tubulin and/or microtubules.

2.2. RP3

The RP3 gene was identified as RPGR,[15,16] and despite being expected to account for 70-90% of XLRP, the mutation frequency was <20%. The reported mutations in the original 19 exons are shown in Figure 3.[13,17-32] Recently, a novel exon, ORF15, was identified between exons 15 and 16 and an additional 50% mutation rate observed.[33] Our analysis of the most highly mutagenic regions of this exon has revealed mutations in only an additional 12% of XLRP patients (data not shown). This discrepancy can be explained in a number of ways. Our patient population is primarily North American while the published mutations are in European families. Therefore, it is possible that the mutation hot spot in our population is in a region of the exon not yet analyzed. The possibility also exists that there is another exon, or even another gene in the region, responsible for the disease in our remaining cohort.

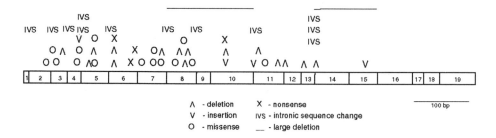

Figure 3. Summary of *RPGR* mutations in the original 19 published exons.

The only clue to the function of RPGR is its homology to the RCC1 protein, a guanine nucleotide exchange factor for Ran-GTPase. In transfected cells, RPGR is detected in the Golgi,[34] while in the retina it is reportedly present in the connecting cilia.[35,36] Two interacting proteins have been identified; the delta subunit of rod cyclic GMP phosphodiesterase[37], involved in solubilizing PDE from the membrane, and the RPGR-interacting protein (RPGRIP1), which has domains homologous to proteins involved in vesicular trafficking.[38-40] The *Rpgr* knockout mouse exhibits slight retinal degeneration beginning at two months, with substantial photoreceptor cell loss by 6 months of age.[35] In addition, there appears to be abnormal distribution of cone opsins. These data suggest that RPGR might be involved in intracellular trafficking. A naturally occuring dog model of *RP3* has been reported, but no mutation published to date.[41]

2.3. *RP15*

The *RP15* locus was mapped to Xp22.13-p22.11 by linkage analysis of a pedigree described as having dominant cone-rod degeneration.[42] After new family members were recruited, we discovered that a female with a critical recombination, described previously as normal, had an affected son. A more recent clinical exam demonstrated that she had an RP phenotype. The disease locus was then remapped to Xp22.1-p11.4, an interval that overlapped *RPGR* and *COD1*. Mutation analysis of *RPGR* revealed a single base pair insertion in ORF15 that is predicted to result in premature truncation of the RPGR protein.[43] As a result, the *RP15* locus has been withdrawn.

2.4. *RP6*

Through linkage mapping and statistical analysis, Ott et al. confirmed the existence of the two major XLRP loci, *RP2* and *RP3* and demonstrated evidence of another locus, distal to Duchenne muscular dystrophy (DMD).[4] We recently began genetic analysis of an apparently X-linked family, characterized earlier by Musarella, as a potential *RP6* family. A detailed haplotype of markers spanning the X-chromosome excluded all XLRP loci with the exception of an approximately 2 cM region between DXS1218 and

DXS1036 corresponding to *RP6* (Figure 4). Because of the small size of the family and the observation of a phenotype in the carrier mother, though less severe than her sons, there is the possibility that the disease is inherited in an autosomal dominant manner. Haplotype analysis was therefore performed across each of the known ADRP loci, and these results are shown in Figure 5. Our data excluded all known ADRP loci, further supporting that the disease locus in this family maps to the *RP6* region. Several genes have been mapped to this region, including a family of four melanoma expressing genes (MAGEs), CKS28, and glycerol kinase. We have excluded the four MAGEs by mutation analysis. ESTs are currently being analyzed as potential positional candidates.

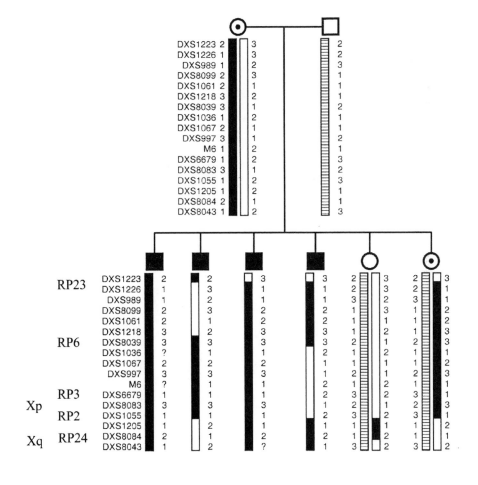

Figure 4. *RP6* pedigree with X-chromosome haplotype. The black bar indicates the disease haplotype.

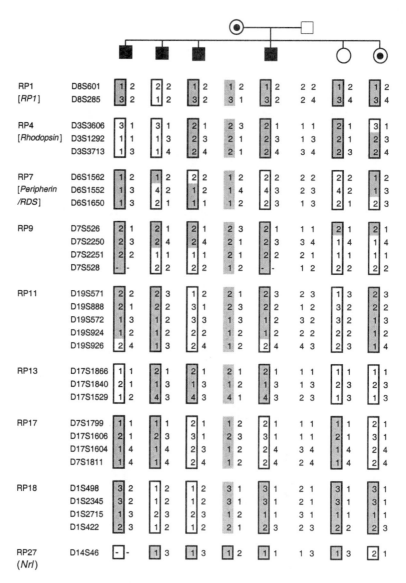

Figure 5. Haplotypes of the *RP6* family members across known ADRP loci. The boxed haplotypes indicate the maternal chromosomes. The shaded markers indicate the likely disease chromosomes.

2.5. *RP24*

Our lab has mapped the *RP24* locus to Xq26-27 by linkage of a large XLRP family.[44] The RP24 family had a different phenotype than other forms of XLRP at an early stage: there was rod photoreceptor dysfunction associated with a negative waveform of the electroretinogram and progressive loss of cone photoreceptor occurring at late disease stages. A transcription map of the region is currently being constructed with five new genes and 12 ESTs accurately mapped.[45] One gene and two ESTs have been screened so far, but no mutations detected. These results are summarized in Table 2.

Table 2. Summary of *RP24* candidate genes

Gene/EST symbol	Expression	Current Status
D16473	Fetal brain	screening in progress...
T89445	Fetal liver, spleen	
D16472	Fetal brain	screening in progress...
AA937500	Bone metastasis	
AA558073	Hematopoietic stem cell	
S5aR	Andrenal	
Factor IX		
MCF.2	Ubiquitous	
WI11452	Brain	
U7snRNA		
SOX3	Brain, stem cell	
NIB1734	Fetal brain	
CDR1	Ubiquitous	
CTp11	Germ cell, testis	
LDOC1	Ubiquitous	Excluded by sequencing
stSG4230	Brain	Excluded by sequencing
W81016	Fetal heart	Excluded by sequencing
SPAN-Xb		
SPAN-Xa		
CT7 (MAGEC1)	Germ, testis, skin	
CT10 (MAGEE1)	Ubiquitous	
HTF2.6		
stSG49622		
L-MYC homologous		
AI283614	Lung, testis, b-cell	
AA332509	Embryo	

2.6. *RP23*

RP23 was recently linked to Xp22 in an interval that overlapped the retinoschisis locus.[46] This is based on a single large family with an atypical RP phenotype exhibiting an unusually early age of onset of disease in affected males. We have not identified any family with a disease locus mapping to the *RP23* region.

3. SUMMARY AND CONCLUSIONS

X-Linked retinitis pigmentosa is a genetically heterogeneous group of severe retinal degenerations with no known treatment or cure. Although several genetic loci have been mapped, *RP3* and *RP2* account for the majority of XLRP. Substantial progress has been made towards the identification of disease-causing mutations in the *RPGR* (for *RP3*) and *RP2* genes. Mutation analysis of *RP2* has yielded mutations in <10% of a North American XLRP population. At this stage, *RPGR* mutations have been identified in <50% of our cohort of over 200 XLRP families. The *RP15* pedigree was remapped to the *RP3* region and a causative mutation identified in the *RPGR* ORF15. Haplotype analysis of an apparently X-linked RP family using markers across the X-chromosome and at all known ADRP loci strongly argues for the existence of the *RP6* locus at Xp distal to *DMD*. Candidate genes are currently being screened for mutations in a single large family in which the disease locus (*RP24*) was mapped to Xq26-27. Further investigations are in progress for the identification of mutations in remaining XLRP families.

4. ACKNOWLEDGEMENTS

We thank the families and their clinicians for contributions to this research. We also thank Dr. Ricardo Fujita for discussions, Drs. Helen Mintz-Hittner, Jacquie Greenberg, Alex Iannacconne, and Marcela Pena for providing some of the families, and Lucia Susani, Jason Cook and Sharyn Ferrara for assistance. This research was supported by grants from NIH (EY07961, EY07003, EY05627), The Foundation Fighting Blindness, the U.S. FDA (FD-R-001232), The Chatlos Foundation, Research to Prevent Blindness (RPB), and Progett Strategico C.N.R. S.G.J. is a recipient of a Senior Scientific Investigator Award and A.S. is a recipient of a Lew Wasserman Merit Award, both from RPB.

REFERENCES

1. JR Heckenlively, *Retinitis Pigmentosa* (Lippincott, Philadelphia, 1988).
2. GA Fishman, Retinitis pigmentosa, *Arch. Ophthalmol.* **96**, 822-826 (1978).
3. JA Boughman, PM Conneally, WE Nance, Population genetic studies of retinitis pigmentosa, A*m. J. Hum. Genet.* **32**,223-235 (1980).
4. J Ott, SS Bhattacharya, JD Chen, MJ Denton, J Donald, C Dubay, GJ Farrar, GA Fishman, D Frey, A Gal, P Humphries, B Jay, M Jay, M Litt, M Machler, M Musarella, M Neugebauer, RL Nussbaum, JD Terwilliger, RG Weleber, B Wirth, F Wong, RG Worton, and AF Wright, Localising multiple X-chromosome-linked retinitis pigmentosa loci using multilocus homogeneity tests, *Proc. Natl. Acad. Sci.* **87**, 701-4 (1990).
5. PW Teague, MA Aldred, M Jay, M Dempster, C Harrison, AD Carothers, LJ Harwick, HJ Evans, L Strain, DJ Brock, et al., Heterogeneity analysis in 40 X-linked retinitis pigmentosa famlies, *Am. J. Hum. Genet.* **55**,105-111 (1994).
6. MA Musarella, L Anson-Cartwright, SM Leal, LD Gilbert, RG Worton, GA Fishman, J Ott, Multipoint linkage analysis and heterogeneity testing in 20 X-linked retinitis pigmentosa families, *Genomics*, **8**,286-296 (1990).

7. U Schwan, S Lenzner, J Dong, S Feil, B Hinzmann, G van Duijnhoven, R Kirschner, M Hemberger, AA Bergen, T Rosenberg, AJ Pinckers, R Fundele, A Rosenthal, FP Cremers, HH Ropers, and W Berger, Positional cloning of the gene for X-linked retinitis pigmentosa 2, *Nat. Genet.* **9**, 327-32 (1998).

8. AJ Mears, L Gieser, D Yan, C Chen, S Fahrner, S Hiriyanna, R Fujita, SG Jacobson, PA Sieving, and A Swaroop, Protein-truncation mutations in the RP2 gene in a North American cohort of families with X-linked retinitis pigmentosa, *Am. J. Hum. Genet.* **64**, 897-900 (1999).

9. AJ Hardcastle, DL Thiselton, L Van Maldergem, BK Saha, M Jay, C Plant, R Taylor, AC Bird, and SS Bhattacharya, Mutations in the RP2 gene cause disease in 10% of families with familial X-linked retinitis pigmentosa assessed in this study, *Am. J. Hum. Genet.* **64**, 1210-5 (1999).

10. Y Mashima, M Saga, Y Hiida, Y Imamura, J Kudoh, and N Shimizu, Novel mutation in RP2 gene in two brothers with X-linked retinitis pigmentosa and mtDNA mutation of leber hereditary optic neuropathy who showed marked differences in clinical severity, *Am. J. Ophthalmol.* **130**, 357-9 (2000).

11. Y Wada, M Nakazawa, T Abe, and M Tamai, A new Leu253Arg mutation in the RP2 gene in a Japanese family with X-linked retinitis pigmentosa, *Invest. Ophthalmol. Vis. Sci.* **41**, 290-3 (2000).

12. DL Thiselton, I Zito, C Plant, M Jay, SV Hodgson, AC Bird, SS Bhattacharya, and AJ Hardcastle, Novel frameshift mutations in the RP2 gene and polymorphic variants, *Hum. Mutat.* **15**, 580 (2000).

13. D Sharon, GAP Bruns, TL McGee, MA Sandberg, EL Berson, and TP Dryja, X-linked retinitis pigmentosa: mutation spectrum of the *RPGR* and *RP2* genes and correlation with visual function, *Invest. Ophthalmol. Vis. Sci.* **41**, 2712-2721 (2000).

14. JP Chapple, AJ Hardcastle, C Grayson, LA Spackman, KR Willison, and ME Cheetham, Mutations in the N-terminus of the X-linked retinitis pigmentosa protein RP2 interfere with the normal targeting of the protein to the plasma membrane, *Hum. Mol. Genet.* **9**, 1919-1926 (2000).

15. A Meindl, K Dry, K Herrmann, F Manson, A Ciccodicola, A Edgar, MRS Carvalho, H Achatz, H Hellebrand, A Lennon, C Migliaccio, K Porter, E Zrenner, A Bird, M Jay, B Lorenz, B Wittwer, M D'Urso, T Meitinger, and AF Wright, A gene (*RPGR*) with homology to the RCC1 guanine nucleotide exchange factor is mutated in X-linked retinitis pigmentosa (RP3), *Nature Genet.* **13**, 35-42 (1996).

16. R Roepman, D Bauer, T Rosenberg, G van Duijnhoven, E van de Vosse, M Platzer, A Rosenthal, H-H Ropers, FPM Cremers, and W Berger, Identification of a gene disrupted by a microdeletion in a patient with X-linked retinitis pigmentosa (XLRP), *Hum. Mol. Genet.* **5**, 827-833 (1996).

17. R Fujita, M Buraczynska, L Gieser, W Wu, P Forsythe, M Abrahamson, SG Jacobson, PA Sieving, S Andréasson, and A Swaroop, Analysis of the RPGR gene in 11 pedigrees with the retinitis pigmentosa type 3 genotype: paucity of mutations in the coding region but splice defects in two families, *Am. J. Hum. Genet.* **61**, 571-580 (1997).

18. M Buraczynska, W Wu, R Fujita, K Buraczynska, E Phelps, S Andreasson, J Bennet, DG Birch, GA Fishman, DR Hoffman, G Inana, SG Jacobson, MA Musarella PA, Sieving, and A Swaroop, Spectrum of mutations in the *RPGR* gene that are identified in 20% of families with X-linked retinitis pigmentosa, *Am. J. Hum. Genet.* **61**, 1287-1292 (1997).

19. RG Weleber, NS Butler, WH Murphey, VC Sheffield, and EM Stone, X-linked retinitis pigmentosa associated with a 2-base pair insertion in codon 99 of the RP3 gene RPGR, *Arch. Ophthalmol.* **115**, 1429-35 (1997).

20. SG Jacobson, M Buraczynska, AH Milam, C Chen, M Jarvalainen, R Fujita, W Wu, Y Huang, AV Cideciyan, and A Swaroop, Disease expression in X-linked retinitis pigmentosa caused by a putative null mutation in the RPGR gene, *Invest. Ophthalmol. Vis. Sci.* **38**, 1983-97 (1997).

21. MG Miano, D Valverde, T Solans, B Grammatico, C Migliaccio, V Cirigliano, C DeBernardo, V Ventruto, T Meitinger, A Wright, G Del Porto, M Baiget, M D'Urso, and A Ciccodicola, Two novel mutations in the retinitis pigmentosa GTPase regulator (RPGR) gene in X-linked retinitis pigmentosa (RP3). *Hum. Mutat.* **12**, 212-3 (1998).

22. GA Fishman, S Grover, SG Jacobson, KR Alexander, DJ Derlacki, W Wu, M Buraczynska, and A Swaroop, X-linked retinitis pigmentosa in two families with a missense mutation in the RPGR gene and putative change of glycine to valine at codon 60, *Ophthalmology* **105**, 2286-96 (1998).

23. S Bauer, R Fujita, M Buraczynska, M Abrahamson, B Ehinger, W Wu, TJ Falls, S Andreasson, and A Swaroop, Phenotype of an X-linked retinitis pigmentosa family with a novel splice defect in the RPGR gene, *Invest. Ophthalmol. Vis. Sci.* **39**, 2470-4 (1998).

24. GA Fishman, S Grover, M Buraczynska, W Wu, and A Swaroop, A new 2-base pair deletion in the RPGR gene in a black family with X-linked retinitis pigmentosa, *Arch. Ophthalmol.* **116**, 213-8 (1998).

25. MG Miano, F Testa, M Strazzullo, M Trjuillo, C De Bernardo, B Grammatico, F Simonelli, M Mangino, I Torrente, G Ruberto, M Beneyto, G Antinolo, E Rinaldi, C Danesino, V Ventruto, M D'Urso, C Ayuso, M Baiget, and A Ciccodicola, Mutation analysis of the RPGR gene reveals novel mutations in south European patients with X-linked retinitis pigmentosa, *Eur. J. Hum. Genet.* **7**, 687-694 (1999).

26. I Zito, DL Thiselton, MB Gorin, JT Stout, C Plant, AC Bird, SS Bhattacharya, and AJ Hardcastle, Identification of novel RPGR (retinitis pigmentosa GTPase regulator) mutations in a subset of retinitis pigmentosa families segregating with the RP3 locus, *Hum. Genet.* **105**, 57-62 (1999).

27. R Kirschner, T Rosenberg, R Schultz-Heienbrok, S Lenzer, S Feil, R Roepman, FP Cremers, HH Ropers, and W Berger, RPGR transcription studies in mouse and human tissues reveal a retina-specific isoform that is deleted in a patient with X-linked retinitis pigmentosa, *Hum. Mol. Genet.* **8**, 1571-8 (1999).

28. KL Dry, FD Manson, A Lennon, AA Bergen, DB Van Dorp, and AF Wright, Identification of a 5'splice mutation in the RPGR gene in a family with X-linked retinitis pigmentosa (RP3), *Hum. Mutat.* **13**, 141-5 (1999).

29. I Zito, A Morris, P Tyson, I Winship, D Sharp, D Gilbert, DL Thiselton, SS Bhattacharya and AJ Hardcastle, Sequence variation within the *RPGR* gene: evidence for a founder complex allele, *Hum. Mut.* MIB #361 (2000).

30. L Liu, L Jin, M Liu, Y Wei, X Wu, Y Liu, H Wang, R Chu, and J Chai, Identification of two novel mutations (E332X and c1536delC) in the RPGR gene in two chinese patients with X-linked retinitis pigmentosa, *Hum. Mutat.* **15**, 584 (2000).

31. I Zito, MB Gorin, C Plant, AC Bird, SS Bhattacharya, and AJ Hardcastle, Novel mutations of the RPGR gene in RP3 families, *Hum. Mutat.* **15**, 386 (2000).

32. M Guevara-Fujita, S Fahrner, K Buraczynska, J Cook, D Wheaton, F Cortes, C Vincencio, M Pena, GA Fishman, H Mintz-Hittner, D Birch, D Hoffman, AJ Mears, R Fujita, and A Swaroop, *Hum. Mutat.* #396 (2000).

33. R Vervoort, A Lennon, AC Bird, B Tulloch, R Axton, MG Miano, A Meindl, T Meitinger, A Ciccodicola, and AF Wright, Mutational hot spot within a new RPGR exon in X-linked retinitis pigmentosa, *Nat Genet.* **25**, 462-466 (2000).

34. D Yan, PK Swain, D Breuer, RM Tucker, W Weiping, R Fujita, A Rehemtulla, D Burke, A Swaroop, Biochemical characterization and subcellular localization of the mouse retinitis pigmentosa GTPase regulator (mRPGR), *JBiol. Chem.* **273**, 19656-19663 (1998).

35. D-H Hong, BS Pawlyk, J Shang, MA Sandberg, EL Berson, and T Li, A retinitis pigmentosa GTPase regulator (RPGR) deficient mouse model for X-linked retinitis pigmentosa (RP3), *Proc. Natl. Acad. Sci.* **97**, 3649-3654 (2000).

36. D-H Hong, G Yue, M Adamian, and T Li, A retinitis pigmentosa GTPase regulator (RPGR) – interacting protein is stably associated with the photoreceptor ciliary axoneme and anchors RPGR to the connecting cilium, *J. Biol. Chem.* [epub ahead of print].

37. M Linari, M Ueffing, F Manson, A Wright, T Meitinger, and J Becker, The retinitis pigmentosa GTPase regulator, RPGR, interacts with the delta subunit of rod cyclic GMP phosphodiesterase, *Proc. Natl. Acad. Sci.* **96**, 1315-1320 (1999).

38. JP Boylan, and AF Wright, Identification of a novel protein interacting with RPGR, *Hum. Mol. Genet.* **9**, 2085-2093 (2000).

39. R Roepman, N Bernoud-Hubac, DE Schick, A Maugeri, W Berger, H-H Ropers, FPM Cremers, and PA Ferreira, The retinitis pigmentosa GTPase regulator (RPGR) interacts with novel transport-like proteins in the outer segments of rod photoreceptors, *Hum. Mol. Genet.* **9**, 2095-2105 (2000).

40. R Roepman, D Schick, and PA Ferreira, Isolation of retinal proteins that interact with retinitis pigmentosa GTPase regulator by interaction screen in yeast, *Methods Enzymol.* **316**, 688-704 (2000).

41. CJ Zeiss, K Ray, GM Acland, and GD Aguirre, Mapping of X-linked progressive retinal atrophy (XLPRA), the canine homolog of retinitis pigmentosa 3 (RP3), *Hum. Mol. Genet.* **9**, 531-537 (2000).

42. RE McGuire, LS Sullivan, SH Blanton, MW Church, JR Heckenlively, and SP Daiger, X-linked dominant cone-rod degeneration: linkage mapping of a new locus for retinitis pigmentosa (RP15) to Xp22.13-p22.11, *Am. J. Hum. Genet.* **57**, 87-94.

43. AJ Mears, S Hiriyanna, R Vervoort, B Yashar, L Gieser, S Fahrner, SP Daiger, JR Heckenlively, PA Sieving, AF Wright, and A Swaroop, Remapping of the RP15 locus for X-linked cone-rod degeneration to Xp11.4-p21.1, and identification of a de novo insertion in the *RPGR* exon ORF15, *Am. J. Hum. Genet.* **67**, 1000-1003 (2000).

44. L Gieser, R Fujita, HH Goring, J Ott, DR Hoffman, AV Cideciyan, DG Birch, SG Jacobson, and A Swaroop, A novel locus (RP24) for X-linked retinitis pigmentosa maps to Xq26-27, *Am. J. Hum. Genet.* **63**, 1439-47 (1998).

45. I. Zucchi, S Mumm, G Pilia, S Mac Millan, R Reinbold, L Susani, J Weissenbach, and D Schlessinger, YAC/STS map across 12 Mb of Xq27 at 25 kb resolution, merging Xq26-qter, *Genomics*, **34**, 42-54 (1996).

46. AJ Hardcastle, DL Thiselton, I Zito, N Ebenezer, TS Mah, MB Gorin, and SS Bhattacharya, Evidence for a new locus for X-linked retinitis pigmentosa (*RP23*), *Invest. Ophthalmol. Vis. Sci.* **41**, 2080-2086 (2000).

CLINICAL VARIABITY OF PATIENTS ASSOCIATED WITH *RDH5* GENE MUTATION

Yuko Wada*, Toshiaki Abe, Miyuki Kawamura, and Makoto Tamai

INTRODUCTION

Fundus albipunctatus is a rare form of congenital stationary night blindness associated white dots in the fundus and extremely slow dark adaptation of the rod photoreceptors. In 1999, it was reported that the RDH5 gene which is localized on chromosome12q13-q14 and encoded 11-cis retinol dehydrogenase, was a causative gene for fundus albipuncataus.[1] In this study, we characterized the clinical features of Japanese patients associated with mutations in the RDH5 gene.

MATERIALS AND METHODS

Subjects and Mutation Analysis

We screened genomic DNA samples from seven patients with fundus albipunctatus including one patient who also had macular dystrophy and 120 unrelated patients with autosomal recessive retinitis pigmentosa (ARRP) in the RDH5 gene. We further screened 200 control chromosomes for these genes.

Genomic DNA was isolated from leukocytes prepared from a sample of each patient's blood (10 to 15 mL), using a protocol previously described in detail.[2] We amplified the entire coding regions of the RDH5 genes by polymerase chain reaction (PCR). For the screenings, we used 8 sets of primer pairs for the RDH5 gene. Products of the PCR were analyzed using non-radioisotopic single-strand conformation polymorphism (SSCP) with a

*Yuko Wada; Department of Opthalmology, Tohoku University, School of Medicine, 1-1 Seiryomachi Aobaku, Sendai Miyagi, 980-8574, Japan. Fax:81-22-717-7298; Tel 81-22-717-7294; e-mailYUKOW@oph.med.tohoku.ac.jp

modification previously described. The DNA fragment that showed abnormal mobility on SSCP was then sequenced to identify the mutation.

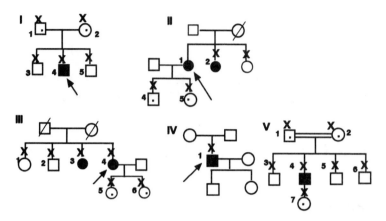

Figure1. Pedigrees of Japanese families with fundus albipunctatus showing affected(solid symbols) and unaffected(open symbols) members. Squares=male members; circles=female members; X= individuals examined in this study; slash=deceased; +=wild type; -=mutated allele; arrows= proband.

Clinical examination

The ophthalmic examinations included best corrected visual acuity, slit-lamp biomicroscopy, kinetic visual field examination, fundus examination, fluorescein angiography, color vision testing, and electroretinography(ERGs). Ophthalmoscopic findings were recorded by color fundus photography. Kinetic visual field examination was performed by a Goldmann perimeter with V-4-e , II-4-e, and I-2-e isopters. The ERG recording were obtained using a single flash or 30-Hz flicker stimulus of red light under light-adapted conditions for cone-isolated responses, a dim blue flash in the dark- adapted condition (30 minutes) for rod-isolated responses, and a bright white flash(20J) in the dark- adapted state for recording maximal responses of both rods and cones. These tests were performed under controlled conditions that conformed to the standard of the International Standardization Committee. [3]

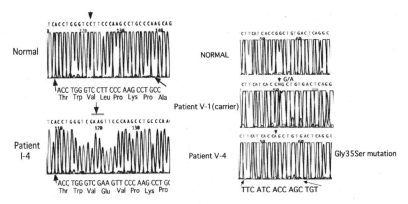

Figure2. Left:Results of DNA sequencing. A normal allele from a healthy control shows C at nucleotide 1085 (top); in the bottom sequence, mutant allele from patients I-4 shows the deletion of one base (C) in nucleotide1085 and the insertion of four bases(GAAG) instead. Arrows indicate the position of the mutation in each allele. Right: The upper sequence, a normal allele from a normal subject shows GCC in codon 35. The middle sequence from the carrier (I-1) shows both normal and mutant alleles. Arrows indicate the position of the mutation.The bottom sequence from patient II-1 shows a homozygous AGC in codon 35, resulting in the substitution of serine for glycine

RESULTS
DNA analysis

The results of nonradioisotopic single-strand conformation polymorphism analysis of exon5b in the RDH5 gene showed abnormal band shifts in all probands from the four unrelated families. The subsequent nucleotide sequencing analysis disclosed an identical homozygous 1085delC/insGAAG mutation in the RDH5 gene. The carriers were heterozygous for this mutation, and no affected member had this mutation. A homozygous Gly35Ser mutation in the RDH5 gene was identified in one affected member with fundus albipunctatus associated with cone dystrophy, and the parents were found to be heterozygous for the Gly35Ser mutation.We confirmed that the 1085delC/insGAAG mutation and the Gly35Ser mutation iin the RDH5 gene were co-segregated with the disease in all pedigrees with fundus albipunctatus and was not detected in the other 120 ARRP patients or 200 control X-chromosomes.

Ophthalmologic Examination

Clinical and molecular genetic findings have been summarized in the Table1. All probands associated with 1085delC/insGAAG mutation had only night blindness from childhood. Results of visual field testing disclosed that these patients had normal to slightly restricted peripheral fields. Fundus examinations of these patients disclosed that many white dots scattered in the retina or at the level of retinal pigment epithelium, without pigmentation and attenuation of retinal vessels.(Figure3) Patient V-4, who had the Gly35Ser mutation experienced a gradual progression of visual impairment including constriction of the visual field, night blindness and photophobia Fundus examination showed atrophic macular lesions in the right eye and sharply-demarcated macular lesions in the left eye with scattered white dots in the retina. (Figure3) The retinal vessels were attenuated. The fluorescein angiogramsof patients I-4 , II-1and IV-1 disclosed diffuse hyperfluorescence mainly in the posterior pole, indicating atrophy in the retinal pigment epithelium. White dots showed neither hyperfluorescence nor hypofluorescence. Electroretinograms of patients with 1085delC/insGAAG mutation showed that scotopic responses were nonrecordable, and single-flash, single flash, standard ERG results showed reduced a and b waves in each patient after 30 minutes of dark adaptation. Regarding cone responses, patients I-4and II-1 showed slightly reduced amplitudes of 30Hz-flicker ERG (right/left: 68/64μV and 73.3/74.6μV, respectively. (Normal range is from 83μV to 175μV). The standard flash ERG of patient V-4disclosed severely decreased a- and b-waves in both eyes.   After 3 hours of dark-adaptation, the standard flash ERG and scotopic ERG showed recovery of the a- and b-waves and scotopic b-waves, respectively.

Table1.   Clinical and molecular genetic findings of patients with fundus albipunctatus.

Cases are indicated by family number (Roman numeral) and patient number(Arabic number).

ND, not done; +, present; NB, night blindness; PB,photophobia; DV, decreased visual acuity

Patient	Age	Symptom	Mutation	Visual acuity OD	OS	Visual field
I-4	36	NB,PB	1085delC/insGAAG	1.2	1.0	Slight constriction
II-1	57	NB	1085delC/insGAAG	1.0	1.0	Slight constriction
II-2	55	NB	1085delC/insGAAG	1.2	1.0	ND
III-3	48	NB	1085delC/insGAAG	1.0	1.0	ND
III-4	45	NB	1085delC/insGAAG	1.2	1.0	Slight constriction
IV-1	40	NB	1085delC/insGAAG	1.0	1.0	Slight constriction
V-4	51	NB,PB,DV	Gly35Ser	0.1	0.08	constriction,ring scotoma

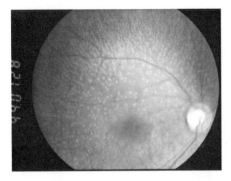 

Figure3.Fundus photographs of patient II-1 (Left) and V-4(Right) with 1085delC/insGAAG and Gly35Sermutation in the RDH5 gene, respectively.

Left; Numerous white dots are scattered in the posterior pole, sparing the macula.

Right: Scattered white dots and macular degeneration are observed.

DISCUSSION

In 1999, Yamamoto et al first reported that mutations of the RDH5 gene caused delayed dark adaptation and fundus albipunctatus.[1] In the present study, we evaluated 7 Japanese patients with fundus albipunctatus and 150 Japanese patients with ARRP. (Figure1)Molecular genetic analysis disclosed that 6 patients with fundus albipunctatus had an identical 1085delC/insGAAG mutation in the RDH5 gene[4], one patient associated with cone dystrophy had a Gly35Ser mutation and that no mutation was detected in the patients with ARRP.(Figure2)Although genealogic studies have not been performed for three or four generations, the1085delC/insGAAG mutation in the RDH5 gene is considered to be the common mutation for Japanese patients with fundus albipunctatus. The affected members with 1085delC/insGAAG mutationshowed similar subjective symptoms, visual acuity, and fundus appearance. (Table1)(Figure3)

On the other hand, the clinical features produced by the Gly35Ser mutation were decreased visual acuity, night blindness, white dots in the retina, progressive retinal degeneration and attenuated retinal vessels.(Figure3) Furthermore, the electroretinograms showed that the recovery of rod function was abnormally delayed and cone responses were severely impaired. These features were different from those of patients associated with 1085delC/insGAAG mutation except for scattered white dots and night blindness.

The clinical expressions associated with the mutation in the RDH5 mutation varied from fundus albipunctatus alone to fundus albipunctatus with cone dystrophy. It has been suggested that there may be a spectrum of gene phenotypes caused by mutations in the RDH5 gene.

In conclusion, we identified a homozygous 1085delC/insGAAG and Gly35Ser mutation in the RDH5 gene in a Japanese patient with fundus albipunctatus and fundus albipunctatus associated with cone dystrophy, respectively. This finding provides evidence that some kind of mutation in the RDH5 gene is related, in part at least, to the pathogenesis of progressive retinal degeneration.

ACKNOWLEDGMENT

This study was supported in part by a grant from the Research Committee on Chorioretinal Degenerations and Optic Atrophy, the Ministry of Health and Welfare of the Japanese Government (Dr. Tamai), Tokyo, Japan, and a Grant-In-Aid for Scientific Resaerch from the Ministry of Education, Science, and Culture of the Japanese Government (Dr.Tamai, A-2-10307041), Tokyo, Japan.

REFERENCES

1.Yamamoto H, Simon A, Eriksson U, et al. Mutation in the gene encoding 11-cis retinol dehydrogenase cause delayed dark adaptation and fundus albipunctatus. *Nature Genetics.* 1999;22:188-191.

2.Nakazawa M, Kikawa-Araki E, Shiono T, Tamai M.   Analysis of rhodopsin gene in patients with retinitis pigmentosa using allele-specific polymerase chain reaction.   *Jpn J Ophthalmol.* 1991; 35:386-393

3.Marmor MF, Arden GB, Nilsson SEG, Zrenner E. Standard for clinical electroretinography. *Arch Ophthalmol.* 1989;107:816-819.

4.Wada Y, Abe T, Fuse N, Tamai M. A frequent 1085delC/insGAAG mutation in the RDH5 gene in Japanese patients with fundus albipunctatus. *Invest Ophthalmol Vis Sci* 2000;41:1894-1897. : 165-170.

JAPANESE PATIENTS WITH FUNDUS ALBIPUNCTATUS CAUSED BY *RDH5* GENE MUTATIONS

Makoto Nakamura, Yoshihiro Hotta, Yozo Miyake[*]

1. ABSTRACT

Fundus albipunctatus is a type of congenital stationary night blindness with an autosomal recessive inheritance pattern. The fundus demonstrates a large number of discrete, small, round or elliptical, yellow-white lesions. Recent studies have demonstrated that mutations of the 11-*cis* retinol dehydrogenase (*RDH5*) gene causes this disease. We analyzed the *RDH5* gene in Japanese families with fundus albipunctatus, and in all, either a homozygous or a compound heterozygous mutation in the *RDH5* gene was identified. We conclude that most Japanese fundus albipunctatus patients are caused by the *RDH5* gene mutation. The mutation of c.928 C deletion/GAAG insertion occurred at a relatively high frequency in Japanese patients, suggesting a founder effect. Most of the elderly fundus albipunctatus patients had cone dystrophy with a depression of visual functions. We suggest that the mutations of the *RDH5* gene causes a cone dystrophy as well as night blindness in Japanese patients.

2. INTRODUCTION

Fundus albipunctatus is a type of congenital stationary night blindness with an autosomal recessive inheritance pattern.[1] The fundus demonstrates a large number of discrete, small, round or elliptical, yellow-white lesions.[2,3] The patients complain of night blindness from early childhood, and the clinical course has been considered to be stationary with normal visual acuity, visual fields and color perception.[4] Electrophysiological examination of patients with fundus albipunctatus revealed normal scotopic and photopic electroretinographic (ERG) responses but a long dark-adaptation

[*] Department of Ophthalmology, Nagoya University School of Medicine, 65 Tsuruma-cho, Showa-ku, Nagoya 466-8560, Japan.

period is necessary to obtain normal scotopic responses.[1,4] Similar yellow-white lesions are observed in patients with retinitis punctata albescense, a kind of retinitis pigmentosa.[5] In such cases, the visual functions are progressively altered with decreased visual acuity and defects of the visual fields. The *peripherin/RDS* gene[6] or the *RLBP1* gene[7,8] are reported as causing genes of retinitis punctata albescense.

The defect in fundus albipunctatus had been supposed to be in the retinal pigment epithelium (RPE) because the white-dots are observed at the RPE level, and because of the features of the ERG and fluorescein angiograms (FAG). This is supported by a recent study that revealed a mutations of the 11-*cis* retinol dehydrogenase (*RDH5*) gene, which is expressed predominantly in RPE, caused this disease.[9]

We have examined 16 patients from 14 separate Japanese families who had been diagnosed as having fundus albipunctatus, and we shall show that a homozygous or a compound heterozygous mutation in the *RDH5* gene was detected in all. Six patients from 6 families also had cone dystrophy.

3. PATIENTS AND METHODS

This study involving human subjects followed the tenets of the Declaration of Helsinki and informed consent was obtained from each patient after an explanation of this study. Sixteen patients with fundus albipunctatus from 14 Japanese families were examined. All individuals examined have been followed in the Departments of Ophthalmology, Nagoya University, Japan. A complete ophthalmologic examination was performed including best-corrected visual acuity, slit-lamp and fundus examination, fundus photography, and electroretinography.

Genomic DNA was extracted from peripheral leukocytes. Exons 2, 3, 4 and 5 of the *RDH5* gene were amplified by polymerase chain reaction (PCR). The PCR conditions and the procedures for direct sequencing were described in detail earlier.[9,10] To search for polymorphisms, 100 alleles from normal, unrelated Japanese individuals were directly sequenced.

Conventional electroretinograms (ERGs) were elicited with Ganzfeld stimuli after 30 minutes of dark-adaptation. The rod (scotopic) ERGs were recorded with a blue light at an intensity of 5.2×10^{-3} cd/m^2/sec. The mixed rod:cone single flash ERGs were recorded with a white stimulus at an intensity of 44.2 cd/m^2/sec. The cone ERGs and the 30 Hz flicker ERGs were elicited with a white stimulus at an intensity of 4 cd/m^2/sec and 0.9 cd/m^2/sec, respectively, under a background illumination of 21 footlamberts.

4. RESULTS

The parents of case 1 were cousins. Case 5's maternal grandfather and father's grandfather were cousins, and in cases 9 and 11, the maternal grandfather and father's grandfather were also cousins. No family history was obtained from the other patients. Molecular genetical examination revealed either a homozygous or a compound heterozygous mutation of the *RDH5* gene in all of the patients examined (Table 1). A C

Table 1. Clinical and genetic findings of patients with fundus albipunctatus

Case	Age (y)	Sex	Visual acuity		Bull's eye	Base change	Amino acid
			OD	OS			
1	74	M	0.02	0.06	+	c.719 G insertion	
						c.841 T to C	Tyr281His
2	65	M	0.07	0.2	+	C.928 Cto GAAG	Leu310GluVal
						c.319 G to C	Gly107Arg
3	63	M	0.1	0.6	+	c.928 C to GAAG	Leu310GluVal
4	59	M	0.05	0.05	+	c.928 C to GAAG	Leu310GluVal
5	58	M	1.2	1.2	+	c.928 C to GAAG	Leu310GluVal
6	53	F	1.2	1.2	-	c.928 C to GAAG	Leu310GluVal
7	53	F	1.2	1.2	-	c.394 G to A	Val132Met
						c.839 G to A	Arg280His
8	48	M	0.03	0.03	+	c.928 C to GAAG	Leu310GluVal
9*	22	M	1.0	1.0	-	c.103 G to A	Gly35Ser
10	20	F	1.2	1.2	-	c.928 C to GAAG	Leu310GluVal
						c.103 G to A	Gly35Ser
11*	19	M	1.2	1.2	-	c.103 G to A	Gly35Ser
12	18	F	1.0	1.0	-	c.928 C to GAAG	Leu310GluVal
						c.839 G to A	Arg280His
13*	18	F	1.0	1.0	-	c.928 C to GAAG	Leu310GluVal
14*	15	M	1.5	1.5	-	c.928 C to GAAG	Leu310GluVal
15	15	M	1.0	1.2	-	c.928 C to GAAG	Leu310GluVal
16	12	M	1.0	1.0	-	c.928 C to GAAG	Leu310GluVal

*Cases 9 and 11, Cases 13 and 14 are brothers, respectively.
Cases 1, 2, 7, 10, and 12 are compound heterozygotes, and others are homozygotes.

deletion at nucleotide 928 with a replacement by a 4-bp insertion of GAAG in exon 5, causing an amino acid substitution of Leu310 into unusual residues of Glu and Val was most frequently found either homozygously or heterozygously. In addition, a missense mutation of G to A at nucleotide 103 in exon 2 predicting a Gly35Ser amino acid substitution; and a missense mutation of G to C at nucleotide 319 in exon 3, predicting a Gly107Arg amino acid substitution were found. A missense mutation of G to A at nucleotide 394 in exon 3 predicting a Val132Met amino acid substitution, a G insertion at nucleotide 719 in exon 4 resulting in a frame shift with 18 unusual amino acid residues and a predicted stop codon at codon 258 were also detected. A missense mutation of G to A at nucleotide 839 in exon 5 predicting an Arg280His amino acid substitution, and a missense mutation of T to C at nucleotide 841 in exon 5, predicting a Tyr281His amino acid substitution were also identified. These mutations were confirmed using both sense and antisense primers.

The sequences of the carriers showed a heterozygous pattern including both the wild and mutant alleles. For example, a daughter of case 2, a son of case 4, the mother of cases 13 and 14, the mother of case 15, and the parents of case 16 showed heterozygous c.928 C to GAAG base change. The mother of case 10 and the mother of cases 9 and 11 showed heterozygous c.103 G to A (Gly35Ser) missense change.

The white punctata were observed in the fundi, especially clearly in young patients (Fig. 1A), and more inconspicuously in elderly patients (Fig. 1B). In most of the elderly patients (cases 1 through 5, and 8), cone dystrophy or macular degeneration was seen (Fig. 1B). Focal chorioretinal atrophy in the mid-periphery or degenerative retinal change in the periphery was observed in some of the elderly patients. The visual acuity in patients with cone dystrophy was reduced to varying degrees with central scotoma or paracentral scotoma (Table 1).

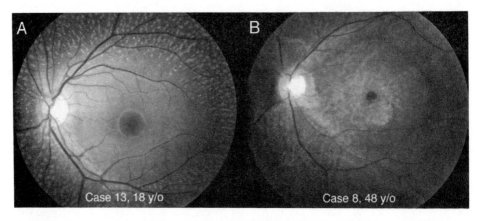

Figure 1. Fundus photographs of fundus albipunctatus patients with mutations of RDH5 gene (c.928 C deletion/GAAG insertion). (**A**) right eye, case 13: numerous dull white-yellow punctations are clearly seen in the midperiphery. (**B**) right eye, case 5: white punctation and Bull's eye. The case number and the age (years) are indicated in each photograph.

The b-wave amplitude of the scotopic ERG was significantly reduced after a short period of dark-adaptation (30 min) in all of the patients. However, after 2 to 3 hours of dark-adaptation, the amplitude improved slightly or to normal levels (Fig. 2). In the fundus albipunctatus patients with cone dystrophy, the photopic b-wave amplitude was also significantly reduced (Fig. 2). In some young patients without cone dystrophy, the photopic ERGs were also reduced.

5. DISCUSSION

Fundus albipunctatus is a rare disease, and to date, only 4 mutations, Ser73Phe,[9] Gly238Trp,[9,11] Arg280His,[11] and Ala294Pro,[11] have been reported in Caucasian patients. This disease is seen with a slightly higher frequency in the Japanese, and other Japanese research groups have also reported the homozygous c.928 C deletion/GAAG insertion in 7 patients,[12,13] a homozygous missense mutation of Val264Gly in one patient,[13] and a compound heterozygous missense mutation of Val177Gly and Arg280His in a patient.[14] Taken together with our results, mutations of the *RDH5* gene have been detected in all Japanese fundus albipunctatus patients, and thus we conclude that mutations of the *RDH5* gene are the cause for most of the fundus albipunctatus in Japanese patients.

Because the mutation of c.928 C deletion/GAAG insertion occurs at a relatively higher frequency in Japanese patients, this is assumed to be a founder effect.[10,12]

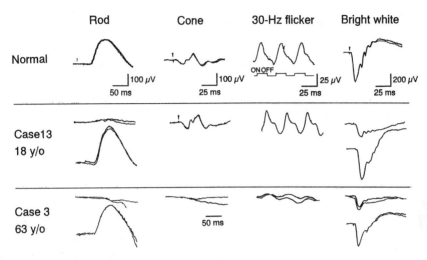

Figure 2. Full-field electroretinograms of a normal subject and fundus albipunctatus patients with mutations of RDH5 gene (c.928 C deletion/GAAG insertion) Case3 was accompanied with cone dystrophy, but case 13 did not. In each case the upper tracing indicate the responses recorded after 30 mimutes of dark adaptation;the lower tracings indicate the responses recorded after prolonged dark adaptation (3 hours). The arrows indicate the stimulus onset. The case number and the age (years) of each patient is noted under the case number in the left column.

Eleven-*cis* retinol dehydrogenase plays an important role in the phototransduction cascade. It oxidize 11-*cis* retinol to 11-*cis* retinal in the RPE for the recycling of retinal to serve as the chromophore of rhodopsin in photoreceptor cells.[15] If the *RDH5* gene is mutated and the coversion to 11-*cis* retinal is delayed, a long dark-adaptation time will be needed for rods to regain sensitivity in darkness to be able to give rise to normal scotopic b wave responses. This is probably the basis for the reduced scotopic b-waves after the conventional dark-adaptation period and also for the night blindness in fundus albipunctatus patients.

It has been believed that fundus albipunctatus was a stationary disease and, other than the night blindness, the visual acuity, visual field, color vision and other visual functions remained normal.[2,4] However, there have been reports that described cases of macular degeneration with fundus albipunctatus,[16-18] and we have also reported several cases of fundus albipunctatus associated with cone dystrophy whose full-field photopic ERGs were severely reduced.[19] At that time, we could not determine whether these cases represented an advanced stage of fundus albipunctatus or a chance combination of fundus albipunctatus and cone dystrophy in the same patient.[19]

Interestingly, mutations of the *RDH5* gene were found in patients associated with cone dystrophy as well as in those with normal cone function.[10] The cone dystrophy with significantly reduced photopic ERG responses was seen frequently in elderly patients, whereas no patient less than 20 years old presented with cone dystrophy. These observations strongly suggest that the association of fundus albipunctatus with cone dystrophy is not due to a chance association of a mutation in two genes.[10] The results of our genetic analysis led us to conclude that cone dysfunction is frequently observed in elderly patients with fundus albipunctatus, and the cone dystrophy with bull's eye maculopathy results in severe loss of visual function. However, it should be noted that all of the elderly patients may not necessarily develop cone dystrophy because cases 6 and 7 maintained good visual acuity without showing cone dystrophy in spite of being 53-years of age. It will be important to follow patients without cone dystrophy for a long time to determine the relationship between genotype and phenotype of fundus albipunctatus and its association with cone dystrophy. Although the exact mechanism whereby mutations of the *RDH5* gene lead to cone dystrophy is unknown, the *RDH5* gene appears to be also important for maintenance of normal cone function.

6. ACKNOWLEDGMENTS

This study was supported in part by grants for Research Committee on Chorioretinal Degenerations from The Ministry of Health and Welfare of Japan, and Grant-in-Aid for Scientific Research (Dr. Miyake, B11470363, Dr. Hotta, C05807166, Dr. Nakamura, C12671703) from Ministry of Education, Science, Sports and Culture, Japan.

REFERENCES

1. Heckenlively J. Congenital Stationary Night Blindness. Genetic diseases of the eye. New York: Oxford University Press 1998;389-396.
2. Krill AE. Hereditary retinal and choroidal diseases. Hagerstown:Harper and Row 1977;2:739-824.
3. Marmor MF. Long-term follow-up of the physiologic abnormalities and fundus changes in fundus albipunctatus. *Ophthalmology.* 1990;97:380-384.
4. Gass JDM. Stereoscopic atlas of macular disease. Diagnosis and treatment. 4th ed. St Louis: Mosby 1997:350-351.
5. Lauber H: Die sogenannte Retinitis punctata albescens. *Klin Monatsbl Augenheilkd.* 1910;48:133-148.
6. Kajiwara K, Sandberg MA, Berson EL, Dryja TP. A null mutation in the human *peripherin/RDS* gene in a family with autosomal dominant retinitis punctata albescens. *Nat Genet.* 1993;3:208-212.
7. Burstedt MSI, Sandgren O, Holmgren G, Forsman-Semb K. Bothnia Dystrophy caused by mutations in the cellular retinaldehyde-binding protein gene (*RLBP1*) on chromosome 15q26. *Invest Ophthalmol Vis Sci.* 1999;40:995-1000.
8. Morimura H, Berson EL, Dryja TP. Recessive mutations in the *RLBP1* gene encoding cellular retinaldehyde-binding protein in a form of retinitis punctata albescens. *Invest Ophthalmol Vis Sci.* 1999;40:1000-1004.
9. Yamamoto H, Simon A, Eriksson U, Harris E, Berson EL, Dryja TP. Mutations in the gene encoding 11-*cis* retinol dehydrogenase cause delayed dark adaptation and fundus albipunctatus. *Nat Genet.* 1999;22:188-191.
10. Nakamura M, Hotta Y, Tanikawa A, Terasaki H, Miyake Y. A high association with cone dystrophy in fundus albipunctatus caused by mutations of the *RDH5* gene. *Invest Ophthalmol Vis Sci.* 2000;41:3925-32.
11. Gonzalez-Fernandez F, Kurz D, Bao Y, Newman S, Conway BP, Young JE, Han DP, Khani SC. 11-*cis* Retinol dehydrogenase mutations as a major cause of the congenital night-blindness disorder known as fundus albipunctatus. *Mol Vision.* 1999;5:41.

12. Wada Y, Abe T, Fuse N, Tamai M. A frequent 1085delC/insGAAG mutation in the *RDH5* gene in Japanese patients with fundus albipunctatus. *Invest Ophthalmol Vis Sci.* 2000;41:1894-7.

13. Hirose E, Inoue Y, Morimura H, Okamoto N, Fukuda M, Yamamoto S, Fujikado T, Tano Y. Mutations in the 11-*cis* retinol dehydrogenase gene in japanese patients with fundus albipunctatus. *Invest Ophthalmol Vis Sci.* 2000;41:3933-5.

14. Kuroiwa S, Kikuchi T, Yoshimura N. A novel compound heterozygous mutation in the *RDH5* gene in a patient with fundus albipunctatus. *Am J Ophthalmol.* 2000;130:672-5.

15. Lion F, Rotmans JP, Daemen FJ, Bonting SL. Biochemical aspects of the visual process. XXVII. Stereospecificity of ocular retinol dehydrogenases and the visual cycle. *Biochim Biophys Acta.* 1975;384:283-92.

16. Scouras J, Chevaleraud J, Papastratigakis B. Macular involvment in retinitis albescens. *J Fr Ophthalmol.* 1979;2:613-617.

17. Sato S, Iijima H. A case of fundus albipunctatus associated with cone dysfunction. *Rinsho Ganka.* 1987;41:1307-1310.

18. Welber RG. Retinitis pigmentosa and allied disorders. In: Ryan SJ, Ogden TE, eds. Retina. St Louis: Mosby 1989;1:346-348.

19. Miyake Y, Shiroyama N, Sugita S, Horiguchi M, Yagasaki K. Fundus albipunctatus associated with cone dystrophy. *Br J Ophthalmol.* 1992;76:375-379.

FUNCTIONAL ANALYSIS OF AIPL1

A novel photoreceptor-pineal-specific protein causing Leber congenital amaurosis and other retinopathies

Melanie M. Sohocki, Dayna L. Tirpak, Cheryl M. Craft, Stephen P. Daiger[1]

1. INTRODUCTION

Leber congenital amaurosis (LCA, OMIM No. 204000) is a severe, early-onset inherited retinopathy that accounts for approximately 5% of all inherited retinal disease. To date, four genes associated with Leber congenital amaurosis have been identified, the most recent of which is the aryl-hydrocarbon interacting-like 1 gene, AIPL1 (OMIM No. 604392, 604393) (Sohocki *et al.*, 2000). Mutations in AIPL1 are the cause of 7 to 9% of cases of autosomal recessive LCA and may cause dominant cone-rod dystrophy (Sohocki *et al.*, 2000a) (Figures 1 and 2).

AIPL1 expression is limited to photoreceptors and the pineal gland, and the gene encodes a protein 384 amino acids in length. The precise functional role of AIPL1 within the photoreceptors remains unknown. However, the human protein sequence contains three tetratricopeptide (TPR) motifs, 34 amino acid motifs that are thought to serve as interfaces for protein-protein interactions (Blatch and Lassle, 1999). TPR motifs are found in proteins that mediate a variety of functions, including protein trafficking or protein folding, and these proteins are usually associated with multiprotein complexes. In addition, a proline-rich region is present at the carboxyl-terminus of the protein in primates, including humans, but not in other mammals. Similar sequences are found in situations requiring rapid recruitment or interchange of several proteins, such as signaling cascades or initiation of transcription.

To improve our understanding of the role of AIPL1 in normal vision and in the pathogenesis of LCA, we have conducted comparative sequencing of AIPL1 orthologs between mammalian species, and yeast two-hybrid analyses to detect AIPL1-binding proteins. This article summarizes our findings in these two areas.

[1]M.M. Sohocki, D.L. Tirpak, S.P. Daiger, Human Genetics Center and Dept. of Ophthalmology and Visual Science, The Univ. of Texas, Houston; C.M. Craft, Doheny Eye Institute and Keck School of Medicine, Univ. of Southern California, Los Angeles.

New Insights Into Retinal Degenerative Diseases
Edited by Anderson *et al.*, Kluwer Academic/Plenum Publishers, 2001

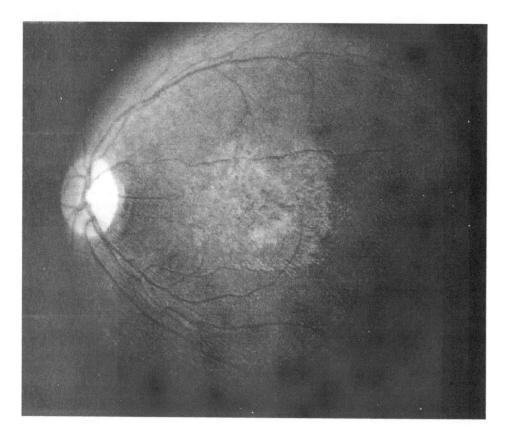

Figure 1. Fundus photograph of Leber congenital amaurosis caused by a homozygous Cys239Arg mutation in AIPL1 (Sohocki *et al.*, 2000).

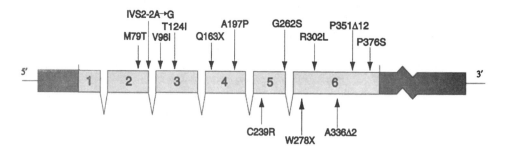

Figure 2. Human AIPL1 protein diagram and spectrum of mutations associated with LCA (Sohocki et al.2000; Sohocki et al. 2000a).

2. METHODS

2.1 Between species comparisons

A zoo blot probed with a 200 bp fragment of AIPL1 indicated likely orthologs in dog, cow and mouse. No signal was present in zebrafish or in chicken. Human, mouse and cow sequences were determined by cDNA sequencing from retinal-cDNA libraries, confirmed, in humans, by genomic sequencing. Chimpanzee, baboon and rhesus monkey sequences were determined from genomic DNA. PCR products for sequencing was generated using "universal" *aipl1* primers, which have high sequence similarity between all mammalian species. Sequencing was conducted using cycle-sequencing PCR with flourescinated nucleotides on a PE Prism 310 DNA Sequencer. Sequence was analyzed using MacVector and the Entrez BLAST server. Deatils are in Sohocki *et al.* 2000b.

2.2 Yeast two-hybrid plasmid construction

A bovine AIPL1 construct for the yeast two-hybrid screen was made by cloning the bovine AIPL1 cDNA into the GAL4 DNA binding domain of a pGBKT7 vector (Clontech). Bovine AIPL1 was used because a human retinal cDNA library appropriate for a yeast two-hybrid screen was not available. Bovine AIPL1 has a 97% amino acid sequence similarity to human AIPL1 (Figure 3). The full-length bovine AIPL1 coding sequence was amplified out of a bovine retinal cDNA library using gene-specific primers. Primers were designed such that *Eco*RI and *Bam*HI restriction sites were incorporated into the 5' and 3' ends of the AIPL1 PCR product respectively. The PCR product was then digested with *Eco*RI and *Bam*HI and ligated in-frame into the pGBKT7 vector and subsequently transformed into E. coli XL1-Blue competent cells (Stratagene). Sequencing of the entire bovine AIPL1 insert was performed using vector and gene specific primers to ensure that the insert was ligated in-frame and would result in a protein with AIPL1 fused to the DNA binding domain (bait plasmid). The bovine retinal cDNA library used for the yeast two-hybrid screen was cloned in-frame into the pGAD10 vector which results in fusion of the library proteins with the activation domain (prey plasmid). The vector-containing bovine AIPL1 (pGBKT7-AIPL1) was used as the bait to screen the bovine retinal cDNA library in the yeast strain AH109 (Clontech). AH109 is a yeast strain capable of reporting protein-protein interactions via activation of three reporter genes: Histidine, Adenine, and lacZ/or Mel1.

2.3 Yeast two-hybrid screen

A GAL4 yeast two-hybrid system was used to identify proteins that interact with AIPL1 (MATCHMAKER 3, Clontech). The pGBKT7-Aipl1 plasmid was first transformed into yeast strain AH109 using the YEASTMAKER lithium acetate method (Clontech). Alpha-galactosidase and growth assays on plates with synthetic medium lacking leucine (Leu), tryptophan (Trp), histidine (His), and adenine (Ade) were performed to ensure that AIPL1 could not autonomously activate the reporters. Next, the pGAD10 bovine retinal library plasmids were transformed into AH109 cells containing the pGBKT7-AIPL1 plasmid. Yeast transformants were incubated at 30°C for 14 days on –Leu/-Trp/-His plates and the single colonies that developed were selected for further analysis. These colonies were then streaked onto –Leu/-Trp/-His + X-α-gal plates to recheck activation of the Mel1 and His reporters. Finally, yeast transformants that indicated activation of the His and Mel1 reporters were streaked onto –Leu/-Trp/-His/-Ade + X-α-gal plates to assay the activation of all three reporters and thereby indicate the presence of a strong protein-protein interaction.

Plasmids containing cDNA inserts that encode potential AIPL1-interacting proteins (pGAD10 prey plasmids) were then isolated following a standard yeast plasmid prep protocol (Hoffman and Winston, 1987). Once isolated from the yeast, the prey plasmids were amplified by transformation into E. coli SURE electroporation competent cells (Stratagene) and purified using a standard miniprep kit (Qiagen). Yeast two-hybrid retests were performed by transforming the prey plasmids into AH109 with either the pGBKT7 vector containing AIPL1, the empty pGBKT7 vector (negative control), or the pGBKT7 vector containing LaminC (negative control). Yeast transformants were incubated at 30°C on –Leu/-Trp (transformation check) and –Leu/ -Trp/-His plates for 4-6 days or until colonies appeared. Prey plasmids that grew only in the presence of the pGBKT7-AIPL1 vector and did not grow with the negative control pGBKT7 vectors were then streaked onto –Leu/-Trp/-His + X-α-gal to recheck for activation of the His and Mel1 reporters. The entire cDNA inserts of these plasmids were sequenced and *in silico* analysis was performed to determine if protein identity is known.

3. RESULTS

3.1 Between species comparisons

Due to the difference in protein length between human and squirrel monkey, the Aipl1 sequences of several primates (chimpanzee, baboon and rhesus monkey) were determined from genomic DNA (Figure 3). Although the chimpanzee and human proteins are identical in length, the difference or similarity in length of the proline-rich region in the other primates tested is inconsistent with evolutionary distance from humans. These data suggest that this region may be susceptible to gain or loss of repeats by slippage, perhaps due to the XXPP repeat-type structure of the proline-rich region. In addition, although several residues in this region are identical across the primates tested, the sequence identity/similarity to the human protein in this region is significantly lower than that within the TPR domains or across the length of the protein (Figure 3).

Sequencing of Aipl1 from cow and mouse retinal cDNA libraries revealed a high level of sequence conservation within mammals, outside of the 56 amino acid proline-rich carboxyl-terminal sequence which is present in human AIPL1 but lacking in non-primates (Figure 3). Genomic sequencing determined that the squirrel monkey Aipl1 proline-rich region is only 44 amino acids in length. In addition, a *Drosophila* protein sequence similar to AIPL1 (GenBank AE003487; 36% identity, 59% similarity) was identified by BLAST. This sequence, however, appears to be more highly related to AIP (39% identity, 59% similarity). As the complete *Drosophila* genomic sequence is known, with no other similar sequences, it is possible that this gene represents an ancestral Aip protein, predating the duplication event leading to separate Aip and Aipl1 in higher organisms.

```
Human            MDAALLLNVEGVKKTILHGGTGELPNFITGSRVIFHFRTMKCDEERTVIDDSRQVGQPMH  60
Chimp            MDAALLLNVEGVKKTILHGGTGELPNFITGSRVIFHFRTMKCDEERTVIDDSRQVGQPMH  60
Baboon           MDAALLLNVEGVKKTILHGGTGELPNFITGSRVIFHFRTMKCDEERTVIDDSRQVDQPMH  60
Rhesus monkey    MDAALLLNVEGVKKTILHGGTGELPNFITGSRVIFHFRTMKCDEERTVIDDSRQVDQPMH  60
Squirrel monkey  MDAALLLNVEGVKKTILHGGTGELPNFITGSRVIFHFRTMKCDEERTVIDDSREVGQPMH  60
Mouse            MDVSLLLNVEGVKKTILHGGTGELPNFITGSRVTFHFRTMKCDEERTVIDDSKQVGQPMS  60
Rat              MDVSLLLNVEGVKKTILHGGTGELPNFITGSRVTFHFRTMKCDEERTVIDDSKQVGQPMN  60
Cow              MDATLLLNVEGIKKTILHGGTGDLPNFITGARVTFHFRTMKCDEERTVIDDSKQVGHPMH  60
                 **  .******.********** .*******.** **************** .  .  **

Human            IIIGNMFKLEVWEILLTSMRVHEVAAEFWCDTIHTGVYPILSRSLRQMAQGKDPTEWHVHT  120
Chimp            IIIGNMFKLEVWEILLTSMRVHEVAAEFWCDTIHTGVYPILSRSLRQMAQGKDPTEWHVHT  120
Baboon           IIIGNMFKLEVWEILLTSMRVHEVAAEFWCDTIHTGVYPILSRSLRQMAQGKDPTEWHVHT  120
Rhesus monkey    IIIGNMFKLEVWEILLTSMRVHEVAAEFWCDTIHTGVYPILSRSLRQMAQGKDPTEWHVHT  120
Squirrel monkey  IIIGNMFKLEVWEILLTSMRVREVAAEFWCDTIHTGVYPILSRSLRQMAQGKDPTEWHVHT  120
Mouse            IIIGNMFKLEVWETLLTSMRLGEVAAEFWCDTIHTGVYPMLSRSLRQVAEGKDPTSWHVHT  120
Rat              IIIGNMFKLEVWETLLTSMRLGEVAAEFWCDTIHTGVYPMLSRSLRQVAEGKDPTSWHVHT  120
Cow              IIIGNMFKLEVWEILLTSMRVSEVAAEFWCDTIHTGVYPILSRSLRQMAEGKDPTEWHVHT  120
                 ************.******.    *************.******.*.*****.*.***
```

TPR I

```
Human            CGLANMFAYHTLGYEDLDELQKEPQPLVFVIELLQVDAPSDYQRETWNLSNHEKMKAVPV  180
Chimp            CGLANMFAYHTLGYEDLDELQKEPQPLVFVIELLQVDAPSDYQRETWNLSNHEKMKAVPV  180
Baboon           CGLANMFAYHTLGYEDLDELQKEPQPLIFVIELLQVDAPSDYQRETWNLSNHEKMKVVPV  180
Rhesus monkey    CGLANMFAYHTLGYEDLDELQKEPQPLIFVIELLQVDAPSDYQRETWNLSNHEKMKVVPV  180
Squirrel monkey  CGLANMFAYHTLGYEDLDELQKEPQPLIFVIELLQVDAPSDYQRETWNLSNHEKMKVVPV  180
Mouse            CGLANMFAYHTLGYEDLDELQKEPQPLVFLYELLQVEAPNEYQRETWNLNNEERMQAVPL  180
Rat              CGLANMFAYHTLGYEDLDELQKEPQPLIFLIELLQVEAPNEYQRETWNLNNEERMQAVPL  180
Cow              CGLANMFAYHTLGYEDLDELQKEPQPLIFIIELLQVEAPSQYQRETWNLNNQEKMQAVPI  180
                 ***********************.    *.***** ** ******** *  * *.  **.
```

TPR II

```
Human            LHGEGNRLFKLGRYEEASSKYQEAIICLRNLQTKEKPWEVQWLKLEKMINTLILNYCQCL  240
Chimp            LHGEGNRLFKLGRYEEASSKYQEAIICLRNLQTKEKPWEVQWLKLEKMINTLILNYCQCL  240
Baboon           LHGEGNRLFKLGRYEEASSKYQEAIICLRNLQTKEKPWEVQWLKLEKMINTLTLNYCQCL  240
Rhesus monkey    LHGEGNRLFKLGRYEEASSKYQEAIICLRNLQTKEKPWEVQWLKLEKMINTLTLNYCQCL  240
Squirrel monkey  LHGEGNRLFKLGRYEEASSKYQEAIICLRNLQTKEKPWEVQWLKLEKMINTLILNYCQCL  240
Mouse            LHGEGNRLYKLGRYDQAATKYQEAIVCLRNLQTKEKPWEVEWLKLEKMINTLILNYCQCL  240
Rat              LHGEGNRLYKLGRYDQAATKYQEAIVCLRNLQTKEKPWEVEWLKLEKMINTLILNYCQCL  240
Cow              LHGEGNRLFKLGRYEEASNKYQEAIVCLRNLQTKEKPWEVQWLKLEKMINTLILNYCQCL  240
                 ********.*****. .*  *** **.*************.***.*.************ **
```

TPR III

```
Human            LKKEEYYEVLEHTSDILRHHPGIVKAYYVRARAHAEVWNEAEAKADLQKVLELEPSMQKA  300
Chimp            LKKEEYYEVLEHTSDILRHHPGIVKAYYVRARAHAEVWNEAEAKADLRKVLELEPSMQKA  300
Baboon           LKKEEYYEVLEHTSDILRHHPGIVKAYYVRARAHAEVWNEAEAKADLQKVLELEPSMQKA  300
Rhesus monkey    LKKEEYYEVLEHTSDILRHHPGIVKAYYVRARAHAEVWNEAEAKADLQKVLELEPSMQKA  300
Squirrel monkey  LKKEEYYEVLEHTSDILRHHPGIVKAYYVRARAHAEVWNEAEAKADLQKVLELEPSMQKA  300
Mouse            LKKEEYYEVLEHTSDILRHHPGIVKAYMRARAHAEVWNAEEAKADLEKVLELEPSMRKA  300
Rat              LKKEEYYEVLEHTSDILRHHPGIVKAYMRARAHAEVWNEAEAKADLEKVLELEPSMRKA  300
Cow              LKKEEYYEVLEHTSDILRHHPGIVKAYYVRARAHAEVWNEAEAKADLEKVLELEPSMRKA  300
                 *********************.*****  **********  ****** *********. **
```

```
Human            VRRELRLLENRMAEKQEEERLRCRNMLSQGATQPPAEPPTEPP--------AQSSTEPPA  352
Chimp            VRRELRLLENRMAEKQEEERLRCRNMLSQGATQPPAEPPTEPP--------AQSSTEPPA  352
Baboon           VRRELRLLENRMAEKQEEERLRCRNMLSQGATQPPTEP---------------------PA  340
Rhesus monkey    VRRELRLLENRMAEKQEEERLRCRNMLSQGATQPPAEPPAQPPTAPPAELSTGPPADPPA  360
Squirrel monkey  VRRELRLLENRMAEKQEEERLRCRNMLSQGATWSPAEP-------------------PA  340
Mouse            VLRELRLLESRLADKQEEERQRCRSMLG                                328
Rat              VLRELRLLESRLADKQEEERQRCRSMLG                                328
Cow              VQRELRLLENRLEEKREEERLRCRNMLG                                328
                 * ******* *. .* **** *** **
```

```
Human            EPPTAPSAELSAGPPAEPATEPPPSPGHSLQH  384
Chimp            EPPPAPSAELSAGPPAETATEPPPSPGHSLQH  384
Baboon           EPHTAPPAELSTGPPAELPLSPGHSLQH  372
Rhesus monkey    EPPTAPPAELSTGPPAELPLSPGHSLQH  392
Squirrel monkey  EPPAESSTEPPAEPPAEPPAELTLTPGHPLQH  372
Mouse                                             328
Rat                                               328
Cow                                               328
```

Figure 3. Between species comparison of AIPL1 cDNA sequences.

3.2 Yeast two-hybrid analysis

Yeast two-hybrids with AIPL1 and the bovine retinal cDNA library were performed twice (Figure 4). Approximately 2×10^6 yeast transformants were screened for AIPL1-interacting proteins. A total of 103 colonies grew after the initial transformation on –Leu/-Trp/-His plates during the 14-day incubation. Of the 103 original colonies 57 regrew on –Leu/-Trp/-His/ +X-α-gal plates and turned blue, indicating the activation of the His and Mel1 reporters. The 57 colonies were then plated onto –Leu/-Trp/-His/-Ade + X-a-gal plates to assay for activation of all three reporters. Twenty-nine of the 57 colonies demonstrated activation of the His, Mel1, and Ade reporters, thereby indicating the presence of a strong protein-protein interaction. Figure 4 summarizes these findings.

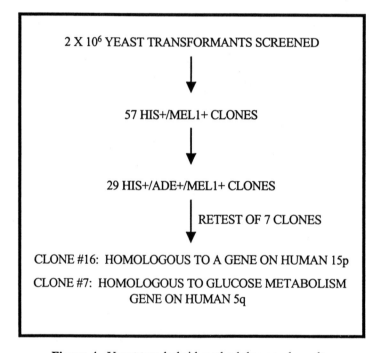

Figure 4. Yeast two-hybrid methodology and results.

To date, 7 of the 57 clones have be subjected to yeast two-hybrid retests. Retests were performed to identify which clones (pGAD10 prey plasmids) are specifically interacting with AIPL1 and which clones are capable of autonomously activating the reporters (a false positive). Five of the 7 clones were either false positives and did not require AIPL1 to activate the reporters, or the transformation failed to produce colonies on the –Leu/-Trp/-His plate. The remaining two clones (clone #7 and clone #16) only grew colonies in the presence of the pGBKT7-AIPL1 plasmid and not with the empty pGBKT7 or pGBKT7-Lamic C plasmids, indicating that these clones contain cDNA inserts which encode proteins that specifically interact with AIPL1 (Figure 4). The entire inserts of both clone #7 and #16 were sequenced and *in silico* analysis was performed. Clone #7 contains an insert size of 1kb and is homologous to a glucose metabolism gene on human chromosome 5q. Clone #16 contains an insert size of 1.2kb in length and is homologous to a novel gene which maps to human chromosome 15p.

4. CONCLUSIONS

The results of this study reveal a high degree of sequence conservation across Aipl1 proteins, at least within mammals, and reinforce the structural relationship of Aipl1 to the Fkbp family of proteins, which function as molecular chaperones in steroid receptor signaling, heat shock responses, and immunosuppression (Miyata *et al.*, 1997). Therefore, AIPL1 is likely performing a similar function within the photoreceptors. The increased sequence conservation within the TPR motifs suggests an important role for these motifs in protein function. In addition, the conservation of certain residues within the proline-rich region suggests that they may be important for primate Aipl1 function.

Two potential AIPL1-interacting proteins have been identified, thus far, using yeast two-hybrid screening. Follow-up yeast two-hybrid studies are currently being performed with the remaining 50 clones. Once it is determined which clones specifically interact with AIPL1, *in vitro* binding assays will be performed to confirm the protein-protein interaction. The presence of the three TPR domains within the AIPL1 protein strongly supports the hypothesis that AIPL1 interacts with at least one protein. Therefore it is likely that these studies will identify proteins that interact with AIPL1, which may allow its placement into a biological pathway leading to a better understanding of its functional role within the visual process, and thereby foster a deeper appreciation of the molecular mechanisms underlying LCA.

5. ACKNOWLEDGMENTS

We are particularly grateful to Dr. Wolfgang Baehr for providing the bovine yeast two-hybrid retinal cDNA library, and to Dr. Gus Aguirre and Dr. Susan Semple-Rowland for providing retinal cDNA libraries. We also thank Dr. M.J. Aivaliotis and Dr. John L. VandeBerg for providing primate genomic DNAs. In addition, we would like to thank Ms. Kimber Malone for expert technical assistance. Supported by grants from the Foundation Fighting Blindness and the George Gund Foundation, the William Stamps Farish Fund, the M.D. Anderson Foundation, and the John S. Dunn Foundation; and by grant EY07142 from the National Eye Institute-National Institutes of Health.

REFERENCES

G. Blatch, M. Lassle, The tetratricopeptide repeat: a structural motif mediating protein-protein interactions, *BioEssays* **21**, 932-939 (1999).

C. Hoffman, F. Winston, A ten-minute DNA preparation from yeast efficiently releases autonomous plasmids for transformation of Escherichia coli, *Gene*, **57**,267-272 (1987).

Y. Miyata, B. Chambraud, C. Radanyi,J. Leclerc, M. Lebeau, J. Renoir, R. Shirai, M. Catelli,L. Yahara, E. Baulieu, Phosphorylation of the immunosuppressant FK506-bindingprotein FKBP52 by casein kinase II: Regulation of HSP90-binding activity of FKBP52. *Proc.Natl. Acad. Sci. USA*. **94**, 14500-14505 (1997).

M.M Sohocki, S.J. Bowne, L.S. Sullivan, S. Blackshaw,C.L. Cepko, A.M. Payne, S.S Bhattacharya, S. Khaliq D.G. Birch, W.R. Harrison, F.F.B. Elder, J.R. Heckenlively, S.P. Daiger, Mutations in a new photoreceptor-pineal gene on 17p cause Leber congenital amaurosis, *Nat. Genet.* **24**:79-83 (2000).

M.M Sohocki, K.A. Malone, L.S. Sullivan, S.P. Daiger, Localization of retina/pineal-specific expressed sequences: identification of novel candidate genes for inherited retinal disorders, *Genomics,* **58**:29-33 (1999).

M.M Sohocki, I. Perrault, B. Leroy, A.M. Payne, S. Dharmaraj,, S.S Bhattacharya, J. Kaplan, I.H. Maumenee, R. Koenekoop, D.G. Birch, J.R. Heckenlively, S.P. Daiger, Prevalence of AIPL1 mutations in inherited retinal degenerative diseases, *Mol. Genet. Metabol.* **70**:142-150 (2000a).

M.M Sohocki, L.S. Sullivan, H.A. Mintz-Hittner, D.G. Birch, J.R. Heckenlively, C.A. Freund, R.R McInnes, S.P. Daiger, A range of clinical phenotypes associated with mutations in CRX, a photoreceptor transcription factor gene, *Am. J. Hum. Genet.* **63**:1307-1315 (1998).

M.M. Sohocki,L.S. Sullivan, D.L. Tirpak, S.P. Daiger, Comparative analysis of aryl-hydrocarbon interacting protein-like 1 (Aipl1), a gene associated with inherited retinal disease in humans. Submitted, *Genomics*. (2000b).

SEARCHING FOR GENOTYPE-PHENOTYPE CORRELATIONS IN X-LINKED JUVENILE RETINOSCHISIS

Kelaginamane T. Hiriyanna, Rita Singh-Parikshak, Eve L. Bingham, Jennifer A. Kemp, Radha Ayyagari, Beverly M. Yashar, and Paul A. Sieving*

1. SUMMARY

We have analyzed 145 RS families for mutations in the XLRS1 gene and have sought clinical correlation of phenotype severity with the type and location of mutation in a subset of these families. Some of the RS families exhibited consistency of either severe or mild clinical phenotypes in multiple affected male members. Intrafamilial variability of clinical phenotypes was encountered in affected males of several other RS families.

2. INTRODUCTION

X-linked juvenile retinoschisis (XLRS; RS; MIM 312700) is a bilateral, highly penetrant recessive degenerative retinopathy. It is one of the more common causes of juvenile macular degeneration in males, with an estimated prevalence of 1:5,000 to 1:25,000 (George et al., 1995; Retinoschisis Consortium, 1998; Sieving, 1999). A comprehensive survey in Finland reported the RS carrier prevalence to be 14 per 10,000 (Forsius et al., 1973). RS presents early in life as intraretinal parafoveal cysts with a characteristic spoke-wheel pattern in the macula that impairs central vision. It is unknown whether the phenotype is always present from birth, but some affected infants suffer massive retinal detachment. Progression of severity is variable and can include splitting of the peripheral retina through the superficial retinal layers, namely the inner limiting membrane (ILM) and nerve fiber layers (NFL) (Yanoff et al., 1968; Condon et al., 1986). Foveal schisis through the outer plexiform layer (OPL) of the retina has been reported in

---

*Department of Ophthalmology and Visual Sciences, Kellogg Eye Center, University of Michigan, Ann Arbor, MI 48105

one case (Ando et al., 2000). Carrier females rarely show fundus changes or have any visual symptoms.

In spite of the wide variations in clinical phenotypes, genetic analyses indicate a single Mendelian locus for the disease. XLRS was mapped to a 2-cM region between DXS418 and DXS999 in Xp22.2 (Weber et al., 1995; George et al., 1996; Pawar et al., 1996; Huopaniemi et al., 1997; Walpole et al., 1997). The RS1 gene was cloned in 1997 (Sauer et al., 1997).

The RS1 gene has six exons and encodes a 224 amino acid protein, processed by N-terminal cleavage into a mature protein with a calculated size of 201 amino acids. The RS1 protein contains a discoidin domain which is highly conserved across slime mold to man and is shared with a number of other proteins (Baumgartner et al., 1998; Retinoschisis Consortium, 1998). Human and murine orthologs show 96% identity in predicted amino acid sequence (Gehrig et al., 1999b). The RS1 protein/retinoschisin (Grayson et al., 2000) is a secretable member of the FA58C/discoidin family that typically has a binding protein in the extracellular matrix or on the cell surface of other cells (Fukuzawa and Ochiai, 1996; Reid et al., 1999). The discoidin domain is implicated in cell-cell adhesion and phospholipid binding functions (Retinoschisis Consortium, 1998), which are in agreement with the observed splitting of the retina in XLRS patients.

The gene transcripts are expressed specifically in the retina (Sauer et al., 1997) and in the photoreceptor cells (Reid et al., 1999) albeit a recent report also indicates their presence in human uterine tissue (Huopaniemi et al., 1999a).   Immunocytochemical analysis with antibodies raised against the protein epitopes indicates that the RS1 protein is localized to multiple layers of the neural retina including photoreceptor inner segments (IS), the inner nuclear layer (INL), the inner plexiform layer (IPL) and the ganglion cells (GCL) (Grayson et al., 2000; Molday et al., 2000; Hiriyanna et al., 2001). Our studies further localize the murine RS1 protein in the IS, the outer plexifom layer (OPL), in a subset of the inner nuclear cells and their axons, IPL, as well as in the Müller cell extensions in the outer nuclear layer (ONL) (Hiriyanna et al., 2001). Staining of Müller cell extensions in the outer nuclear layer indicates that the Müller cells may be involved in translocation of the protein across different layers. RS1 is synthesized and secreted from photoreceptors and transported to the target cells (Grayson et al., 2000). This transport is critical to RS1 biological function as indicated by the mutant phenotypes in RS families with a missense mutation in the signal peptide (Hiriyanna et al., 1999b). The gene expression and protein distribution patterns in mouse retina (Reid et al., 1999; Grayson et al., 2000; Livesey et al., 2000) suggest a role for the RS1 gene in the development and distribution of certain interneuronal connections in the retina (Molday et al., 2000; Hiriyanna et al., 2001).

More than 350 RS families worldwide have been identified with 120 different pathological RS1 gene mutations (Sauer et al., 1997; Retinoschisis Consortium, 1998; Gehrig et al., 1999a; Hiriyanna et al., 1999b; Huopaniemi et al., 1999b; Mashima et al., 1999; Mendoza-Londono et al., 1999; Shastry et al., 1999; Shinoda et al., 1999; Eksandh et al., 2000; Inoue et al., 2000) and others (see www.dmd.nl/rsl/html). At the University of Michigan we have screened patients from 145 apparently unrelated RS families and sporadic cases, and identified XLRS1 mutations in 121 of them. Most mutations are missense mutations localized to exons 4-6 of the RS1 gene, a region of the gene that encodes the discoidin domain. This emphasizes the critical role of the discoidin domain in RS1 physiological function (Retinoschisis Consortium, 1998). Clinical indications of RS in carrier females are quite rare, although there is a report that carrier females may have abnormal rod-cone interactions (Arden et al., 1988). Some of the unusual cases of females presenting RS clinical phenotypes as in their affected male siblings are due to mutations on both the alleles (Mendoza-Londono et al., 1999). RS1 is located on Xp22 where many of the genes have been shown to escape X-inactivation (Sudbrak et al., 2001). This may be the case with the RS1 gene also as X-inactivation patterns are similar in RS carrier and non-carrier females and the mutation bearing X chromosome is not preferentially subject to X-inactivation (Huopaniemi et al., 1999a). A skewed secondary sex ratio in the offspring of carriers of the 214G>A mutation of the RS1 gene, however, has been reported for some families in Finland (Huopaniemi et al., 1999a).

Traditionally, fundus examination and electroretinography (ERG) accomplish the clinical diagnosis of X-linked juvenile retinoschisis. RS typically presents a characteristic 'electronegative' ERG with a relatively intact a-wave, but reduced or absent b-wave. This ERG pattern, along with the histopathological evidence, was considered to suggest Müller cell pathology. However, there is evidence in some species, particularly in cat, that the b-wave may originate directly from the bipolar cells (Robson and Frishman 1996). The ERG however, is not an infallible indicator of X-linked retinoschisis as demonstrated in some molecularly confirmed cases of RS patients (Bradshaw et al., 1999; Sieving et al., 1999a). However, the identification of the XLRS gene, as well as advances in sequencing technology, has made the molecular diagnostics more effective and straight forward since it complements other existing clinical skills for early detection of at-risk males. Further, molecular genetic analysis is probably the only means of confirming carrier status in females with familial history. The strength of molecular screening lies in its high sensitivity, specificity, and reliability (Sieving et al., 1999b).

No clear correlations between the gene mutations and the clinical phenotype have been established thus far. A wide variability of clinical phenotypes is often encountered even for affected members belonging to the same family (Retinoschisis Consortium, 1998; Roesch et al., 1998; Bradshaw et al., 1999; Hiriyanna et al., 1999a; Sieving et al., 1999a; Eksandh et al., 2000; Shinoda et al., 2000; Hotta et al., 2001). The purpose of our study was to identify RS families presenting minimal intrafamilial variability, if any, in the clinical features of its affected members and to correlate these phenotypes with the associated RS genotype. We studied 15 apparently unrelated RS families each with multiple affected members, and report here some of the families that suggest an intrafamilial consistency in disease severity.

3. PATIENTS AND METHODS

We examined the genotypes in 145 apparently unrelated XLRS families. Sixty-four patients in fifteen (A-O) of these families, each with multiple affected members were further studied for correlation in severity of clinical phenotypes with the associated gene mutation. Informed consent was obtained from participating individuals. Ophthalmologic examinations included the best-corrected visual acuity measurement, fundus examination, fluorescein angiography, and full field electroretinography (ERG).

3.1. Genotyping and Mutation Analysis

DNA from the lymphocytes of the peripheral blood samples was isolated by standard procedures. Molecular analysis included mapping of the disease to RS locus (Pawar et al., 1996) and mutation screening for the RS1 gene (Retinoschisis Consortium, 1998; Hiriyanna et al., 1999b). The six exons and the flanking sequences were PCR amplified using primers described in Sauer et al. 1997 (Sauer et al., 1997) and sequenced using Thermosequanase cycle sequencing (Amersham, Arlington Heights, Ill) protocols. PCR of the 5' upstream region of the gene containing the putative promoter was carried out using a set of primers derived from the database sequences of the region (Eksandh et al., 2000). One hundred normal chromosomes were screened to identify the polymorphisms and the putative pathological mutations in the RS gene.

3.2. Fundus Examination

XLRS fundus presents characteristic features that are generally bilateral with >95% of cases exhibiting stellate/spokewheel cystic maculopathy. In mild cases, the parafoveal spoke-wheel intraretinal cysts are observed only by careful ophthalmoscopy and are rarely evident on fluorescein angiography (FA). However, some older affected males exhibit macular atrophy of the retinal pigment epithelium, causing hyperfluorescent defects on FA. Visual fields were determined with a Goldman perimeter. Schisis of the peripheral retina disrupts transmission of visual signals and causes a scotoma in the involved areas. Large cystic inner layer elevation with holes and vitreous veils can occur, most commonly involving the inferotemporal. Other complications include mottled RPE atrophy, macular hole, retinal detachment and vitreous hemorrhage, optic nerve neovascularization and neovascular glaucoma.

3.3. Electroretinography

Clinical Ganzfeld electroretinograms were recorded for many but not all affected males using full-field xenon flashes . Pupils were dilated (phenylepherine hydrochloride 10% and tropicamide 1%), and Burian-Allen bipolar corneal electrodes (Hansen Ophthalmic Instruments, Iowa City, Iowa) were placed after topical corneal anesthesia (proparcaine hydrochloride 0.5%). ERGs were recorded with scotopic white flash, scotopic blue flash, photopic white flash and 30Hz flicker beginning after 45 minutes of dark adaptation. Responses were evaluated against normals. The rod-isolated b-wave was elicited with a blue flash of $-0.82 \log \text{cd-s/m}^2$ in the dark-adapted state; this response has very little or nearly no rod a-wave contamination and thus reflects proximal retinal activity.

3.4. Grading Scheme

We classified XLRS disease qualitatively as mild, moderate, or severe based on three factors of central involvement, peripheral involvement, and ERG responses. Variations from normal visual acuity (VA) were taken as one measure of the extent of foveal dysfunction. Peripheral involvement was judged by fundus appearance and visual field abnormalities. The dark-adapted rod-isolated b-wave was measured. Severe disease characteristics included poor visual acuity of 20/200 or worse, extensive peripheral retinoschisis or retinal detachment, and rod ERG b-wave less than 25% of normal amplitude. Moderate disease characteristics included VA of 20/50-20/100, moderate foveal and peripheral schisis, and rod ERG b-wave between 25-80% of normal amplitude. Mild disease had VA of 20/40 or better, mild foveal schisis, and rod ERG b-wave at least 80% of normal.

4. RESULTS

The general vision morbidity profile of our patients included VA of 20/40-20/80 in 64% and VA of $\leq$20/200 in 20%. Foveal changes were seen in 97%, and peripheral schisis occurred in 47%. Retinal detachment was observed in 10% of eyes. An electronegative ERG was found in 92%. The rod ERG amplitudes indicated mild phenotype in 12% of the affected members, moderate in 60%, and severe phenotype in 28% of the affected members. In the following sections, case families are presented in order of most severe to mild disease.

4.1. Families with Severe Disease

Family G has a large deletion involving exons 2-3 and intron 3. All four affected males examined in this family have a severe RS phenotype. Patient #1 by age 5 months developed extensive peripheral schisis that also involved the macula OD and caused vitreous traction and intraretinal hemorrhage OS. Patient #2 at age 17 had VA of 20/400, 20/200 and had had three surgical procedures for retinal detachment repair OS. Patient #3 by age 42 had VA of 20/200, 20/60 with a macular hole OD and severe inferior schisis OS. Lastly, patient #4 at age 72 had VA of 20/200, 20/200 and previous bilateral retinal detachment repair.

Family F has a deletion involving exon 2. The proband had an acuity of 20/200 bilaterally by age five with moderate to extensive parafoveal spoke-wheel cysts and inferior schisis (severe phenotype). Three other affected males from this family were studied subsequently. Their VA ranged from 20/80 to no light-perception (NLP). On average VA was 20/200 and presented a moderate to severe retinal pathology.

Family E has the Arg182Cys mutation and presented a severe phenotype with VA of 20/200 and 20/400, severe foveal cysts, inferior schisis, and a major ERG reduction. Cysteine residues are frequently important for the three-dimensional structure of proteins. There are 10 cysteine residues in XLRS1. Several of them are localized in the evolutionarily conserved discoidin domain whose integrity seems to be critical for the protein function. Creation of new residues by mutational changes, as in this family, also seems to be equally detrimental for protein function (Retinoschisis Consortium, 1998; Hiriyanna et al., 1999b).

Family D has a Glu72Lys mutation. Two affected brothers presented with a severe disease phenotype at a young age. One had extensive schisis by age six months, and the other had vitreous hemorrhage by age six years. This particular mutation is the most frequently found RS1 mutation and has been reported in forty-five families (http://www.dmdnl/rs/rshome), across wide ethnic and geographical boundaries. The prevalence of this mutation suggests that it may have multiple origins. This is further supported by its occurrence in different haplotypes (Retinoschisis Consortium, 1998; Hiriyanna et al., 1999b).

Family C has an IVS4+1G→A nucleotide change causing a splice site mutation. Splice site mutations may result in a truncated protein and thus might cause severe disease. Two affected brothers in this family have early onset severe disease, one had extensive schisis and vitreous veils at four months of age, and the other had 20/400 bilateral acuity by age five.

Family M has a Leu127Pro mutation, a change that is associated with moderate to severe disease, in two of its affected members whom we examined.

Patients in two of our families (A and B) exhibited severe disease but neither a point mutation nor a deletion was found in the exonic and the splice site sequences. We had previously mapped these families to the RS locus (Pawar et al., 1996). Family A has five affected males with moderate to severe acuity loss from foveal schisis and showed extensive inferior schisis, including a three-year-old who had bilateral retinal detachment. Similarly, Family B, which included four affected males, presented a severe phenotype with visual acuities of 20/200 to NLP, foveal atrophy, inferior schisis, and vitreous veils. These two families are potentially important, as they indicate that mutations other than those in the protein coding or splice site sequences can cause RS disease. The 5' upstream regions of the gene including the promoter are now being sequenced in an attempt to identify the gene lesion.

4.2. Families with Mild to Moderate Disease

Families H and I have a Gly70Ser mutation, and family J has a Gly70Arg mutation. Patients from these families had an average visual acuity of 20/80 with mild to moderate foveal schisis without any associated peripheral schisis. The Gly70Ser mutation has previously been reported in 11 apparently unrelated families from the United States of America (Retinoschisis Consortium, 1998; Hiriyanna et al., 1999b). These two mutations affect the same amino acid residue that is localized at the amino terminal region of the functionally critical discoidin domain.

Families K and L also have missense mutations in the discoidin domain and presented a mild to moderate disease phenotype. Family K has the Trp96Arg mutation. Three affected males who were examined had an average VA of 20/60, mild foveal changes, and inferior schisis. Family L with the Thr213Trp mutation has four XLRS males with an average VA of 20/70, mild foveal schisis, and inferior schisis.

4.3. Families with Variable Phenotype

Families N and O also have missense mutations resulting primarily in mild disease. Family N has a mutation (Gly140Arg), which changes a non-polar residue to a positively charged residue in the discoidin domain. In this family with four affected males examined two were found to have mild disease (average VA of 20/50 and subtle foveal changes) while the other two had severe phenotypes with one teenager exhibiting retinal detachment and a 67 year old with late foveal atrophy. Family O has an Ile199Thr mutation in the discoidin domain. There are four affected males with average VA 20/40 and minimal foveal changes with one exception, a 15-month-old with foveal cystic changes and large inferior schisis.

5. DISCUSSION

In the course of this study we have come across a few RS families in which several affected male members had consistently either a severe or a mild phenotype. In family G with an exon 2-3 deletion, for example, all five affected members exhibited a severe form of the phenotype and the youngest member had severe retinal detachment. Similarly a severe phenotype was shown to be associated with a 33-bp deletion at the exon 3-intron 3 boundary (Shinoda et al., 1999; Shinoda et al., 2000). All nine affected members in families A and B also presented a severe clinical phenotype. The disease in these families has been mapped to the RS locus (Pawar et al., 1996), although no pathological genetic lesion was seen in the coding and splice site sequences.   These families are being analyzed for possible mutations in 5'UTR sequences including the putative promoter.

Promoter mutations and deletions may result in non-expression (null allele) or abnormal expression of the gene. Although the effects of abnormal expressions of the RS gene has so far not been documented, the possible complete suppression of the RS1 expression by partial deletions and premature stop codons have not always been associated with severe phenotypes (Retinoschisis Consortium, 1998).   A severe phenotype was associated with a 20 kb deletion involving exon-1 and the promoter region in one study (Shinoda et al., 2000), but a wide variability in RS phenotype was seen in other families reported with a similar mutation (Eksandh et al., 2000; Huopaniemi et al., 2000). The largest reported gene deletion is 136 kb in length and disrupts two other flanking genes, STK9 and PPEF-1, along with exons 1-3 and the promoter region of the RS1 gene. Yet patients in the families with this mutation exhibited a range of mild to moderate retinoschisis phenotype, although one patient in this group is possibly syndromic with mild neurological disorders (Huopaniemi et al., 2000).   A deletion involving exon 2 (Family F) also presented a moderate to severe retinal pathology. Conversely, severe retinoschisis has been reported for some RS families with small deletions (Hiriyanna et al., 1999a; Shinoda et al., 1999;Shinoda et al., 2000). A severe phenotype was also observed in family C with a splice site mutation on intron 4.

Both affected members in family D (Glu72Lys) had a severe phenotype that started at an early age. The prevalence of this mutation in the population is presumably due to its multiple origins (Retinoschisis Consortium, 1998). Several cases have been reported which are associated with either the severe phenotypes or mild phenotypes (Retinoschisis Consortium, 1998; Shinoda et al., 2000; Hotta et al., 2001). In one study a severe phenotype with multiple white flecks in the temporal posterior pole of the eye was observed in two RS patients presenting Glu72Lys mutation (Hotta et al., 2001). In our study a severe phenotype was also found associated with a family presenting the Arg182Cys mutation in affected males. A similar observation was made in two RS families in Japan presenting the same mutation (Shinoda et al., 2000). Our families with Gly70Ser/Arg (H and I), as well as Trp96Arg (K), Gly140Arg (N), and Thr213Trp (L), presented in general mild to moderate phenotypes.

Although there is a considerable variation in the clinical expression of the disease, there is no evidence of genetic heterogeneity. 85%-95% of the tested patients who are clinically suspect for XLRS subsequently are found to have RS1 coding or splice site mutations. In the patients who have clear linkage to XLRS1 but for whom mutations have not yet been found, we are continuing to search for possible pathological mutations or deletions involving the 5' upstream region of the gene.   In our experience, it is not common to find RS families, which present a consistent clinical phenotype in their affected members. However, searching for and studying such families is important, as this may provide clues to the determinants, in addition to RS1 gene mutations, which contribute to clinical severity of this disease.

6. ACKNOWLEDGMENTS

Supported in part by grants from NIH RO1EY 10259 and Foundation Fighting Blindness (PAS), P30 EY07003 (CORE) and Midwest Eye Banks Transplantation Center grants (KTH). The authors thank Ms. Austra Liepa for help in proofreading this manuscript.

REFERENCES

Ando, A., Takahashi, K., Sho, K., Matsushima, M., Okamura, A., and Uyama, M. 2000, Histopathological findings of X-linked retinoschisis with neovascular glaucoma, *Graefes Arch. Clin. Exp.Ophthalmol.,* **238**(1): 1-7.

Arden, G. B., Gorin, M. B., Polkinghorne, P. J., Jay, M., and Bird, A. C. 1988, Detection of the carrier state of X-linked retinoschisis, *Am. J. Ophthalmol.,* **105**(6): 590-595.

Baumgartner, S., Hofmann, K., Chiquet-Ehrismann, R., and Bucher, P. 1998, The discoidin domain family revisited: new members from prokaryotes and a homology-based fold prediction, *Protein Sci.,* **7**(7): 1626-1631.

Bradshaw, K., George, N., Moore, A., and Trump, D. 1999, Mutations of the XLRS1 gene cause abnormalities of photoreceptor as well as inner retinal responses of the ERG, *Doc. Ophthalmol.,* **98**(2): 153-173.

Condon, G. P., Brownstein, S., Wang, N. S., Kearns, J. A., and Ewing, C. C. 1986, Congenital hereditary (juvenile X-linked) retinoschisis. Histopathologic and ultrastructural findings in three eyes, *Arch. Ophthalmol.,* **104**(4): 576-583.

Eksandh, L. C., Ponjavic, V., Ayyagari, R., Bingham, E. L., Hiriyanna, K. T., Andreasson, S., Ehinger, B., and Sieving, P. A. 2000, Phenotypic expression of juvenile X-linked retinoschisis in Swedish families with different mutations in the XLRS1 gene, *Arch. Ophthalmol.,* **118**(8): 1098-1104.

Forsius, H., Krause, U., Helve, J., Vuopala, V., Mustonen, E., Vainio-Mattila, B., Fellman, J., and Eriksson, A. W. 1973, Visual acuity in 183 cases of X-chromosomal retinoschisis, *Can. J. Ophthalmol.,* **8**(3): 385-393.

Fukuzawa, M., and Ochiai, H. 1996, Molecular cloning and characterization of the cDNA for discoidin II of Dictyostelium discoideum, *Plant Cell Physiol.,* **37**(4): 505-514.

Gehrig, A., Weber, B. H., Lorenz, B., and Andrassi, M. 1999a, First molecular evidence for a de novo mutation in RS1 (XLRS1) associated with X linked juvenile retinoschisis [letter], *J. Med. Genet.,* **36**(12): 932-934.

Gehrig, A. E., Warneke-Wittstock, R., Sauer, C. G., and Weber, B. H. 1999b, Isolation and characterization of the murine X-linked juvenile retinoschisis (Rs1h) gene, *Mamm. Genome.,* **10**(3): 303-307.

George, N. D., Payne, S. J., Bill, R. M., Barton, D. E., Moore, A. T., and Yates, J. R. 1996, Improved genetic mapping of X linked retinoschisis, *J. Med. Genet.,* **33**(11): 919-922.

George, N. D., Yates, J. R., and Moore, A. T. 1995, X linked retinoschisis, *Br. J. Ophthalmol.,* **79**(7): 697-702.

Grayson, C., Reid, N. M. S., Ellis, J. A., and al., e. 2000, Retinoschisin, the X-linked retinoschisis protein, is a secreted photoreceptor protein, and is expressed and released by Weri-Rb1 cells., *Hum. Mol. Genet.,* **9**: 1873-1879.

Hiriyanna, K. T., Bingham, E. L., Yashar, B., Richards, J., and Sieving, P. A. 1999a, XLRS1 gene mutation spectrum and correlation to clinical phenotypes (Abstract)., *Invest. Ophthalmol. Vis. Sci.,* **40** (Suppl): S471.

Hiriyanna, K. T., Bingham, E. L., Yashar, B. M., Ayyagari, R., Fishman, G., Small, K. W., Weinberg, D. V., Weleber, R. G., Lewis, R. A., Andreasson, S., Richards, J. E., and Sieving, P. A. 1999b, Novel mutations in XLRS1 causing retinoschisis, including first evidence of putative leader sequence change, *Hum. Mutat.,* **14**(5): 423-427.

Hiriyanna, K. T., Takada, Y., Kondo, M., Bingham, E. L., and Sieving, P. A. 2001, X-linked retinoschisis: RS1 Functional genomics and physiological dysfunction., *Invest. Ophthalmol. Vis. Sci.,* **42** (Suppl): Abs.

Hotta, Y., Makoto, N., Okamoto, Y., Nomura, R., Teraski, H., and Miyake, Y. 2001, Different mutation of the XLRS1 gene causes juvenile retinoschisis with white flecks, *Br. J. Ophthalmol.,* **85**: 238-248.

Huopaniemi, J., Fellman, A., Rantala, A., and et al. 1999a, Skewed secondary sex ratio in the offspring of carriers of the 214G>A mutation of the RS1 gene., *Ann. Hum. Genet.,* **63**: 521-533.

Huopaniemi, L., Rantala, A., Forsius, H., Somer, M., de la Chapelle, A., and Alitalo, T. 1999b, Three widespread founder mutations contribute to high incidence of X-linked juvenile retinoschisis in Finland, *Eur. J. Hum. Genet.,* **7**(3): 368-376.

Huopaniemi, L., Rantala, A., Tahvanainen, E., de la Chapelle, A., and Alitalo, T. 1997, Linkage disequilibrium and physical mapping of X-linked juvenile retinoschisis, *Am. J. Hum. Genet.,* **60**(5): 1139-1149.

Huopaniemi, L., Tyynisma, H., Rantala, A., Rosenberg, T., Alitalo, T. 2000, Characterization of two unusual RS1 gene deletions segregating in Danish retinoschisis families., *Hum. Mutat.,* **16**: 307-314.

Inoue, Y., Yamamoto, S., Okada, M., Tsujikawa, M., Inoue, T., Okada, A. A., Kusaka, S., Saito, Y., Wakabayashi, K., Miyake, Y., Fujikado, T., and Tano, Y. 2000, X-linked retinoschisis with point mutations in the XLRS1 gene, *Arch. Ophthalmol.,* **118**(1): 93-96.

Livesey, F. J., Furukawa, T., Steffen, M. A., Church, G. M., and Cepko, C. L. 2000, Microarray analysis of the transcriptional network controlled by the photoreceptor homeobox gene Crx, *Curr. Biol.,* **10**(6): 301-310.

Mashima, Y., Shinoda, K., Ishida, S., Ozawa, Y., Kudoh, J., Iwata, T., Oguchi, Y., and Shimizu, N. 1999, Identification of four novel mutations of the XLRS1 gene in Japanese patients with X-linked juvenile retinoschisis. *Hum. Mutat.,* **13**(4): 338, http://journals.wiley.com/1059-7794/pdf/mutation/234.pdf.

Mendoza-Londono, R., Hiriyanna, K. T., Bingham, E. L., Rodriguez, F., Shastry, B. S., Rodriguez, A., Sieving, P. A., and Tamayo, M. L. 1999, A Colombian family with X-linked juvenile retinoschisis with three affected females finding of a frameshift mutation, *Ophthalmic Genet.,* **20**(1): 37-43.

Molday, R., Molday, L. L., Sauer, C. G., Hicks, D., and Weber, B. H. F. 2000, Characterization and localization of RS1, the protein encoded by the gene for X-linked juvenile retinoschisis, *Invest. Ophthalmol. Vis. Sci.,* **41** (Suppl)(4): 1736.

Pawar, H., Bingham, E. L., Hiriyanna, K., Segal, M., Richards, J. E., and Sieving, P. A. 1996, X-linked juvenile retinoschisis: localization between (DXS1195, DXS418) and AFM291wf5 on a single YAC, *Hum. Hered.,* **46**(6): 329-335.

Reid, S. N., Akhmedov, N. B., Piriev, N. I., Kozak, C. A., Danciger, M., and Farber, D. B. 1999, The mouse X-linked juvenile retinoschisis cDNA: expression in photoreceptors, *Gene,* **227**(2): 257-266.

Retinoschisis Consortium. 1998, Functional implications of the spectrum of mutations found in 234 cases with X-linked juvenile retinoschisis. The Retinoschisis Consortium, *Hum. Mol. Genet.,* **7**(7): 1185-1192.

Roesch, M. T., Ewing, C. C., Gibson, A. E., and Weber, B. H. 1998, The natural history of X-linked retinoschisis, *Can. J. Ophthalmol.,* **33**(3): 149-158.

Sauer, C. G., Gehrig, A., Warneke-Wittstock, R., Marquardt, A., Ewing, C. C., Gibson, A., Lorenz, B., Jurklies, B., and Weber, B. H. 1997, Positional cloning of the gene associated with X-linked juvenile retinoschisis, *Nat. Genet.,* **17**(2): 164-170.

Shastry, B. S., Hejtmancik, F. J., and Trese, M. T. 1999, Recurrent missense (R197C) and nonsense (Y89X) mutations in the XLRS1 gene in families with X-linked retinoschisis, *Biochem. Biophys. Res. Commun.,* **256**(2): 317-319.

Shinoda, K., Ishida, S., Oguchi, Y., and Mashima, Y. 2000, Clinical characteristics of 14 Japanese patients with X-linked juvenile retinoschisis associated with XLRS1 mutation., *Ophthalmic Genet.,* **21**: 171-180.

Shinoda, K., Mashima, Y., Ishida, S., and Oguchi, Y. 1999, Severe juvenile retinoschisis associated with a 33-bps deletion in XLRS1 gene, *Ophthalmic Genet.,* **20**(1): 57-61.

Sieving, P. A., 1999, Juvenile retinoschisis. in: *Genetic Diseases of the Eye,* E. I. Trabolusi ed., Oxford University Press, New York, pp. 347-355.

Sieving, P. A., Bingham, E. L., Kemp, J., Richards, J., and Hiriyanna, K. 1999a, Juvenile X-linked retinoschisis from XLRS1 Arg213Trp mutation with preservation of the electroretinogram scotopic b-wave, *Am. J. Ophthalmol.,* **128**(2): 179-184.

Sieving, P. A., Yashar, B. M., and Ayyagari, R. 1999b, Juvenile retinoschisis: a model for molecular diagnostic testing of X-linked ophthalmic disease, *Trans. Am. Ophthalmol. Soc.,* **97**: 451-469.

Sudbrak, R., Wieczorek, G., Nuber, U. A., Mann, W., Kirchner, R., Erdogan, F., Brown, C. J., Wohrle, D., Sterk, P., and et al. 2001, X chromosome-specific cDNA arrays: Identification of genes that escape from X-inactivation and other applications, *Hum. Mol. Genet.,* **10**: 77-83.

Walpole, S. M., Nicolaou, A., Howell, G. R., Whittaker, A., Bentley, D. R., Ross, M. T., Yates, J. R., and Trump, D. 1997, High-resolution physical map of the X-linked retinoschisis interval in Xp22, *Genomics,* **44**(3): 300-308.

Weber, B. H., Janocha, S., Vogt, G., Sander, S., Ewing, C. C., Roesch, M., and Gibson, A. 1995, X-linked juvenile retinoschisis (RS) maps between DXS987 and DXS443, *Cytogenet. Cell Genet.,* **69**(1-2): 35-37.

Yanoff, M., Kertesz Rahn, E., and Zimmerman, L. E. 1968, Histopathology of juvenile retinoschisis, *Arch. Ophthalmol.,* **79**(1): 49-53.

RP1 MUTATION ANALYSIS

Sara J. Bowne[1], Stephen P. Daiger[1,2], Kimberly A. Malone[1], Jin Zuo[2,3], Kyeongmi Cheon[3], David G. Birch[4], Dianna Hughbanks-Wheaton[4], John R. Heckenlively[2,5], Debora B. Farber[2,5], Eric A. Pierce[2,6], Shomi S. Bhattacharya[7], Chris F. Inglehearn[8], Lori S. Sullivan[1,2]

1. INTRODUCTION

Retinitis pigmentosa (RP) is a heterogeneous group of inherited retinal diseases which affects approximately 1 in 3500 people worldwide. RP is initially characterized by night blindness followed by progressive degeneration of the retina, often culminating in legal or complete blindness in the later decades of life (Heckenlively et al., 1997). To date 11 autosomal dominant (adRP), 13 autosomal recessive (arRP), 5 X-linked (xlRP), and one digenic form of retinitis pigmentosa have been identified (RetNet). While several disease-associated genes have been identified, the majority of genes remain unknown. Recently we identified the gene and mutations responsible for the RP1 form of autosomal dominant retinitis pigmentosa located on chromosome 8q (Sullivan et al., 1999; Pierce et al., 1999, Guillonneau et al., 1999).

The RP1 locus was first identified using a large nine-generation American family and was subsequently refined to a 4cM interval between D8S601 and D8S285 using a second family (Fields et al., 1982, Heckenlively et al., 1982; Blanton et al., 1991; Xu et al., 1996). The RP1 gene was discovered using positional candidate cloning and differential display analysis. It is arranged in four exons with the initiation codon being present in exon 2. The RP1 mRNA appears to be retinal specific, is approximately 7000bp long, and encodes a protein 2156 amino acids in length with a calculated

[1]Human Genetics Center and Dept. of Ophthalmology, Univ. of Texas-Houston; [2]The RP1 Consortium; [3]St. Jude Children's Research Hospital, Memphis, TN; [4]Retina Foundation of the Southwest, Dallas, TX; [5]Jules Stein Eye Inst. UCLA; [6]Scheie Eye Inst., Univ. of Penn.; [7] Inst. of Ophthalmology, Univ. College, London, UK; [8]Molecular Medicine Unit, Leeds Univ., UK.

molecular weight of 240kD. The gene product shows partial sequence similarity to the human doublecortin gene (DCX) and its expression is believed to be regulated by retinal oxygen levels; however, exact functional information is unknown at this time (Sullivan et al., 1999; Pierce et al., 1999, Guillonneau et al., 1999).

The three initial studies reported a total of four different mutations in RP1 that cause adRP. One mutation, Arg677X, was found in the both of the original RP1 families that mapped to 8q. An additional nonsense mutation, Gln679X, and two small deletions were also identified in these initial studies (Sullivan et al., 1999; Pierce et al., 1999; Guillonneau et al., 1999). The purpose of this study was to determine the types and frequencies of mutations in RP1 that are responsible for inherited retinal degenerations.

2. MATERIALS AND RESULTS

Our laboratory tested 250 probands with adRP and 409 probands with other forms of retinopathy for mutations in RP1. Initially 56 probands were tested by a combination of SSCP and sequencing for mutations in the entire RP1 gene. All 56 probands were members of American families with adRP who tested negative for mutations in rhodopsin, peripherin/RDS, and CRX. This analysis identified two different RP1 mutations in three families. These two mutations, Arg677X and a 5-bp deletion of bases 2285-2289, were also identified in other families in the initial RP1 studies.

After this preliminary analysis indicated that mutations seemed to cluster in a small region of exon 4, we tested the remaining probands by SSCP for mutations in this small region only. These probands were comprised of individuals from Europe and America with cone or cone-rod dystrophy, autosomal dominant RP, autosomal recessive RP, isolated RP, or other forms of retinopathy. The segment tested spans 442 bp from nucleotide 1947 to 2388 (Genbank AF143222). All SSCP variants were sequenced using an ABI Prism 310 Genetic Analyzer (Perkin Elmer) to determine the underlying variant or mutation.

In total we identified 8 different nonsense and frameshift mutations in 17 of the 250 adRP probands tested (Table 1). No mutations were identified in probands with other forms of retinopathy. All the mutations identified result in a truncated *RP1* protein about one-third the size of the wild-type protein. Two missense variants, Leu1808Pro (5423T→C) and Lys663Asn (1989G→T), were also detected in two different probands. Neither of these variants was seen in unaffected controls. Recent testing of additional affected family members determined that the Leu1808Pro variant does not segregate with disease. No additional family members were available to test segregation of the Lys663Asn variant with disease.

Subsequent to our analysis, several other laboratories have conducted RP1 mutation screens in different cohorts of patients. The mutations identified in these studies are also summarized in table 1. In addition to the disease-causing mutations, a large number of polymorphic or non-disease-causing variants in RP1 have been identified. These variants and their frequencies are described in table 2 (Grimsby et al., 2000; Jacobson et al., 2000; Payne et al., 2000; Tam et al., 2000).

Table 1. Summary of RP1 mutations.

cDNA change	Protein Change	Phenotype	Number of probands	References
1498-1499insGT	Frameshift Met500→termination codon at 532	adRP	1	Payne et al., 2000.
1 bp insertion at unknown position	Frameshift Pro658 → termination	adRP	1	Jacobsen et al., 2000.
2029del[*]	Frameshift Arg677→ termination codon at 681	adRP	1	Bowne et al., 1999.
2029C→T (CGA→TGA)[*]	Arg677X	adRP	40	Jacobsen et al., 2000; Bowne et al., 1999; Sullivan et al., 1999; Guillonneau et al., 1999; Pierce et al., 1999; Grimsby et al., 2000; Payne et al., 2000.
2035C→T (CAA→TAA)	Gln679X	adRP	2	Sullivan et al., 1999; Grimsby et al., 2000.
2098G→T (GAG→TAG)[*]	Glu700X	adRP	1	Bowne et al., 1999.
2168-2181del[*]	Frameshift Gly724 → termination codon at 730	adRP	2	Bowne et al., 1999.
2167G→T (GGA→TGA)	Gly723X	adRP	2	Grimsby et al., 2000.
2168-2181del	Frameshift Ile725 →termination codon at 730	adRP	2	Payne et al., 2000.
2169-2170insG[*]	Frameshift Ile725 → termination codon at 728	adRP	1	Bowne et al., 1999.
1 bp deletion at unknown position	Frameshift Glu729 → termination codon at 737	adRP	1	Grimsby et al., 2000.
2206-2207insT	Frameshift Thr736→termination codon at 739	adRP	1	Payne et al., 2000.
2232T→A (TGT→TGA)[*]	Cys744X	adRP	1	Bowne et al., 1999.
2239del	Frameshift Ser747 → termination codon at 762	adRP	1	Jacobsen et al., 2000.
2285-2289del[*]	Frameshift Leu762 →termination codon at778	adRP	12	Jacobsen et al., 2000; Bowne et al, 1999; Pierce et al., 1999; Grimsby et al., 2000; Payne et al., 2000.

Table 1 (continued). Summary of RP1 mutations.

cDNA change	Protein Change	Phenotype	Number of probands	References
2287-2290	Frameshift Asn763→termination codon at 773	adRP	2	Pierce et al., 1999; Grimsby et al., 2000.
2303del[*]	Frameshift Lys769 →termination codon at 774	adRP	1	Bowne et al., 1999.
2608-2609insA	Frameshift Arg872 →termination codon at 873	adRP	1	Payne et al., 2000.
3157del	Frameshift Tyr1053→termination codon at 1056	adRP	1	Jacobsen et al., 2000.
652G→A	Ala218Thr	? adRP	1	Grimsby et al., 2000.
1989G→T[*]	Lys663Asn	? adRP	1	Bowne et al., 1999.
5378C →T	P1793S	? adRP	1	Payne et al., 2000.
6338C →A	T2113N	? adRP	1	Payne et al., 2000.

[*]Mutations identified in this study.

3. CONCLUSIONS

To date a total of 20 different RP1 mutations have been reported in 74 adRP probands. All of these are nonsense mutations or indels that result in a truncated RP1 protein 1/3 the size of the wildtype. Mutations have only been identified in individuals with adRP indicating that mutations in RP1 are not a significant cause of other forms of retinopathy. The most common RP1mutation, Arg677X, was found in 40 probands and accounts for 54% of the mutations identified. The second most common mutation, a 5-bp deletion of bases 2285-2289, was found in 12 of the probands and accounts for 16% of the RP1 mutations identified. All of the remaining 18 mutations were only seen in 1 or 2 probands each.

In addition to the 20 disease-causing mutations, 4 potential disease-causing mutations have been identified in adRP samples (Ala218Thr, Lys663Asn, Pro1793Ser, Thr2113Asn). In each instance, there are no additional family members available for testing to determine if the mutation segregated with disease. Each of these residues is conserved in baboon, squirrel monkey, and cow and all the but threonine at residue 2113 is conserved in mouse (Malone et al., 2000). The conservation of these bases in other species may indicate that they play an important role in RP1 protein function. Further investigation in needed to determine conclusively if these mutations are responsible for retinal degeneration or if they are just rare variants.

The RP1 gene appears to tolerate a large degree of variation since 25 different benign variants have been identified (Table 2). Seven of these variants alter the amino acid sequence of RP1 and occur at polymorphic levels in the population. It is possible that these amino acid variants play a modifier role in the wide range of clinical phenotypes seen among adRP patients with RP1 mutations. Further research is needed to test this hypothesis.

Collectively RP1 mutations cause 7-10% of adRP and are the second most common known cause of adRP next to mutations in rhodopsin. Data summarized here demonstrates that the RP1 protein must be significantly affected to cause disease and that most missense mutations in RP1 are not a likely cause of disease. Further study is needed to determine the exact function of RP1 in both the normal and disease states.

4. ACKNOWLEDGEMENTS

This work was supported by NEI-NIH grant EY07142; and by grants from the Foundation Fighting Blindness, the George Gunn Foudation, the William Stamps Farish Fund, the M.D. Anderson Foundation, the John S. Dunn Foundation and Mr. Alfred Lasher, III.

Table 2. Benign variants found in RP1.

CDNA change	Protein change	Number of Probands or Frequency	References
-315T→C	none	<.01	Bowne et al., 1999
228C→?	Leu76Leu	<.01	Jacobsen et al., 2000
279G→?	Thr93Thr	<.01	Jacobsen et al., 2000
502C→G	Arg168Gly	1	Grimsby et al., 2000
616-6T→C	none	1	Payne et al., 2000; Jacobsen et al., 2000
788-68T→C	none	.43	Payne et al., 2000
1118C→T	T373I	2	Grimsby et al., 2000; Payne et al., 2000
1510T→G	Ser504Ala	<.01	Jacobsen et al., 2000
2379A→G	Arg793Arg	.03	Unpublished
2615G→A	Arg872His	.25	Sullivan et al., 1999
2953A→T	Asn985Tyr	.46	Sullivan et al., 1999
3012C→G	Leu1004Leu	<.01	Unpublished
3215A→G	Asp1072Gly	1	Grimsby et al., 2000
4249C→G	Leu1417Val	<.01	Jacobsen et al., 2000
4250T→C	Leu1417Pro	2	Grimsby et al., 2000
4784G→A	Arg1595Gln	.01	Jacobsen et al., 2000
5008G→A	Ala1670Thr	.22	Sullivan et al., 1999
5071T→C	Ser1691Pro	.23	Sullivan et al., 1999
5195A→G	Gln1725Gln	.20	Sullivan et al., 1999
5377C→T	Pro1793Ser	1	Grimsby et al., 2000
5423T→C	Leu1808Pro	<.01	Bowne et al., 1999
5797C→T	Arg1933X	1	Tam et al., 2000
6098G→A	Cys2033Tyr	.54	Sullivan et al., 1999
6196G→A	Asp2066Asn	1	Grimsby et al., 2000
6366T→C	Ile2122Ile	<.01	Bowne et al., 1999

REFERENCES

Blanton, S.H., Heckenlively, J.R., Cottingham, A.W., Friedman, J., Sadler, L.A., Wagner, M., Friedman, L.H., Daiger, S.P., 1991, Linkage mapping of autosomal dominant retinitis pigmentosa (RP1) to the pericentric region of human chromosome 8, *Genomics* 11:857-869.

Bowne, S.J., Daiger, S.P., Hims, M.M., Sohocki, M.M., Malone, K.A., McKie, A.B., Heckenlively,J.R., Birch, D.G., Inglehearn, C.F., Bhattacharya, S.S., Bird, A., Sullivan, L.S., 1999. Mutations in the RP1 gene causing autosomal dominant retinitis pigmentosa, *Hum. Mol. Genet.* 11:2121-2128.

Field, L.L., Heckenlively, J.R., Sparkes, R.S., Garcia, C.A., Farson, C., Zedalis, D., Sparkes, M.C., Crist, M., Tideman, S., Spence, M.A., 1982, Linkage analysis of five pedigrees affected with typical autosomal dominant retinitis pigmentosa, *J. Med. Genet.* 19:266-70.

Grimsby, J.L., Adams, S.M., McGee, T.L., Pierce, E.A., Berson, E.L., Dryja, T.P., 2000, Mutations in the RP1 gene causing dominant retinitis pigmentosa, *Invest. Ophthalmol. Vis. Sci.* 41:A1004.

Guillonneau, X., Piriev, N.I., Danciger, M., Kozak,C.A., Cideciyan, A.V., Jacobson, S.G., Farber, D.B., 1999, A nonsense mutation in a novel gene is associated with retinitis pigmentosa in a family linked to the RP1 locus, *Hum. Mol. Genet.* 8:1541-1546.

Heckenlively, J.R., and Daiger, S.P., 1997, *Hereditary Retinal and Choroidal Degenerations*, Rimon, D.L., Conner, J.M. and Pyeritz, R.E. eds., Churchill Livingston, New York, pp. 2555-2576.

Heckenlively, J.R., 1988, *Retinitis Pigmentosa*, Lippincott , Philadelphia.

Jacobson, S.G., Cideciyan, A.V., Iannaccone, A., Weleber, R.G., Fishman, G.A., Maguire, A.M., Affatigato, L.M., Bennett, J., Pierce, E.A., Danciger, M., Farber, D.B., Stone E.M., 2000, Disease expression of RP1 mutations causing autosomal dominant retinitis pigmentosa, *Invest. Ophthalmol. Vis. Sci.* 41:1898-1908.

Malone, K.A., Zuo, J., Farber, D.B., Heckenlively, J.R., Pierce, E.A., Inglehearn, C.F., Daiger, S.P., Sullivan, L.S., 2000, Comparative sequencing of RP1 to identify evolutionarily conserved region of the protein, *Am. J. Hum. Genet.* 67:A1061.

Payne, A., Vithana, E., Khhaliq, S., Hammed, A., Jane, D., Abu-Safieh, L., Kermani, S., Leroy, B.P., Mehdi, S. Q., Moore, A.T., Bird, A.C., Bhattacharya S., 2000, RP1 protein truncating mutations predominate at the RP1 adRP locus, *Invest. Ophthalmol. Vis. Sci.* 41:4069-4073.

Pierce, E.A., Quinn, T., Meehan, T., McGee, T.L., Berson, E.L., Dryja, T.P., 1999, Mutations in a gene encoding a new oxygen-regulated photoreceptor protein cause dominant retinitis pigmentosa, *Nat. Genet.* 22:248-254.

RetNet; http://www.sph.uth.tmc.edu/retnet

Sullivan, L.S., Heckenlively, J.R., Bowne, S.J., Zuo, J., Hide, W.A., Gal, A., Denton, M., Inglehearn, C.F., Blanton, S.H., Daiger, S.P., 1999, Mutations in a novel retina-specific gene cause autosomal dominant retinitis pigmentosa, *Nat. Genet.* 22:255-259.

Tam, P.O.S., Zhang, X.L., Lam, D.S.C., Yeung, K.Y., Fan, D.S.P., Pang, C.P., 2000, RP1 mutations in Chinese retinitis pigmentosa patients, *Invest. Ophthalmol. Vis. Sci.* 41:A1003.

Xu, S.Y., Denton, M., Sullivan, L., Daiger, S.P., Gal, A., 1996, Genetic mapping of RP1 on 8q11-q21 in an Australian family with autosomal dominant retinitis pigmentosa reduces the critical region to 4 cM between D8S601 and D8S285, *Hum. Genet.* 98:741-743.

THE MOLECULAR BASIS OF ACHROMATOPSIA

Susanne Kohl, Herbert Jägle, Eberhart Zrenner, Lindsay T. Sharpe, and Bernd Wissinger[*]

1. SUMMARY

Achromatopsia is a rare genetic heterogenic disorder with known loci on chromosome 2q11 (*ACHM2*) and chromosome 8q21 (*ACHM3*). These loci encode the genes for the channel-forming α- (*CNGA3*) and the modulatory β-subunit (*CNGB3*) of the cone photoreceptor cGMP gated channel – the final component of the cone photoreceptor transduction cascade. Candidate gene screening of these genes in patients affect by Achromatopsia resulted in the identification of a large number of predominantly missense mutations in the *CNGA3* gene and a discrete number of mostly nonsense mutations in the *CNGB3* gene. Mutations in both genes result in Achromatopsia with clinically indistinguishable phenotypes.

2. INTRODUCTION

Complete Achromatopsia (syn.: Rod Monochromacy, Total Colorblindness; OMIM 262300, 216900) is an autosomal recessively inherited congenital disorder with an incidence of 1:30.000 (Francois, 1961). It is characterized by the absence of color discrimination, low visual acuity (<0.2), photophobia and nystagmus. In electroretinographic recordings cone function cannot be established (for review see Hess et al., 1990).

We have reported recently that mutations in the *CNGA3* gene encoding the α-subunit of the cone photoreceptor cGMP gated channel on chromosome 2q11 (*ACHM2*) are responsible for this disease (Kohl et al., 1998). Now we were able to show that mutations in the *CNGB3* gene encoding the putative modulatory β-subunit of the cone

[*] Susanne Kohl, Bernd Wissinger, Molecular Genetics Laboratory, University Eye Hospital, Auf der Morgenstelle 15, 72076 Tuebingen, Germany. Herbert Jägle, Lindsay T. Sharpe, Psychophysics Laboratory, University Eye Hospital, Calwerstr. 7/7, 72076 Tuebingen, Germany. Eberhart Zrenner, University Eye Hospital, Schleichstr. 12-16, 72076 Tuebingen, Germany.

photoreceptor cGMP gated channel cause Achromatopsia linked to chromosome 8q21 (*ACHM3*) (Kohl et al., 2000).

CNG channels are expressed as heterotetramers which consist of channel-forming α- and modulatory β-subunits (MacKinnon, 1991; Liman et al., 1992). The cone photoreceptor cGMP gated cation channel (CNG) is located in the membrane of the outer segments of the cone photoreceptor. In the dark it is kept open by high intracellular levels of cGMP and is permeable for mono- and divalent cations (dark current). Upon illumination the cone pigments are excited and activate transducin which in turn activates the photoreceptor phosphodiesterase. This enzyme catalyses the hydrolytic cleavage of cGMP to 5'GMP and decreases the intracellular cGMP-level. The channel misses its ligand and closes. This generates a membrane hyperpolarization signal which is transmitted to the inner segment where it decreases the steady-level of glutamate release at the cone synapse (Müller &Kaupp, 1998).

Daylight and color vision in human depends on the presence of three different cone types (trichromatic vision). Their spectral specificity is determined by the expression of different opsins which convey their distinct spectral sensitivity: short, middle and long wavelength sensitive cone pigments – also referred to as blue, green and red, respectively.

The functional loss of all three types of cones in patients affected by Achromatopsia suggests that both the *CNGA3* gene and the *CNGB3* gene are commonly expressed and functional in red and green cones as well as in the evolutionary distant blue cones.

Here, we report the results of the mutation screening of the *CNGA3* and the *CNGB3* gene in Achromatopsia patients.

3. MATERIAL AND METHODS

3.1. Patients And Clinical Examination

Patients were recruited and clinically examined in several clinical centers in Germany, the Netherlands, Denmark, Italy, Sweden, France, Turkey and the United States. The ophthalmologic examination included color vision testing, visual acuity testing, electroretinography (scotopic and photopic ERG recordings), funduscopy, dark adaptation curve, presence/absence of nystagmus and photophobia, as well as onset and course of symptoms.

Venous blood was taken from patients and available family members after informed consent. Total genomic DNA was extracted according to standard procedures.

3.2. Mutation Screening

Exons and flanking intronic sequences of the *CNGA3* and the *CNGB3* gene were amplified by PCR from total genomic DNA (Kohl et al., 1998; Kohl et al., 2000) and subjected to direct DNA sequencing or SSCP analysis. For SSCP analysis, PCR products were separated on 10% PAA gels (with or without 10% glycerol) in 1x TBE at 4°C or room temperature followed by silver-staining. For direct sequencing PCR products were purified by Centricon-100 ultrafiltration (Amicon, Bedford, USA) or PCR Purification columns (Qiagen, Hilden, Germany; GibcoBRL Lifetechnologies, Paisley, GB) and subjected to DNA sequencing employing the AmpliTaq FS Dye Terminator or in later stages the BigDye Terminator Chemistry (Applied Biosystems, Weiterstadt, Germany). Samples were separated on an ABI 373A or an ABI 377 DNA sequencer (Applied Biosystems, Weiterstadt, Germany). Sequences were analyzed for mutations by visual inspection and computer-aided alignments employing Lasergene software (DNAStar, Madison, WI). Screening for particular mutations in family members and exclusion of all mutations in 100 healthy control individuals (=200 chromosomes) was performed either by RFLP with appropriate restriction enzymes, by SSCP or by direct sequencing of PCR products.

4. RESULTS

4.1. Mutations In The *CNGA3* Gene

We have screened over 250 patients with a clinical presentation compatible with Complete or Incomplete Achromatopsia, Rod Monochromacy, stationary Cone Dysfunction as well as 45 cases with Cone Dystrophy for mutations in the *CNGA3* encoding the channel-forming α-subunit of the cone photoreceptor cGMP gated channel.

Sequencing analysis of the *CNGA3* gene resulted in the identification of 35 new *CNGA3* mutations in 51 independent pedigrees in addition to the eight mutations in five independent families reported previously (Kohl et al., 1998) (Table 1). Mutations on both mutant chromosomes were identified in 45 patients including fifteen cases with homozygous mutations and 30 cases with compound-heterozygous mutations. Only single mutations were identified in six additional families.

The vast majority of all known *CNGA3* mutations (36/43) are simple amino acid substitutions in comparison with only four stop codon mutations, two 1bp insertions and one 3bp in-frame deletion.

All missense mutations affect highly conserved amino acid residues and were excluded in 100 healthy controls (200 chromosomes). Segregation analysis was performed in all available families and was found consistent with the autosomal recessive mode of inheritance and proved the independent segregation of the mutant alleles.

All mutations but one (144insG) are located in the last three coding exons. These exons encode for almost 2/3 of the CNGA3 polypeptide and contain all functional important domains of the CNG channel protein, i.e. transmembrane helices 1-6, the pore and the cGMP-binding site. A clustering of mutations can be observed at the carboxyterminal part of transmembrane helix S1, transmembrane helix S4 and parts of the cGMP-binding domain.

Several mutations have been observed recurrently in independent families. These are R277C (9 chromosomes), R283W (17 chromosomes), R436W (6 chromosomes) and F547L (12 chromosomes).

Table 1. Spectrum of *CNGA3* mutations.

Location	Alteration Nucleotide sequence[1]	Alteration Polypeptide[2]	No. Chromosomes
Exon 2	144insG	G49fs	1
Exon 5	604A>T	D162V	1
Exon 5	607C>T	P163L	3
Exon 5	542A>G	Y181C	2
Exon 5	544A>T	N182Y	1
Exon 5	556C>T	L186F	1
Exon 6	572G>A	C191Y	3
Exon 6	580G>A	E194K	1
Exon 6	667C>T	R223W	3
Exon 6	671C>G	T224R	1
Exon 7	778G>A	D260N	1
Exon 7	800G>A	G267D	1
Exon 7	829C>T	R277C	9
Exon 7	830G>A	R277H	2
Exon 7	847C>T	R283W	17
Exon 7	848G>A	R283Q	2
Exon 7	872C>G	T291R	1
Exon 7	934-936delATC	I312del	3
Exon 7	947G>A	W316X	1
Exon 7	1021T>C	S341P	1
Exon 7	1106C>G	T369S	1
Exon 7	1114C>T	P372S	3
Exon 7	1139T>C	F380S	1
Exon 7	1217T>C	M406T	1
Exon 7	1228C>T	R410W	3
Exon 7	1279C>T	R427C	3
Exon 7	1306C>T	R436W	6
Exon 7	1320G>A	W440X	1
Exon 7	1350insG	V451fs	1
Exon 7	1412A>G	N471S	1
Exon 7	1454A>T	D485V	1
Exon 7	1547G>A	G516E	1
Exon 7	1565T>C	I522T	1
Exon 7	1574G>A	G525D	1
Exon 7	1585G>A	V529M	1
Exon 7	1609C>T	Q537X	1
Exon 7	1641C>A	F547L	12
Exon 7	1669G>A	G557R	2
Exon 7	1688G>A	R563H	3
Exon 7	1694C>T	T565M	1
Exon 7	1718A>G	Y573C	1
Exon 7	1777G>A	E593K	1
Exon 7	1963C>T	Q655X	1

[1] Sequence position within the *CNGA3* cDNA with nt1 denoting the first nucleotide of the ATG start codon of Genbank entry AF065314. [2] fs - frameshift.

4.2. Mutations In The *CNGB3* Gene

Complete sequencing analysis of the eighteen exons of the *CNGB3* gene was performed in eleven independent families with Complete Achromatopsia linked to chromosome 8q21. We identified six mutations including one missense mutation, two stop codon mutations, a 1bp and a 8bp deletion, as well as a putative splice site mutation of intron 13 (Kohl et al., 2000) (Table 2).

Mutations on both mutant chromosomes were identified in nine out of the eleven families. In addition, only a single heterozygous mutation – the 1bp deletion – was observed in one family. Homozygous mutations were found in seven cases and compound-heterozygosity in the remaining two families. One patient did not show any mutation in the *CNGB3* gene. Since this patient had already been excluded in the *CNGA3* gene, at least one additional Achromatopsia locus and further genetic heterogeneity has to exist.

In contrast to *CNGA3* mutations the vast majority of *CNGB3* mutations are nonsense mutations (two stop codons, two frame-shifts, one splice defect) generating truncated proteins with probably no residual function. Only a single homozygous missense mutations has been observed.

Surprisingly, two recurrent, probably very frequent mutations were observed. The 1bp deletion T383fs constituted eleven out of nineteen identified mutant alleles and was therefore responsible for 58% of all disease chromosomes. The other frame-shift mutation P273fs accounted for seven disease chromosomes and was found in 37% of all mutant alleles.

All mutations were excluded in 100 healthy control individuals (200 chromosomes). Segregation analysis was performed in all families and was found consistent with the autosomal recessive mode of inheritance and proved the independent segregation of the mutant alleles.

Table 2. Spectrum of *CNGB3* mutations.

Location	Alteration Nucleotide Sequence[1]	Alteration Polypeptide[2]	No. Chromosomes
Exon 5	607C>T	R203X	1
Exon 6	819-826del	P273fs	7
Exon 9	1006G>T	E336X	2
Exon 10	1148delC	T383fs	11
Exon 11	1304C>T	S435F	2
Intron 13	1578+1G>A	Splice defect	2

[1] Sequence position within the *CNGB3* cDNA with nt1 denoting the first nucleotide of the ATG start codon in Genbank entry AF272900. [2] fs - frameshift.

5. DISCUSSION

We have recently shown that mutations in the *CNGA3* and the *CNGB3* genes encoding the α- and the β-subunit of the cone photoreceptor cGMP gated channel cause

Achromatopsia linked to chromosome 2q11 (*ACHM2*) (Kohl et al., 1998) and chromosome 8q21 (*ACHM3*) (Kohl et al., 2000; Sundin et al., 2000), respectively.

It has been suggested that native CNG channels represent heterotetramers of two α- and two β-subunits (Liu et al., 1998). In heterologous expression systems the α-subunit is able to form functional cation channels (Wissinger et al., 1997). In contrast, the β-subunit alone cannot conduct measurable ion currents although the secondary structure of the polypeptide is conserved between α- and β-subunits, and relevant functional domains – six transmembrane helices, pore and cyclic nucleotide binding domain – are also present in the β-subunit (Chen et al., 1993; Gerstner et al., 2000). Since mutations in either of the two genes give rise to a clinically indistinguishable phenotype of Achromatopsia, it has to be assumed that the β-subunit confers some essential function to the native channel complex within the context of phototransduction.

Here we present a first overview about mutations in *CNGA3* and *CNGB3* involved in Achromatopsia.

Forty-three different mutations were identified in the *CNGA3* gene. The vast majority of *CNGA3* mutations (36/43) are amino acid substitutions compared to only four stop codon mutations, two 1bp insertions and one 3bp in-frame deletion.

All mutations but one were observed at the aminoterminal 2/3 part of the CNGA3 polypeptide. A clustering of mutations can be observed in the evolutionary conserved structural and functional domains of the channel, i.e. transmembrane helix S4, pore and cGMP binding site. The S4 transmembrane helix contains homologous sequence motifs to the voltage sensor motif in the closely related voltage-dependent K^+-channels.

Several mutations in the *CNGA3* gene were identified recurrently, i.e. R277C, R283W, R436W and F547L. The high prevalence of the R283W mutation is in part due to homozygosity, whereas the other three mutations were almost exclusively identified compound-heterozygously. Interestingly the majority of patients with R283W mutations originate from Scandinavian countries and Northern Italy.

In contrast, all but one mutations identified in the *CNGB3* gene are nonsense mutations including two stop-mutations, two frame-shift mutations and one splice defect.

The T383fs mutation in the *CNGB3* gene was found recurrently. T383fs was present homozygously in 4 out of 10 families and compound-heterozygously in two cases in combination with another mutation. In one family this mutation could only be identified as a single heterozygous mutation. All in all the T383fs allele constituted 11 out of 22 disease chromosomes in our patient sample. This surprisingly high frequency of this one mutant allele can be explained by two different mechanisms. One possible explanation could be the presence of a mutational hot spot giving rise to this mutation independently in several instances. The other hypothesis is that the high prevalence of this very mutation is due to a founder effect. Preliminary haplotype analysis suggests the presence of a founder effect implying that most T383fs alleles derive from one common ancestral mutant allele (data not shown).

The *ACHM3* locus was originally mapped to chromosome 8q21 in pedigrees from a Western Pacific Island named Pingelap (Winnick et al., 1999). Therefore this condition has also been acknowledged as Pingelapese blindness (OMIM 262300). Achromatopsia is highly frequent in this Pingelap Islander population due to genetic drift after a devastating storm in the 18th century. These instances have come to public attention by Oliver Sacks book "The Island of the Colorblind" and a recent BBC television documentary. We were able to analyse one family originating from this population. Only a sole homozygous alteration S435F was identified in the *CNGB3* gene. Another study on families from Pingelap also identified this very mutation (Sundin et al., 2000). Therefore we argue that this is the particular mutation causing the Pingelapese blindness.

Apart from the description of the mutation spectrums for the *CNGA3* and the *CNGB3* genes in Achromatopsia, this study also provided evidence that there is further genetic heterogeneity in Achromatopsia. Neither *CNGA3* nor the *CNGB3* gene mutations could be identified in one Achromatopsia family. Furthermore we have excluded both known loci in several families by linkage analysis (data not shown). These findings suggest that there is at least one further locus responsible for this retinal disorder.

In conclusion, our data show that mutations in the genes for the α- and the β-subunit of the cone photoreceptor cGMP gated channel, the final component of the phototransduction cascade in cone photoreceptors, can result in a common phenotype of Complete Colorblindness with indistinguishable phenotypes.

6. ACKNOWLEDGEMENTS

This work was supported by grants from the Bundesministerium für Bildung und Forschung (01KS9602: IB2, Q1) and the Interdisziplinäres Zentrum für Klinische Forschung (IZKF) Tuebingen, the Deutsche Forschungsgemeinschaft (SFB430/A5), and the fortüne program (grant no. 583) of the Medical Faculty at the University Tuebingen

REFERENCES

Chen, T.Y., Peng, Y.W., Dhallan, R.S., Ahamed, B., Reed, R.R., and Yau, K.W., 1993, A new subunit of the cyclic nucleotide-gated cation channel in retinal rods,. *Nature* **362**:764-767.

François, J., 1961, Heredity in ophthalmology, Mosby, St. Louis.

Gerstner, A., Zong, X., Hofmann, F., and Biel, M., 2000, Molecular cloning and functional characterization of a new modulatory cyclic nucleotide-gated channel subunit from mouse retina, *J. Neurosci.* **20**:1324-1332.

Hess, R.F., Sharpe, L.T., and Nordby, K., 1990, Night vision: basic, clinical and applied aspects, eds., Cambridge University Press, Cambridge.

Kohl, S., Marx, T., Giddings, I., Jägle, H., Jacobson, S. G., Apfelstedt-Sylla, E., Zrenner, E., Sharpe, L.T., and Wissinger B., 1998, Total colorblindness is caused by mutations in the gene encoding the α-subunit of the cone photoreceptor cGMP-gated cation channel, *Nature Genet.* **19**:257-259.

Kohl, S., Baumann, B., Broghammer, M., Jägle, H., Sieving, P., Kellner, U., Spegal, R., Anastasi, M., Zrenner, E., Sharpe, L.T., and Wissinger B., 2000, Mutations in the *CNGB3* gene encoding the b-subunit of the cone photoreceptor cGMP-gated channel are responsible for achromatopsia (ACHM3) linked to chromosome 8q21, *Hum. Mol. Genet.* **9**:2107-2116.

Liman, E.R., Tytgat, J., and Hess, P., 1992, Subunit stoichiometry of a mammalian K+ channel determined by construction of multimeric cDNAs, *Neuron* **9**:861-871.

Liu, D.T., Tibbs, G.R., Paoletti, P., and Siegelbaum, S.A., 1998, Constraining ligand-binding site stoichiometry suggests that a cyclic nucleotide-gated channel is composed of two functional dimers, *Neuron* **21**:235-248.

MacKinnon, R., 1991, Determination of the subunit stoichiometry of a voltage-activated potassium channel, *Nature* **350**:232-235.

Müller, F., and Kaupp, U.B., 1998, Signaltransduktion in Sehzellen, *Naturwissenschaften* **85**:49-61.

Sacks O., 1997, The island of the colorblind, Vintage books, Random Hause, New York, NY.

Sundin, O.H., Yang, J.-M., Li, Y., Zhu, D., Hurd, J.N., Mitchell, T.N., Silva, E.D., and Maumenee, I.H., Genetic basis of total colourblindness among the pingelapese islanders, *Nature Genet.* **25**:289-293.

Winnick, J.D., Blundell, M.L., Galke, B.L., Salam, A.A., Leal., S.M., and Karayiorgou, M., 1999, Homozygosity mapping of the achromatopsia locus in the pingelapese. *Am. J. Hum. Genet.* **64**:1679-1685.

Wissinger, B., Müller, F., Weyand, I., Schuffenhauer, S., Thanos, S., Kaupp, U.B., and Zrenner, E., 1997, Cloning, chromosomal localization and functional expression of the gene encoding the α-subunit of the cGMP-gated channel in human cone photoreceptors, *Eur. J. Neurosci.* **9**:2512-252.

Wissinger, B., Jägle, H., Kohl, S., Broghammer, M., Baumann, B., Hanna, D.B., Hedels, C., Apfelstedt-Sylla, E., Randazzo, G., Jacobson, S.G., Zrenner, E., and Sharpe, L.T., 1998, Human rod monochromacy: linkage analysis and mapping of a cone photoreceptor expressed candidate gene on chromosome 2q11, *Genomics* **51**:325-331.

CHARACTERISATION OF THE *CRX* GENE; IDENTIFICATION OF ALTERNATIVELY SPLICED 5' EXONS AND 3' SEQUENCE

Matthew D. Hodges, Kevin Gregory-Evans and Cheryl Y. Gregory-Evans[*]

1. INTRODUCTION

The cone-rod homeobox gene (*CRX*) is a *paired-like* transcription factor belonging to the *otd/Otx* family that is essential for the maintenance and development of photoreceptors[1-5]. The gene has been mapped to chromosome 19q13.3 and has been implicated in autosomal dominant cone-rod dystrophy (adCRD)[1, 2, 6] and Leber's congenital amaurosis (LCA)[7, 8, 9]. adCRD is characterised by an initial loss of colour vision and visual acuity followed by nyctlopia, progressive loss of peripheral vision and advancing retinal pigmentation[10], LCA is defined as total blindness at birth or soon afterwards, involving retinal dystrophy and is nearly always recessively inherited[11]. There is distinct heterogeneity between adCRD and LCA both clinically and genetically. Mutations in *RETGC-1*,[12] *RPE65*,[13] and *AIPLI*[14] have been attributed to LCA, with linkage to a further three loci and at least four loci have been linked to adCRD[1]. The CRX protein itself has been shown to interact with NRL,[15] p300/CBP,[16] RX,[17] and phosducin[18]. This implies that mutations in multiple loci may therefore cause adCRD or LCA.

The *CRX* gene encodes two mRNA species of approximately 4.5 kb and 3.0 kb based on northern analysis[1], and encode for a protein of 299 amino acids[1-4]. Apart from the highly conserved homeodomain, the protein has a nuclear localisation signal, basic region, WSP and OTX motifs [1-4, 19]. Deletion analysis of the protein has ascertained that a complete HD is required for normal DNA binding, and that the WSP and OTX motifs are necessary for controlling positive regulatory function[20].

A review of the current literature has revealed fifteen well characterised mutations within the *CRX* gene that lead to adCRD or LCA as summarised in Table 1.

[*] Matthew. D. Hodges, Cheryl Y. Gregory-Evans, Section of Cell and Molecular Biology and Kevin Gregory-Evans, Department of Ophthalmology, Imperial College School of Medicine, Exhibition Road, London SW7 2AZ

Table 1. Mutations within the *CRX* gene leading to cone rod dystrophy or Leber's congenital amaurosis

Phenotype	Mutated codon	Base change	cDNA[a]	Mutation type	Comments	Fraction correct a.a.	Reference
LCA	Pro 9	ins G	24	Frameshift	Loss of HD, WSP, OTX tail and C terminus	9/69	9
adCRD, RP	Arg41Trp	CGG-TGG	120	Missense	Alters HD sequence	298/299	2
adCRD	Arg41Gln	CGG-CAG	121	Missense	Alters HD sequence	298/299	2, 21
adCRD	Glu80Lys	GAG-AAG	238	Missense	Alters HD sequence	298/299	6
adCRD	Glu80Ala	GAG-GCG	239	Missense	Alters HD sequence	298/299	1, 21
LCA	Arg90Trp	CGG-TGG	262	Missense	Alters HD sequence	298/299	8
LCA	Leu 146	del 12bp	436	In-frame deletion	Codons 147-150 deleted	295/295	21
adCRD	Ala158Thr	GCC-ACC	472	Missense	WSP altered. Detected in unaffected controls	298/299	2
LCA	Glu 168	del AG	503	Frameshift	Partial loss of WSP, loss of OTX tail and C terminus	168/171	7
adCRD	Glu 168	del G	503	Frameshift	Partial loss of WSP, loss of OTX tail and C terminus	168/185	1
LCA	Ala 196	ins C	585	Frameshift	Loss of OTX tail and C terminus	196/234	22
adCRD	Ala 196, Pro 197	del CCCC	586	Frameshift	Loss of OTX tail and C terminus	195/216	2
LCA	Gly 217	del G	649	Frameshift	Loss of OTX tail and C terminus	216/217	7
LCA	Lys 237	del C	709	Frameshift	Loss of OTX tail and C terminus	236/369	9
LCA	Val242Met	GTG-ATG	724	Missense	Not in conserved region	298/299	2

[a] Sequence refers to accession number AF024711 counting from the first base of the initiation codon

1.1. Homeodomain Mutations

Five *CRX* mutations specifically affect the homeodomain (HD): Arg41Trp, Arg41Gln, Glu80Ala, Glu80Lys and Arg90Trp[1, 2, 6, 8, 21]. These mutations are all of the missense variety, implying that the phenotype is directly related to the sequence change seen in the HD. Pro-9 ins 1bp[9] is a special case and will be discussed in Section 1.3.

Arg41Trp gives rise to adCRD with patients having normal vision with late onset (35-50yrs) progressive vision loss[2]. Arg41Gln patients also diagnosed as adCRD, were asyptomatic with late-onset loss of visual acuity through cone and rod degeneration. Glu80Ala was similarly diagnosed as adCRD. The visual loss was early onset (5-30yrs), with central visual loss and macular pigmentation[21]. The Glu80Lys phenotype was clinically similar to the Arg41Trp phenotype: adCRD, with early onset, progressive retinal degeneration (childhood to early teens)[6]. The Arg90Trp patients were clinically determined to have autosomal recessively inherited LCA (arLCA), with the proband having severe vision loss from birth[9].

Arg41 and Glu80 are within regions completely conserved between all members of the *paired* HD class, implying high importance to these residues in structural/functional integrity. Arg90 is conserved through all members of the *otd/Otx* family[1], within the *paired* class, there is a basic residue (Lys or Arg) at this position[23, 24].

Using the crystal structures of a typical paired like homeodomain and *engrailed* a member of this family, it is apparent that Arg41 is equivalent to Arg3, Glu80 is equivalent to Glu42 and Arg90 is equivalent to Lys52[23, 24]. Arg3 and Glu42 form a charged hydrogen bond that is disrupted when either of these residues are mutated, Glu42 also forms a salt bridge with Arg31, that may be important in DNA recognition[1].

Lys52 does not seem to have specific importance in the HD, but the surrounding residues Asn51 and Arg53 are important for binding with the DNA recognition sequence[24]. The replacement with Trp at this site may cause some loss of DNA recognition. Arg41Trp, Glu80Ala and Arg90Trp have all been shown to have reduced DNA binding activity with *CRX* target sequences, further substantiating these mutations as disease causative[2, 8, 21, 25].

The Arg41Trp/Gln mutations have a similar phenotype, as do the Glu80Ala/Lys mutations, which is unsurprising as the same residues are affected. The latter pair of mutations are associated with earlier onset than the former, this may be due to the greater importance of the Glu80 residue in the HD, as detailed above.

1.2. Truncated Proteins

It is known that at least four other genes interact with CRX, but it is as yet undetermined where these proteins bind on the CRX protein. Thus any conclusions drawn from a truncated CRX protein in relation to an adCRD or LCA patient's phenotype would be premature until more is known about how CRX-associated proteins interact. Six *CRX* mutations are predicted to truncate the nascent protein through frameshift mutations: Glu 168del 1bp, Glu 168 del 2bp, Ala 196/Pro 197 del 4bp, Gly 217 del 1bp and Lys 237 del C[7, 9,11, 22]. Additional complexity in the characterisation of these mutations is aquired through the extra out-of- frame residues predicted at the C terminus, each of differing length and frame. It is also uncertain if the mutant protein is stable, or even if the mutant mRNA is stable in each case, (Table 1).

The Leu 146 del 12bp mutation causes an in frame deletion, resulting in the removal of four residues[21]. Clinically the affected members of this family have severe visual disturbances from birth accompanied with nystagmus, hence the LCA diagnosis.

The Glu 168del 1bp, adCRD, and Glu 168 del 2bp, autosomal dominant LCA (adLCA), mutations[1, 7] correctly encode the first 168 residues, with a partial loss of the WSP, OTX and C terminus, and have additional residues at the C terminus; eight in the case of Glu 168del 1bp and 5 in the case of Glu 168 del 2bp. Glu 168del 1bp was clinically determined to be adCRD, the mutation being associated with maculopathy of an early onset and progressive type. The Glu 168 del 2bp mutation was associated with LCA, with nystagmus and visual inattention at 1 month. Interestingly the patient was heterozygous for the mutation and the disease was therefore described as a dominant form of LCA. It is however clear that the mutation occurred *de novo*, as both parents were unaffected[26].

Ala 196 ins 1bp, Ala 196/Pro 197 del 4bp, Gly 217 del 1bp and Lys 237 del 1bp[2, 7, 9, 22] all cause frameshifts with the resulting protein lacking the OTX motif and C terminus. The Ala 196/Pro 197 del 4 bp mutation is associated with adCRD whereas the other three are all associated with LCA, and of these Gly 217 del G and Lys 237 del 1bp were defined as adLCA. As previously seen for the Glu 168 del 2bp mutation, these mutations also occurred *de novo* since the mutation was only observed in the proband.

1.3. Special Cases

Pro 9 ins G was clinically diagnosed as adLCA and caused by a frameshift at base 24 resulting in the loss of all functional domains after amino acid 9 and replacement with an additional 60 residues out of frame[9]. This mutation can therefore be regarded as a null allele of *CRX,* implying that haploinsufficiency of CRX is not responsible for retinal disease in this case. The proband was clinically diagnosed with adLCA but as with the Glu 168 del 2bp, Gly 217 del G and Lys 237 del 1bp mutations, this mutation also appears to be *de novo* and so contradicts the diagnosis of adLCA. Interestingly the father, like his affected daughter was also heterozygous for the same mutation but is himself unaffected. The most obvious explanation for this is that the proband has a *de novo* mutation in a *CRX* interacting gene (digenic inheritance) or that the proband has also inherited a maternal mutation in a different part of the *CRX* gene[9]. Although mutation screening of *RPE65* and *RETGC-1* was carried out (without the identification of any mutations) screening in *AIPLI*[14] was not, so digenic inheritance can not be excluded. Neither can a mutation in the mother's *CRX* allele since the entire coding sequence has yet to be determined[1, 2].

Ala158Thr and Val242Met have both been associated with adCRD. It was therefore surprising to note that the Ala158Thr mutation was observed in 1% of control chromosomes, and that the Val242Met mutation was not in a conserved recognisable motif[2]. Additional *CRX* mutation screening in these families will be necessary once the full sequence of *CRX* is available. Indeed, either of these families are candidates to display digenic inheritance with a mutated CRX interacting gene.

2. MATERIALS AND METHODS

Human Y-79 retinoblastoma cells (American Type Culture Collection HTB18) were cultured in RPMI 1640 medium (Sigma) supplemented with 15% fetal bovine serum (Gibco BRL) at 37°C with 5% CO_2. RNA was extracted using the Trizol reagent (Gibco BRL) following the manufacturers instructions and cDNA was synthesised from 1-2µg of RNA using the Superscript Preamplification system for first strand cDNA synthesis (Gibco BRL).

RT-PCR was carried out using 1 µl of the final cDNA synthesis mixture in a 50 µl reaction consisting of 10 µM each primer, 10 mM Tris-HCl (pH 8.3.), 50 mM KCl, 1.5 mM $MgCl_2$, 200 µM dNTP and 0.5U Taq DNA polymerase (Bioline). RT-PCR was carried out on a Biometra Personal cycler under the following conditions 4 minutes at 94°C and then 35 cycles of 93°C for 1 minute, 55°C for 1 minute and 72°C for 1 minute, followed by 10 minutes at 72°C.

5' and 3' RACE was carried out using Human retina Marathon-Ready cDNA (Clontech) and the 5' RACE system for rapid amplification of cDNA ends version 2 (Gibco BRL).

Accession number AC008745 was identified from the HTGS database through a BLAST search with *CRX* exon 1. It contains approximately 150 kb of genomic sequence from human chromosome 19 including all published sequence from the *CRX* locus.

DNA Sequencing was carried out using ABI prism DNA sequencing and ABI 377 genetic analyser kit (Perkin-Elmer).

Oligonucleotide Primers (5'-3') 3'RACEc: G G T C A G A A T C A C C G T G - C C C T T G A A. 3'RACEd: G C A C A A T A T A T G C T T A C G A G T T G G - T A. 3'RACEe: C G T G G A C A C T T C T T T A C A T A T G G T T A. CRX1AN: A G G A C A G G T C T C A T T A G C. CRX1AR : C T C C T G A T G G G A G A - A T C C A G T C C. 1aaF: G A G C A C T G T A C A C T T C T T T C T G. 1aaF1: T A G C T G A A G C T G G A G G A A T A. CRXexon1r: C C T G G T G C A T C - A G G T C C A C A T. 1AAF2: G A G G G G A T T A A G C A G A C G G. 1AAF3: C T G A G G A G T A G G G G C T T A T. 1AAF4: G A C C A T C T G A C A G T - G C T. 1AR6: G A T C T G A T A G A A G A G A T G A C A G. 1AR7: C A G T G - C T T T A T A T T G T T T C A G. Exon1AR: C C T G A G C C A C T G G G T G - G A CRXexon1R: C C T G G T G C A T C A G A T C C A A T G. S1AR2: C T T C - C A G C G G - G A C A G A G A A

3. RESULTS

We set out to determine the sequence the full length *CRX* transcript(s). We report here on the identification of two different 3' UTR sequences and two different alternatively spliced 5' exons from *CRX*. One of the exons is itself alternatively spliced to give three further variants. We have also identified a C-A polymorphism in one of the new 5' exons. We also report the identification of the mouse homologous sequence to one of the 5'exons.

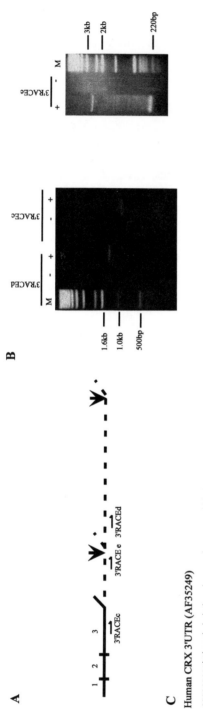

C

Human CRX 3'UTR (AF35249)

```
aggacgagtctccatctctccatcggcctggaacccttttcttttctgaatctgcttcctgcagtttagatcccggatggcattcctgagaaagcaaccgaaccagctgtccttctgacagctcgtgttcagcttacagagaccac
Cccttcctccacaggagaggctccctcctccctggacagctcacagtctgcctagtgatctctccaacccacacgtgtgcagcggacagatctctcaaccctgacaagcaggtaggggggttgataacttaacttccacgtgacagaattttttttttgttttgtttgttttgcaacacag
agaagaagtcacaaactgcagttgattgggcatgttagtttggcatcgttttcaccgtttcctgacaagacgcaggtaggggggcttgataacttaacttccacgtgacagaattttttttttgttttgtttgttttgcaacacag
tctagctcgtgtcccagctggagtgacatcctcaaccgtccccagttcacgccattcctccccggttcacgccattcctccccggttcacgccattcctccccaaagtcacccgatgcccgccctttttttttttttttttttttgtattt
tctcactctttgcttaggtcggagtgcagtggtgtgatctcagctcactgcaaccctccccgttgaacctcccaatctccaatagctggattacaggtgtgctccatcgcccgctcaattttttcatt
attagtagacaggtttaccatgttgggcagtatgtccgcgctcgcctggcctccccaaagtgcggattacagacatgtgccttgcctggcctacaggacacttctttagcatatgtt
aggacccttctagaattccaaagacagacttaagaaggcccctcggaaacctcctcggttgacctctccccaaagtgctttgtcactctagttctttccttcatgttttgttctttcttttaaagacatttatcacctcattcgctcattagcacattggaatttcacagactccaaaatatattattaatttaaatttaatttctttaaagacatttatcacctcattcgctcattagcacattggaatttcacagactccaaaatatattattaatttaaattt
agagacaaagatctacattcatatattagctatacatatgtactctttctcttttagtttcttttaaagacatttatcacctcattcgctcattagcacattggaatttcacagactccaaaatatattaatttaaattt
tctctactcctttttctttgttttagatttctaaagtgagacaccggtagctccttagttctcaaccaagtgctattcagactcaccatagtgtgcattccagagaccacactagtgtcacctctgtgcattccagagaccacactagtgtcacct
ggaagagggggaaatagatttggaggaggcagcccatcttcaaacatacatttctttttttagttttttttttttttttttttttttgttttcagtttgatgtggattggaagaagttggctcagcattttcctggggtggcacc
attatttctgaaatagaatcttactatgtgaaccctgaagacgacttgatgctcagaagttgatgtcagagtcccaagaaagggaaaggtt gttgtcagttcctggtttcagtttccggtggt
ggttccaagacccttcattggtggttcaaagcaagtgtaaaaaaaagtaggccagaagacagtggtgtgctgcaagtccaacatgggctccaagaagaaggtgttgtggagttgttgtcagttcctggttttgtgtgagttgtggttcagcatattgtacgagttggtacacacgtcagcaggtcaggaagagctggta
gggagtatcagatgaccttttcagcactgacagagaggtgcacacgccagaagaggtacaccctgagggtacaccccctgagggtacccctgagaccctaatcccacatggtgtgctcccacacgctgagccccacacgctgaggaccgtgtc
tctggacaataagccaggatggccgggtcggtgctgctcacgcctaatccacagcttggaaggcagcaggaaattgctgaaggcagcaggaaattctcttattcaccggagggctgtgtcttgtgcaaacaagtcatagaaa
aaattagccggcatggtggcacgcgtctcagtagtcccagctacgtgggatgaaaataaaccaggagacccacatggtggcacaaattcacaacgggaggctgaggatgcaaaattcacaacgggaggctgag
ataataacaacacaaataaccaggtggtcaaagtcattcttcagagacacggagactcctcagctgaagcccggagcccgggaccaaggctcagctgaagcccccaaatgaaatataaacaggagaatataaacaggaga
ctgatgggagtttggggaggactgaaggatgccaaggtgtggatgaaaataaaccaggagagccgtgaggagttggtcttccggctgttcttccggctgtttcctgagtgttgttcaaccctctaaggctgaggggttcttccggctgtttcctgagtgttgttcaaccctctaaggctgacctcaaccct
gagaggcaggagcccaaatccgactaggaggttacagattgctttccatatttatctgtcctcctccctccccacacactgccttatctgtcctcctctcccacacactgccttatctggctttccgtttcaaccctgggac
aagaagcaggagcccctgagaagatgctcacctatctgtcctcctctcccacacactgccttatctggctttccgtttcaccctgggac*caattaaa*atgttgatctcatttaaal
```

Figure 1. Sequence of the *CRX* transcript. Solid line indicates the published region, dashed line indicates 3'UTR. Primers are shown below the sequence. **B.** 3' RACE using primers indicated in panel A, +/– indicates presence/absence of template in reaction. **C.** The 3' UTR of *CRX*. Bold star indicates an additional 200bp sequence previously reported at the 3' terminus of AF024711. The sequence primers used in 3'RACE experiments have been underlined. Polyadenylation signals are in bold . The termination point of the shorter/longer transcript are indicated with a bold vertical bar.

3.1. Identification Of Two Polyadenylation Sites At The *CRX* 3' End

We identified accession number AC008745 as detailed in Section 2. Using the NIX computer programme at the HGMP[27] we were able to characterise this sequence. Excellent matches to pineal and retinal EST sequences were identified downstream of the published *CRX* terminus. The sequence included four putative polyadenylation signals, which were not present in the original *CRX* published transcript[1]. Five of the EST clones were acquired from the HGMP; IMAGE clones 220333, 275861 and 381301 from retina and clones 230860 and 397262 from pineal, all these clones terminated near to polyadenylation signals 1/2 or 4, it was impossible to distinguish signals 1 and 2 from each other due to their close proximity to each other (Figure 1). The clones were sequenced from both ends, confirming that clone 220333 terminated after polyadenylation signal 1/2 and the remainder terminated after signal 4. No clones identified by the NIX searches appeared to terminate after signal 3.

To confirm that the two termination sites were attributable to the *CRX* 3' UTR we were able to successfully carry out 3' RACE. Primer 3'RACEc was designed from accession number AF024711 (the cDNA sequence of *CRX*), and a band approximately 900 bp was amplified, predicting the end of the cDNA to terminate at polyadenylation signal 1/2. Primer 3'RACEd was designed from accession number AC008745, this amplified a product of approximately 1.5 kb, predicting the end of the cDNA to be after signal 4. A further primer 3'RACEe was designed approximately 200 bp upstream of polyadenylation signal 1/2. This amplified a band of 220 bp, inferring termination at signal 1/2 and a band of approximately 2.2 kb inferring termination at signal 4. The 3' RACE products were cloned and sequenced and had identical sequence to the equivalent region of accession number AC008745. The sequence differed to that of AF024711 however. The sequences are identical from nucleotide 1253 (the 5' end of 3'RACEc) to nucleotide 1324 counting from AF024711, but the last 86 bp of AF024711 are not seen in the 3'RACE sequences. Analysis of the 3' end of AF024711 identified a positive match to an *Alu* repeat over the additional sequence. We propose that the terminal sequence in AF024711 has arisen through chimarism between *CRX* and an *Alu* sequence and therefore explains why there is not a polyadenylation signal in the published *CRX* sequence.

3.2. Identification Of Exon1A

Three primers were designed from accession number AF024711 (CRX exon 1R, CRX 1AN and CRX 1AR). cDNA was synthesised using primers 1R, Poly-C tailed and two rounds of nested PCR were carried out using primer CRX 1AN and CRX 1AR, with the abridged primer and AUAP primer respectively (Gibco BRL).

Approximately 300 bp of novel sequence was identified upstream of the published *CRX* sequence. A BLAST search with the 400 bp fragment against accession number AC008745 gave an exact match to a region approximately 12 kb upstream of exon1. Putative splice sites at the 5' end of exon 1 and the 3' end of the upstream sequence were also identified (Table 2). To confirm the validity of these results we designed an RT-PCR primer from the new sequence and were able to amplify a correctly sized product with CRX 1R primer (exon1) from Y-79 derived cDNA (data not shown), showing that

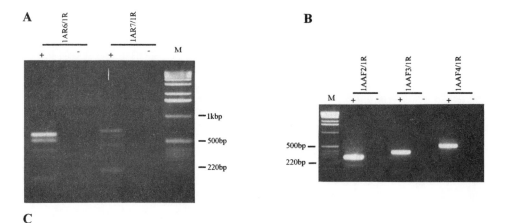

C

Human exon 1Aα (559 bp) (AF335246)

```
cagtgctttatattgtttcagtaggataatgatctgatagaagagatgacaggacttatggatggattggatgtaagtggtgagg
agaaagggttaggggccaggctgagtgcccagatttctgcctgagccactgggtggagatgagaaggaaagcaagttgcctcag
aagacctagctcagagaagaagtgcttggatttgatttcagcttacttgatcccagggccatgtgcaggaccagctgaaacccac
ctctctgtaccccacccctggaaggactctggatccaggagccaaccagtacccagattccccaggcaggtgatttaagagtt
gctgacacccttctctgtcccgctggaagttattttctcctccctctccctgtacagctgacttccttccctgctgcagggccag
gggtctcgttccaggactggattctcccatcaggagcctctggactggtcgccgagggtcagacggcccctccctctcttgctgt
catccctggctcttcaagctaatgagacctgtcctgattcctcagccag
```

Human exon1Aβ (322 bp) (AF335593)

```
cagtgctttatattgtttcagtaggataatgatctgatagaagagatgacaggacttatggatggattggatgtaagtggtgagg
agaaagggttaggggccaggctgagtgcccagatttctgcctgagccactgggtggagatgagaaggaaagcaagttgcctcag
aagacctagctcagagaagaagtgcttggatttgatttcagcttacttgatcccagggccatgtgcaggaccagctgaaacccac
ctctctgtaccccacccctggaaggactctggatccaggagccaaccagtacccagattccccagg
```

Human exon 1Aγ (136 bp) (AF335594)

```
ggccaggggtctgttccaggactggattctcccatcaggagcctctggactggtcgccgagggtcagacggcccctccctctct
tgctgtcatccctggctcttcaagctaatgagacctgtcctgattcctcag
```

Human exon 1Aδ (72 bp) (AF335595)

```
cagtgctttatattgtttcagtaggataatgatctgatagaagagatgacaggacttatggatggattggat
```

D

Human exon 1AA (428 bp) (AF335247)

```
gaccatctgacagtgctctccttcctctttgatgcctctctcttgggatactagctgaagctggagagacaccagttagacctaa
ggaaggacttccctgaggagtaggggcttatggtcaccggcaggagctggggcctcccttccccatcagccctaattgccaagat
gtcatggggggaagaggagggggattaagcagacgggtgcccctcccccctcccagccaatgtcacctcctggtgcccagtcgagtc
cccaccttggccgggattaccctccgagttccaggccatgacaaatgacatcactcccgcccaggcttaaaatctccccatgt
gaggggatgtgtttccttcagcctctgctgtctggccgctctgtctaggtcctgggccacgggagagccccgtccctcctttctg
aag
```

Mouse exon 1aa (427 bp) (AF335248)

```
gagcactgtacacttctttctgtgcttttgaagacatttcttttcagatagtagctgaagctggaggaatacccctttagacccaa
ggaaagacttccctgagaagtgggggctcctggttgcaggcaggagttgggctttccctcccttcttcatcaaccctaattgcca
agatgtcactgtgggggtggtgggattaagcagacggtgcccttccccccagccaatgtcacctcctgttgcccagtgagtcccc
caccttccctgggattacccttccagatccaggccacaatagatgacatcactcccagcccaggcttaaagtcgccccgtgtgag
gggacctatttcctctagcctctgctgtctgatgacaccatccagtacctgaacatccaggagagtccccatttctcctttccaa
ag
```

Figure 2. A. RT-PCR from Y-79 derived cDNA primed with oligonucleotide 1AR6 and 1R, 1AR7 and 1R. +/- denotes presence/absence of RT in PCR mix. M, kb ladder (Gibco BRL) **B.** As before but with primers 1AAF2, 1AAF3 and 1AAF4, primers are underlined. **C.** Sequence of exon 1Aα and 3 variants put into the same frame as the published sequence of *CRX*. Start codons are underlined, stop codons are in bold and position of 1AR6 and 1AR7 are indicated. **D.** As before but sequence of exon 1AA and 1aa

exon 1 is spliced to exon 1A (the new exon) in *CRX* cDNA. Further, we have identified multiple ESTs derived from exon 1A and exon 1 (data not shown).

3.2.1. Exon 1A Is Itself Alternatively Spliced

We were able to extend the sequence of exon 1A by designing primers for RT-PCR from sequence upstream of the 5' RACE product from accession number AC008745. Primers 1AR6 and 1AR7 primed with an exon 1 primer (exon1R) amplified four bands from Y-79 derived cDNA (Figure 2, panel A). No amplified products were detected in the no reverse transcriptase, confirming that these were bona fide *CRX* transcripts. The PCR products were cloned and sequenced. The largest fragment corresponded to the predicted 559 bp PCR product. The three smaller products appeared to be alternately spliced variants of exon 1A. To distinguish theses, exon 1A was renamed exon1Aα, and the three variants as: 1Aβ (322bp), 1Aγ (136bp) and 1Aδ (72bp) (Figure 2, panel C). All the alternate transcripts have recognisable splice sites at their 3' and 5' ends respectively, (Table 2).

Exons 1Aα, 1Aβ and 1Aγ all remain in the same relative reading frame when placed in front of the previously reported exon 1 sequence of *CRX*[1, 2]. Using the published start codon of *CRX* and working upstream, exon 1Aα does not appear to be translated as there are numerous stop codons in-frame in the new sequence. Exon 1Aβ similarly has the same stop codons when put in frame with the rest of the *CRX* transcript. Exon 1Aβ has two in-frame start codons that are not followed by additional stop codons, although neither fit the Kosak consensus well[28]. Exon1Aγ does not have start or stop codons in frame with the rest of the *CRX* transcript. Exon 1Aδ is not in the same relative frame as the other 5' exons. It has a start codon in-frame with the rest of the *CRX* transcript but has three stop codons immediately downstream of it, implying that it is also untranslated.

The most 5' region of *CRX* is shared between exons 1Aα, 1Aβ and 1Aγ. When this region was identified in genomic DNA an excellent match with acceptor splice-site was found (Table 2), implying that there are additional *CRX* exon(s) upstream. Further evidence of this has come from the inability to perform RT-PCR with primers designed upstream of this putative splice site and from the unsuccessful search for promoter like elements in the genomic sequence adjacent to exon 1A.

3.2.2. Exon 1A Has A C-A Polymorphism

To date, no mutation has been identified in the original adCRD family that defined the *CORD2* locus (MIM 120970)[1, 10, 29]. In the course of mutation screening the new sequences reported here we identified an A-C sequence polymorphism, that is detectable in exon 1Aα and 1Aβ (A205C), (Figure 2). This variation generates an *Mse*I site, that we were able to utilise in an examination of the adCRD family. We amplified genomic DNA using primers exon 1AR and S1AR2 and then digested with *Mse*I. The A variant was detected in 10/12 chromosomes in the adCRD family (the polymorphism was not linked to adCRD) and 4/6 chromosomes in wild type controls. The C-A polymorphism

Table 2. Exon-Intron boundaries of the Human *CRX* gene

Exon	Exon size (bp)	Acceptor splice site	Donor Splice site	Intron size (bp)
?[a]	?[a]	?[a]	?[a]	?[a]
1Aα	559	tttcttccagCAGTGCTT	TCAGCCAGgcctgtagc	~12000[b]
1Aβ	322	tttcttccagCAGTGCTT	TTCCCAGGgcaggtgat	~12000[b] or 97[c]
1Aγ	136	cctgctgcagGGCCAGGG	TCAGCCAGgcctgtagc	~12000[b]
1Aδ	72	tttcttccagCAGTGCTT	GATTGGATgtaagtggt	~12500[b]
1AA	428	gatctctcagGACCATCT	TTCTGAAGgtgagcgtc	~10500[b]
1	136	tctcttgcagGCCCCCTG	CTACCCAAgtgagtaca	1700
2	152	cccaccccagGCGCCCCC	GGGTTCAGgtggggtgg	2925
3[d]	1859	tatccccagGTTTGGTT	–	–
3[d]	4044	tatccccagGTTTGGTT	–	–

[a] Exon(s) yet to be identified.
[b] Exons are all spliced to exon 1
[c] If exon 1Aβ is spliced to exon 1Aγ there is a 97bp intron otherwise exon 1Aβ is spliced to exon 1 and exon 1Aγ is skipped.
[d] Size depends on which polyadenylation signal is used

introduces a stop codon into exon 1Aα and 1Aβ if both sequences are read in the same frame as the published sequence of *CRX*.

3.3. Identification Of Exon1AA

Searching the EST database for additional *CRX* sequence, we identified accession number BE254470, that had additional 5' sequence to the published *CRX* sequence but was not exon 1A. We predicted that this would be an additional 5' exon. A search of accession number AC008745, identified the additional sequence, approximately 10.5 kb upstream of exon 1. There were excellent splice sites at the 3' end of this sequence further suggesting that this was a exon of *CRX*. We examined the genomic sequence and identified matches to 5' splice sites in the genomic sequence. It was then possible to amplify portions of the sequence using RT-PCR (Figure 2, panel B). We were unable to amplify products with primers designed upstream of the 5' splice site, implying that we had identified the entire exon. The entire exon is predicted to be 428 bp and we termed it exon 1AA. Unlike exon 1A it does not appear to be alternatively spliced (Figure 2, panel D). We have been unable to amplify between exon 1A and exon 1AA by RT-PCR, suggesting these are two alternate 5' exons..

3.3.1. Identification Of Exon 1aa The Mouse Homologue Of Exon 1AA

In a BLAST search of the HTGS database with the sequence of exon1AA we identified accession number AC080167, an unfinished genomic mouse sequence. We also identified accession number BB606342 from the mouse EST database, an eyeball derived cDNA. Analysis of the mouse genomic sequence showed that the 3' and 5' splice

A

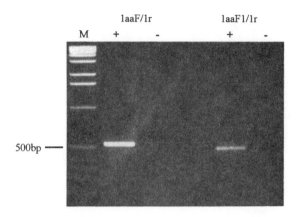

B

```
1AA    1   cctctctttt ttcagacgat ctctcagGAC CATCTG-ACA GTGCTCTCCT
           |||  ||||||  |          |  ||| ||||| ||  ||| ||| | || ||
1aa    1   cctttctttt ctg------t ctccagGAG CA-CTGTACA CTTCTTTCTG

1AA   50   TCCTCTTTGA TGCCTC--TC TCTTGGGATA CTAGCTGAAG CTGGAGAGAC
           | ||  ||||| | |      || | ||  ||||  ||||||||| ||
1aa   44   TGCT-TTTGA AGACATTTTC TTTTCAGATA GTAGCTGAAG CTGGAGGAAT

1AA   98   ACCAGTTAGA CCTAAGGAAG GACTTCCCTG AGGAGTAGGG GCTTATGGTC
           |||  ||| || | |||||  |||||||| | || ||| |||  |||   ||||
1aa   93   ACCCTTTAAC CCAA-GAAAG -ACTTCCC-G AGAAGTGGGG GCTCCTGGTT

1AA  148   ACCGGCAGGA GCTGGGG--- -CCTCCCTTC CCCATCAGCC CTAATTGCCA
           |  ||||||| | ||||      |||||||||  ||||| ||  |||||  ||||
1aa  140   GCAGGCAGGA GTTGGGCTTT CCCTCCCTTC TTCATCAACC CTAAT-GCCA

1AA  194   AGATGTCA-- -TGGGGGGAA GAGGAGGGGA TTAAGCAGAC GGGTGCCCCT
           ||||||||    |||||||          |||||||||| || |||||  ||
1aa  189   AGATGTCACT GTGGGGGTGG T-----GGGA TTAAGCAGAC GG-TGCCCTT

1AA  241   CCCCCTCCCA GCCAATGTCA CCTCCTGGTG CCCAGTCGAG TCCCCCACCT
           ||    ||||  ||||||||||||  |||||||| || |||||| ||| |||||||||||
1aa  233   CC----CCCA GCCAATGTCA CCTCCTGTTG CCCAGT-GAG TCCCCCACCT

1AA  291   TGGCCGGGAT TACCCTCCGA GTTCCAGGCC ATGACAAATG ACATCACTCC
           | | |||||| ||||||||| | | |||||| | | ||| |||| |||||||||
1aa  278   TCCCTGGGAT TACCCTTCCA GATCCAGGCC ACAATAGATG ACATCACTCC

1AA  341   CGGCCCAGGC TTAAAATCTC CCCATGTGAG GGGATGTGTT TCCTTCAGCC
           | ||||||||| ||||| || | ||| |||||| ||||  | || |||| ||||
1aa  328   CAGCCCAGGC TTAAAGTCGC CCCGTGTGAG GGGACCTATT TCCTCTAGCC

1AA  391   TCTGCTGTCT GGCCGCTCTG TCTAGGTCCT GGGCCACG-GG AGAG-CCCC
           |||||||||||  |     |  || || ||| | | | | || |||| ||||
1aa  378   TCTGCTGTCT GATGACACCA TCCAGTACCT GAACATCCAGG AGAGTCCCC

1AA  439   GTCCCTCCTT TCTGAAGgtg ag
           |  ||||||| || |||||| ||
1aa  428   ATTTCTCCTT TCCAAAGgtg ag
```

Figure 3. A. RT-PCR of *Crx* exon 1aa from C57Bl6/6J retinal cDNA, +/- denotes presence/absence of reverse transcriptase in PCR mix, M is Kb ladder (Gibco BRL). **B.** Alignment of *CRX* exon 1AA and *Crx* exon 1aa genomic DNA. Uppercase exon DNA, bold indicates splice sites. There is 75% sequence similarity between the two homologous exons. Underlined sequence, primers used for RT-PCR used in panel A.

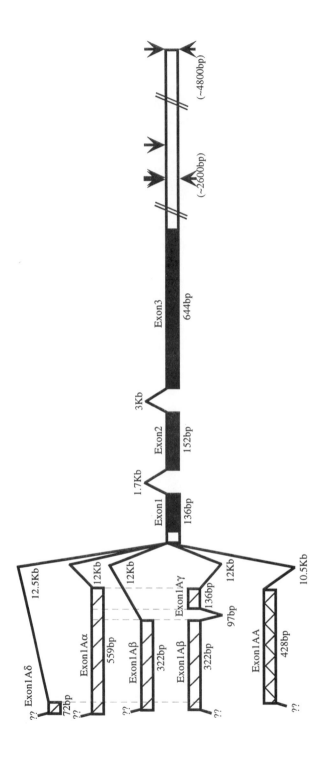

Figure 4 The human *CRX* transcript. Solid boxes represent coding exons and open boxes represent non-coding regions of these exons, with the sizes of exons depicted underneath. Introns have been indicated, not to scale, by open triangles. At the 5' end two alternative untranslated exons are depicted with open hatched boxes. Exon 1Aα is itself alternatively spliced to generate exons 1Aβ, 1Aγ, and 1Aδ. Double hatched box represents exon 1AA, an unrelated exon to exon1Aα. Question marks at the 5' ends indicate that there are further upstream exon(s) yet to be identified. At the 3' end, potential polyadenylation signals are represented by vertical arrows above the open boxes. The actual polyadenylation signals that appear to be used are shown below the boxes. The first of these polyadenylation signals gives a transcript of approximately 2600 bp and the second generates a transcript of approximately 4800 bp. The 3' region is not drawn to the scale of the rest of the transcript, bars are used to indicate this.

sites were conserved with the human sequence. We designed RT-PCR primers (1aaF, 1aaF1 and 1r) and were able to amplify products of the predicted size from cDNA derived from C57Bl/6J mice (Figure 3). Murine exon 1aa is 427bp in size and has 75% sequence similarity to the homologous human sequence. Exon 1aa, like exon 1AA does not seem to be translated as there are stop codons in all three reading frames (data not shown).

4. DISCUSSION

We have identified extensive 3' sequence from the *CRX* 3'UTR and have shown that the *CRX* gene utilises two polyadenylation sites. We have also identified exon 1Aα (524bp) and exon 1AA (428bp), two alternative 5' untranslated exons of *CRX*. Exon 1Aα is alternatively spliced to generate exon 1Aβ (302bp), 1Aγ (131bp) and 1Aδ (72bp). A diagram detailing the complexity of the CRX transcript is shown in Figure 4. Exon 1AA is conserved in the mouse, however, exon 1Aα and its multiple species do not appear to be conserved.

The sequence that we have identified accounts for approximately 95% of the *CRX* transcript based on northern blot experiments[1]. However, the 5' end of exon 1Aα, including its alternate forms and exon 1AA/1aa have excellent matches to the acceptor splice site sequence, suggesting that there is additional unidentified sequence upstream.

The novel exons we report here are probably non-coding when in frame with the rest of the *CRX* gene. The function of the 5' untranslated exons remains to be elucidated and will be a key issue in understanding the regulation of the gene. Possible functions for these 5' non-coding exons include the binding of *trans* elements, enhancement of translation of the CRX protein and thermodynamic stability of the *CRX* mRNA. Alternatively, the untranslated exons may direct tissue-specific expression. For example, the human PKR protein kinase gene has an untranslated exon 1, followed by 3 alternate untranslated exon 2 sequences, which together with part of exon 3 constitute the 5'-untranslated region of the PKR mRNA. It has been shown that expression levels of the alternate PKR transcripts vary in different cell types and that the sequence of the 5'UTR is responsible for this[30]. Similar findings have been attributed to the CC chemokine receptor 3 gene[31]. In a similar manner this mechanism may be used to control expression of alternate *CRX* transcripts in the pineal gland and retina.

Multiple species of mRNA are generated from the *CRX* locus, through use of alternate splicing and use of different polyadenylation signals. Upon identification of the additional exon(s) upstream, we will be able to determine if the mRNAs generated at this locus are under the control of one, or multiple promoters, adding further complexity to the gene structure. It is also important to note that exon 1Aα and the additional variants are not conserved in mouse. This may signify a significant physiological difference between the *CRX* gene in these two species.

The recently identified *RINX* gene[32], also a *paired*-like transcription factor expressed in the retina, has multiple splice variants, including alternate 5' and 3' ends. It is interesting to note that these variants include alternate splicing of the protein coding region, and in one variant the HD is missing altogether. One of our next goals will be to investigate whether the 5' exons of *CRX* are similarly involved with the generation of alternate proteins from the locus.

We have not identified any mutations in the new exons or splice sites and polyadenylation sites in the *CORD2* family. It may be the case that a mutation remains to be found in the unidentified, but predicted 5' exon(s) and promoter. The *CRX* sequence we report here is now available for further mutational screening of adCRD and LCA patients. This may be particularly beneficial to the *de novo* adLCA mutations and *CRX* mutations that are not immediately explainable by the identified mutation, such as Ala158Thr[2] and Val242Met[2].

5. ACKNOWLEDGEMENTS

This work was supported by the British Retinitis Pigmentosa Society, Grant GR 518

REFERENCES

1. C. L. Freund, C. Y. Gregory Evans, T. Furukawa, M. Papaioannou, J. Looser, L. Ploder, J. Bellingham, D. Ng, J. A. Herbrick, A. Duncan, S. W. Scherer, L. C.Tsui, A. Loutradis-Anagnostou, S.G. Jacobson, C. L.Cepko, S.S. Bhattacharya and R. R. McInnes, Cone-rod dystrophy due to mutations in a novel photoreceptor-specific homeobox gene (*CRX*) essential for maintenance of the photoreceptor, *Cell.* **91**, 543-553 (1997).
2. P. K. Swain, S. Chen, Q.L.Wang, L.M. Affatigato, C.L. Coats, K.D. Brady, G.A. Fishman, S.G. Jacobson, A. Swaroop, E. Stone, P.A. Sieving, and D.J. Zack, Mutations in the cone-rod homeobox gene are associated with the cone-rod dystrophy photoreceptor degeneration, *Neuron.* **19**, 1329-1336 (1997).
3. S. Chen, Q. Wang, Z. Nie, H. Sun, G. Lennon, N.G. Copeland, D.J. Gilbert, N.A. Jenkins and D.J. Zack, Crx, a novel Otx-like paired-homeodomain protein, binds to and transactivates photoreceptor cell-specific genes, *Neuron.* **19**, 1017-1030 (1997).
4. T. Furukawa, E.M. Morrow and C. Cepko, Crx, a novel *otx*-like homeobox gene, shows photoreceptor-specific expression and regulates photoreceptor differentiation, *Cell.* **91**, 531-541 (1997).
5. T. Furukawa, E.M. Morrow, T. Li, F.C. Davis and C.L.Cepko, Retinopathy and attenuated circadian entrainment in *Crx*-deficient mice, *Nat. Genet.* **23**(4), 466-70 (1999).
6. E. M. Sankila, T. H. Joensuu, R. H. Hamalainen, N. Raitanen, O.Valle, J. Ignatius and B. Cormand, A *CRX* mutation in a Finnish family with dominant cone-rod retinal dystrophy, *Hum. Mutat.* **1**, 94 (2000).
7. C.L.Freund, Q.L.Wang, S.Chen, B.L.Muskat, C.D.Wiles, V.C.Sheffield, S.G. Jacobson, R.R.McInnes, D.J.Zack and E.M.Stone, *De novo* mutations in the *CRX* homeobox gene associated with Leber congenital amaurosis, *Nat.Genet.* **18** 311-312 (1998).
8. A. Swaroop, Q. Wang, W. Wu, J. Cook, C. Coats, Xu. S, S. Chen, D. J. Zack and P. A. Sieving, Leber congenital amaurosis caused by a homozygous mutation (R90W) in the homeodomain of the retinal transcription factor CRX: direct evidence for the involvement of CRX in the development of photoreceptor function, *Hum. Mol. Genet.* **8** (8), 299-305 (1999).
9. E. Silva, J. Yang, Y. Li, S. Dharmaraj, O. H. Sundin and I. H. Maumemenee, A *CRX* null mutation is associated with both Leber congenital amaurosis and a normal ocular phenotype, *Invest. Opthalmol. Vis. Sci.* **41**(8), 2076-2079 (2000).
10. K. Evans, A. Fryer, C. Inglehearn, J. Duvall-Young, J. L. Whittaker, C. Y. Gregory, R. Butler, N. Ebenezer, D. M. Hunt and S. Bhattacharya, Genetic linkage of cone-rod retinal dystrophy to chromosome 19q and evidence for segregation distortion. *Nat.Genet.* **9**, 210-213 (1994).
11. R. Schroeder, M. B. Mets and I.H. Maumenee, Leber's congenital amaurosis. *Arch Ophthalmol.* **105**, 356-359 (1987).
12. I. Perrault, J.M. Rozet, P. Calvas, S. Gerber, A. Camuzat, H. Dollfus, S. Chatelin, E.Souied, I. Ghazi, C. Leowski, M. Bonnemaison, D.L. Pralier, J. Frezal, J.L-. Dufier, S. Ipttler, A. Munnich and J. Kaplan, Retinal-specific guanylate cyclase gene mutations in Leber's congenital amaurosis. *Nat. Gen.* **14**, 461-141 (1996).
13. F. Marlhens, C. Bareil, J.-M. Griffoin, E. Zrenner, E. Amalric, P. Eliaou, S.-Y. Liu, E. Harris, T. M. Redmond, B. Arnaud, M. Claustress and C.P. Hamel, Mutations in RPE65 cause Leber's congenital amaurosis. *Nat.Gen.* **17**, 139-141 (1997).

14. M.M. Sohocki, I. Perrault, B.P.Leroy, A.M. Payne, S.Dharmaraj, S.S. Bhattacharya, J. Kaplan, I.H. Maumenee, R. Koenekoop, F.M. Meire, D.G. Birch, J.R. Heckenlively and S.P.Daiger, Prevalence of AIPL1 mutations in inherited retinal disease, *Mol. Genet. Metab.* **70**(2), 142-50 (2000).

15. K.P. Mitton, P.K. Swain, S. Chen S. Xu, D.J. Zack A. Swaroop, The leucine zipper of NRL interacts with the CRX homeodomain. a possible mechanism of transcriptional synergy in rhodopsin. *J. Biol. Chem.* **275**(38), 29794-9 (2000).

16. Y. Yanagi, Y. Masuhiro, M. Mori, J.Yanagisawa and S.Kato, p300/CBP acts as a coactivator of the cone-rod homeobox transcription factor, *Biophys. Res. Commun.* **269**(2), 410-4 (2000).

17. A. Kimura, D.Singh, E.F. Wawrousek, M. Kikuchi, M. Nakamura and T. Shinohara, Both PCE-1/RX and OTX/CRX interactions are necessary for photoreceptor-specific gene expression, *J. Biol. Chem.* **275**(2),1152-60 (2000).

18. X. Zhu and C.M. Craft, Modulation of CRX transactivation activity by phosducin isoforms. *Mol. Cell. Biol.* **20**(14), 5216-26 (2000).

19. Y. Fei and T.E. Hughes, Nuclear trafficking of photoreceptor protein *crx*: the targeting sequence and pathologic implications. *Invest. Ophthalmol. Vis. Sci.* **41**(10), 2849-56 (2000).

20. K.Y. Chau, S. Chen, D.J. Zack and S.J. Ono, Functional domains of the cone-rod homeobox (CRX) transcription factor, *J. Biol.Chem.* **275**(47), 37264-70 (2000).

21. M. M. Sohocki, L.S. Sullivan, H.A. Mintz-Hittner, D. Birch, J.R. Heckenlively, C.L. Freund, R.R. McInnes and S.P. Daiger, A range of clinical phenotypes associated with mutations in *CRX*, a photoreceptor transcription-factor gene, *Am. J. Hum. Genet.* **63**, 1307-1315 (1998).

22. R.T. Tzekov, M.M. Sohocki, S.P. Daiger and D.G. Birch, Visual phenotype in patients with Arg41Gln and Ala 196+1bp mutations in the *CRX* gene, *Ophth. Genet.* **21**(2), 89-90 (2000).

23. D.S. Wilson, B. Guenther, C. Despain and J. Kuriyan, High resolution crystal structure of a paired (Pax) class cooperative homeodomain dimer on DNA, *Cell.* **82**, 709-719 (1995).

24. C.R. Kissinger, B. liu, E. Martin-Blanco, T. B. Kornberg and C. O. Pabo, Crystal structure of an engrailed homeodmain-DNA complex at 2.8Å resolution: A framework for understanding homeodomain- DNA interactions, *Cell.* **63**, 579-590 (1990).

25. M. Hodges, K. Gregory-Evans, C.Y and Gregory-Evans (unpublished results).

26. S.G. Jacobson, A.V. Cideciyan. Y. Huang, D. B. Hanna, C.L. Freund, L.M. Affatigato, R.E. Carr, D.J. Zack, E.M. Stone and R.R. McInnes, Retinal degenerations with truncation mutations in the cone-rod homeobox (*CRX*) gene, *Invest. Opthalmol. Vis. Sci.* **39**(12), 2417-2426 (1998).

27. United Kingdom MRC Human Genome Project Resource Centre; http://www.hgmp.mrc.ac.uk/

28. M. Kosak, Interpreting cDNA sequences: some insights from studies on translation. *Mamm. Genome.* **7** 563-574 (1996).

29. J. Bellingham, C.Y. Gregory-Evans and K. Gregory-Evans, Microsatellite markers for the cone-rod retinal dystrophy gene, on 19q13.3. *J. Med. Genet.* **35**(6), 527 (1998).

30. K. Kawakubo, K.L. Kuhen, J.W.Vessey, C.X.George and C.E. Samuel, alternate splice variants of the PKR protein kinase possessing different 5'-untranslated regions: Expression in untreated and interferon-treated cells and translational activity. *Virology.* **264** 106-114 (1999).

31. N. Zimmermann, B.L Daugherty, J.L. Kavanaugh, F.Y. El-Awar, E.A. Moulton and M.E. Rothenberg, Analysis of the CC chemokine receptor 3 gene reveals a complex 5' exon organization, a functional role for untranslated exon 1, and a broadly active promoter with eosinophil-selective elements. *Blood.* **96**(7) 2346-2354.

32. T. Hayashi, J. Huang and S.S. Deeb, *RINX(VSX1)*, a novel homeobox gene expressed in the inner nuclear layer of the adult retina. *Genomics.* **67** 128-139 9(2000).

RHODOPSIN MUTATIONS IN SECTORIAL RETINITIS PIGMENTOSA

S. Kaushal, D. A. R. Bessant, A. M. Payne, S. M. Downes, G. E. Holder, S. S. Bhattacharya, and A. C. Bird[1]

1. SUMMARY

Nearly 100 mutations in the rhodopsin gene have been described in association with autosomal dominant retinitis pigmentosa (ADRP), autosomal recessive retinitis pigmentosa (ARRP) and congenital stationary night blindness (CSNB), a nonprogressive disease. Additionally, mutations in the rhodopsin gene have been observed in patients with sectorial retinitis pigmentosa (SRP). We set out to determine the prevalence of rhodopsin mutations in sectorial RP, and the nature of the ocular phenotype associated with each mutation. Separately, we have also studied the biochemical phenotype of some of these mutant opsin by expression studies in COS cells.

2. INTRODUCTION

Retinitis pigmentosa (RP) (MIM 268000 - *Online Mendelian Inheritance in Man*: http:www.ncbi.nlm.nih.gov/omim) is the term applied to a clinically and genetically heterogeneous group of retinal degenerations which primarily affect the rod photoreceptors, and have an overall prevalence of about 1:3000[2]. RP is characterised by progressive loss of vision, initially manifesting as night blindness, with subsequent loss of the peripheral visual fields, and later frequently involving deterioration of central vision[2]. RP may be inherited as an autosomal recessive, autosomal dominant, digenic, or X-linked trait. Autosomal dominant RP (ADRP) accounts for 20-25% of all cases of RP[4].

Patients affected by the sectorial form of retinitis pigmentosa have predominant involvement of the inferior retinal with corresponding loss of the superior visual field. This is associated with a less severe attenuation of electroretinographic (ERG) responses than is observed in patients with diffuse RP. The visual prognosis in sectorial RP is relatively good with many patients retaining 20/30 (6/9) or better Snellen visual acuity throughout life.

New Insights Into Retinal Degenerative Diseases
Edited by Anderson *et al.*, Kluwer Academic/Plenum Publishers, 2001

Over the last eight years many rhodopsin mutants have been linked to autosomal dominant retinitis pigmentosa – the most common hereditary retinal degeneration that leads to photoreceptor death and subsequent severe loss of peripheral and night vision. Some of these mutant proteins have been purified and studied biochemically and biophysically. The mutant phenotypes fall into three classes[5]: Class I mutants are expressed at wild-type levels, form the normal rhodopsin chromophore with 11-*cis* retinal, and are transported to the cell surface. However, on illumination, they inefficiently activate transducin, the cognate G-protein of the photoreceptor cell. Class II mutants remain in the ER and do not bind 11-*cis* retinal to form the chromophore. Class III mutants are expressed at low levels and form the chromophore poorly. They also remain in the ER and, as expected, show high mannose glycosylation. Most of these mutants show abnormal bleaching kinetics and activate transducin inefficiently e.g. P23H. Greater than 80% of the mutants are either misfolded (Type II) or slowly folding (Type III) proteins that are retained in the ER. These results from heterologous expression of the opsin mutants in mammalian cells correlates with the findings in transgenic rhodopsin animals[6].

A number of mutations in rhodopsin have been described in autosomal dominant pedigrees with the sectorial RP phenotype (Asn-15-Ser, Thr-17-Met, Pro-23-His, Thr-58-Arg, Gly-106-Arg, Gly-182-Ser, and several mutations of codons 190 and 267)[2-4, 7-12]. With the exception of Pro-23-His these mutations have each been described in only a very small number of families.

We set out to determine the prevalence and clinical effects of rhodopsin mutations amongst patients with the sectorial RP phenotype in the U.K. Furthermore we have initially characterized some of the mutant opsin proteins associated with this form of RP.

3. MATERIALS AND METHODS

3.1. Mutation screening

After gaining informed consent from subjects, genomic DNA was prepared from peripheral blood. Oligonucleotide primers to the five exons of the *rhodopsin* gene (Genbank: U49742) were designed from the published genomic sequence[13]. PCR was performed in a 50μl reaction with 1μg of genomic DNA, 20 pmoles of each primer, 200μM dNTPs, 1.5mM MgCl$_2$ and 1 unit of Taq DNA polymerase. A three stage PCR consisting of 35 cycles of 94°C, 58°C, and 72°C, each for 1 minute, was used. The resulting product was allowed to cool slowly to room temperature to maximise the formation of heteroduplexes[14]. The PCR products generated were then electrophoresed overnight at 1600-2000Vh on MDE (Flowgen) polyacrylamide gels cast in 24 cm long, 1mm thick vertical gel tanks (Hoefer). Gels were then stained with ethidium bromide and photographed.

Once detected sequence variants were directly sequenced using an automated fluorescent sequencer (ABI Biosystems model 373). PCR products were purified and then reamplified using the FS kit (Perkin Elmer). Variants were sequenced in both the forward and reverse directions to confirm the location of sequence changes.

In addition to heteroduplex screening all DNA samples were subjected to two restriction enzyme digests (*DdeI* and *ApaI*) known to identify the mutant sequence of the Thr-58-Arg and Gly-106-Arg mutations respectively. Restriction digests were performed directly on 20μl of PCR product and incubated overnight at 37°C to ensure complete digestion. The resultant bands were photographed after running on a 2% agarose gel and staining with ethidium bromide.

3.2. Electrodiagnostic Assessment

Electrophysiologic assessment was undertaken on 6 patients. Electroretinography (ERG) was performed according to the guidelines for standard electrophysiology published by the International Society for Clinical Electrophysiology of Vision[15] (including rod ERG, bright white flash mixed rod-cone ERG, and 30-Hz cone-derived flicker ERG). Pattern ERG (PERG) was also performed according to standard guidelines.[16]

3.3. Mutant Opsin Expression and Initial Characterization

The mutant opsins T58R, Δ68-71 and G106R were constructed by standard mutagenesis techniques in the vector pMT4 which contains the wild-type opsin gene. Transfection of COS-7 cells with the wild-type and mutant vectors was as previously described.[17] The opsin from the transfected cells was reconstitution with 11-*cis* retinal and immunoaffinity purification by well-established methods.[18] Samples of the eluate were quantitated by UV/visible absorption spectroscopy.

4. RESULTS

The results of the mutation screening and the visual acuities of some patients are shown in Figure 1. The location of these mutations in a secondary structure model of human rhodopsin is shown in Figure 2.

No.	Mutation	Age	Acuity R	Acuity L
1	58	38	6/5	6/5
2	58	69	6/6	6/6
3	58	76	6/5	6/12
4	58	48	6/9	6/9
5	58	57	6/9	6/9
6	58	55	6/9	6/9
7	58	26	6/9	6/12
8	58	24	6/9	6/12
9	58	42	6/36	6/18
10	58	29	6/5	6/6
11	68-71del	56	6/6	6/5
12	106	46	6/9	6/9
13	190	34	6/12	6/9
14	269del	48	6/6	6/6
15	269del	42	6/5	6/6

Figure 1. Results of genetic screening of families with sectorial retinitis pigmentosa.

4.1. Thr-58-Arg mutation families

Twelve affected individuals (aged between 26 and 76 years) from the five families with the Thr-58-Arg mutation were examined.

These patients typically complained of nyctalopia from around the age of 35 years and loss of side vision within 5 years of the onset of nyctalopia. Visual field loss forced the older subjects to stop driving at around 55 years of age.

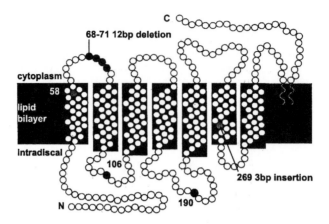

Figure 2. Secondary structure model of human rhodopsin. The dark colored circles represent mutation sites.

11 of these 12 patients had a Snellen visual acuity of 20/30 or better in each eye. A mild posterior subcapsular cataract was found in one eye of only one individual (aged 69 years). Fundus examination of 26 and 29 year old subjects revealed a small area of bone spicule pigmentation in the inferior retina. In older patients a larger area of pigmentation and atrophy was observed in the inferior and inferonasal retina in association with attenuation of inferior retinal vessels and moderate disc pallor (Figure 3).

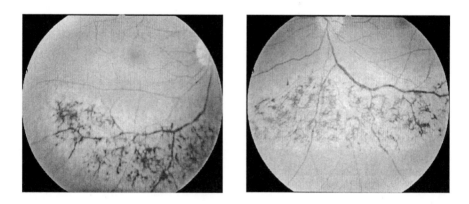

Figure 3. Fundus photographs illustrating the restricted region of clinically observable retinal changes.

In all subjects visual field loss predominantly affected the superior field, but with considerable inter- and intra-familial variation in severity. One 69 year old patient (3-II:1) exhibited field loss entirely confined to the upper fields), whilst another 26 year old subject (1-III:5) had significant loss of inferior field under both photopic and scotopic conditions.

The results of electrodiagnostic testing in patients (3-II:1 and 3-III:1) are shown in Figure 4. These demonstrate that there may be no unequivocal abnormality in affected individuals under the age of 40 years, whilst a marked reduction in the amplitude of the b wave of both the photopic and scotopic ERGs, and a reduction in the light rise of the EOG, occurs in older patients. Significant reductions in the amplitude of both the P50 and N95 components of the Pattern ERG confirm involvement of the macula in the later stages of the disease process. Cone b-wave implicit times remained normal in younger patients, but were prolonged in some patients over the age of 50 years.

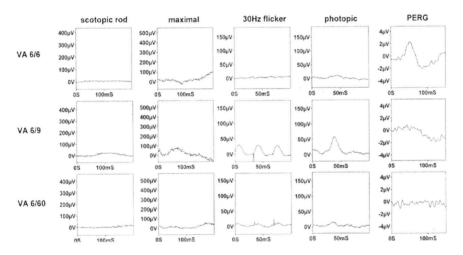

Figure 4. Electroretinographic characteristics of two different patients with rhodopsin mutations associated with sectorial retinitis pigmentosa.

4.2. Gly-106-Arg mutation families

Three affected members of this family, aged 36 to 74 years, were examined. Their fundi and visual fields appeared very similar to those of age matched members of the Thr-58-Arg families. However, patients with Gly-106-Arg typically had better preservation of ERG a and b wave amplitudes (dark adapted bright white flash, blue light, and 30Hz flicker) than those with Thr-58-Arg. 30Hz flicker b wave implicit times were normal in younger subjects, but were markedly delayed in the 74 year old patient.

4.3. Codon 269 three base pair insertion family

The 48 year old proband from this family was completely asymptomatic and was referred following a routine examination by her optometrist. Her visual acuity was

20/20 in each eye. Fundal examination revealed an area of bone spicule pigmentation inferiorly in both retinas that was smaller and more discrete than that typically seen in age matched individuals from the other families. Her remaining retina appeared healthy.

The amplitudes and implicit times of her photopic and scotopic ERGs and the Arden ratio of her EOG were all within normal limits.

4.4. Mutant Protein Spectra

Spectra of the immunoaffinity purified wild-type, T58R, Δ68-71 and G106Rmutant proteins are shown in Figure 5. Note that the total amount of mutant chromophore is less than that of wid-type (A_{500}).

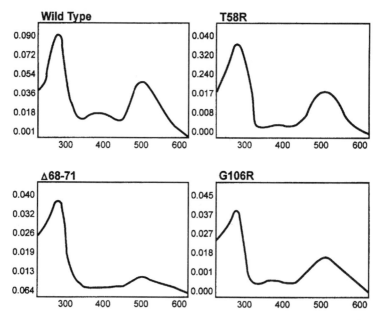

Figure 6. UV/visible spectrum of wild-type and 3 of the mutants associated with sectorial retinitis pigmentosa. Measurements on the X-axis are nm, on the Y-axis OD.

5. DISCUSSION

Rhodopsin mutations were identified in 11 out of 21 apparently unrelated families with sectored retinitis. From our data, only 5 different mutations accounted for all the patients. This may be, at least in part, due to a founder effect which is being investigated. One intriguing unresolved question is why such mutations that are presumably found in all rod photoreceptors is clinically manifest in only a restricted region of the retina.

Most patients in our study maintained good central vision, even in those greater than 50. Electrophysiologically most of patients had a significant loss of rod function with some degree of cone dysfunction.

The three studied *in vitro*, T58R, Δ68-71 and G106R are found in different regions of the rhodopsin protein – the first transmembrane helix, furst cytoplasmic loop and third transmembrane helix respectively.

All three mutant opsins could be expressed in COS-7 cells but at levels much less than the wild-type protein. Furthermore, they can bind 11-*cis* retinal in the dark but the spectral ratio is much greater than that of the wild-type protein. In general, the spectral ratio of rhodopsin (A_{280}/A_{500}) reflects the extent of regeneration i.e., the higher the ratio the greater the unregenerated fraction. Pure rhodopsin has a spectral ratio of 1.6. T58R has a near wild-type spectral ratio while Δ68-71 has a high ratio, 6.5, when the proteins were eluted from an immunoaffinity column. Such variation in the spectral ratio and low levels of protein recovered from the cell would imply at least four possibilities. One, the rate and extent of folding is different in for the members of the class III mutants. Indeed, it is known that proteins which fold slowly or are misfolded are retained in the ER where they may eventually be targeted for degradation[19-21]. Second, the rate of translation could be different in the mutants. Third, the kinetics of degradation could be accelerated for the mutants. Fourth, there are non-opsin proteins that associate with the mutants and are co-eluted during the immunoaffinity purification. We are investigating these possibilities to further elucidate the molecular mechanisms of rhodopsin related RP.

Regardless, one important conclusion of this preliminary work is that only a fraction of the expressed mutant opsin can bind to 11-*cis* retinal. This is similar to other class III mutant opsins that are misfolded. Such misfolded proteins that are retained in the endoplasmic reticulum have been associated with other inherited disorders including α_1-antitrypsin disease, cystic fibrosis, hypercholesterolemia and diabetes insipidus.[22] The retention of such proteins has been associated with the demise of the cell.

6. ACKNOWLEDGEMENTS

Funding for this research was provided by grants from the Medical Research Council and the Foundation Fighting Blindness.

REFERENCES

1. S. K., Department of Ophthalmology, University of Minnesota. D.A.R.B., Department of Molecular Genetics. A.M.P., Institute of Ophthalmology. S.M.D., Moorfield Eye Hospital, London. G.E.H., Department of Molecular Genetics and Institute of Ophthalmology. A.C.B., Moorfield Eye Hospital, London and Institute of Ophthalmology.
2. Bundey S, Crews SJ. (1984) A study of retinitis pigmentosa in the city of Birmingham - I. Prevalence. *J. Med. Genet.* **21**: 417-420
3. Bird AC. (1995) Retinal photoreceptor dystrophies. *Am. J. Ophthalmol.* **119**: 543-562
4. Jay M. (1982) On the heredity of retinitis pigmentosa. *Br. J. Ophthalmol.* **66**: 405-416
5. Kaushal, S. & Khorana, G. Structure and function in rhodopsin. 7. Point mutations associated with autosomal dominant retinitis pigmentosa. *Biochemistry* **33**, 6121-6128 (1994).
6. Roof, D.J., Adamian, M. & Hayes, A. Rhodopsin accumulation at abnormal sites in retinas of mice with a human P23H rhodopsin transgene. *Invest Ophthalmol Vis Sci* **35**, 4049-62 (1994).

7. Sullivan L, Makris M, Dickinson P, Mulhall M, Forrest S, Cotton R, Loughnan M. A new codon 15 rhodopsin gene mutation in autosomal dominant retinitis pigmentosa is associated with sectorial disease. *Arch. Ophthalmol.* **111**: 1512-7 (1993)

8. Hayakawa M., Y. Hotta, Y. Imai, K. Fujiki, A. Nakamura, K. Yanashima & A. Kanai. Clinical features of autosomal dominant retinitis pigmentosa with rhodopsin gene codon 17 mutation and retinal neovascularization in a Japanese patient. *Am J Ophthalmol* **115**: 168-73 (1993)

9. Stone E.M., A. E. Kimura, B. E. Nichols, P. Khadivi, G. A. Fishman & V. C. Sheffield. Regional distribution of retinal degeneration in patients with the proline to histidine mutation in codon 23 of the rhodopsin gene. *Ophthalmology* **98**: 1806-13 (1991)

10. Berson E.L., B. Rosner, M. A. Sandberg & T. P. Dryja. Ocular findings in patients with autosomal dominant retinitis pigmentosa and a rhodopsin gene defect (Pro-23-His). *Arch Ophthalmol* **109**: 92-101 (1991)

11. Fishman G.A., E. M. Stone, L. D. Gilbert, P. Kenna & V. C. Sheffield. Ocular findings associated with a rhodopsin gene codon 58 transversion mutation in autosomal dominant retinitis pigmentosa. *Arch Ophthalmol* **109**: 1387-93 (1991)

12. Fishman, E. M. Stone, L. D. Gilbert & V. C. Sheffield. Ocular findings associated with a rhodopsin gene codon 106 mutation. Glycine-to-arginine change in autosomal dominant retinitis pigmentosa. *Arch Ophthalmol* **110**: 646-53 (1992)

13. Nathans J, Hogness DS. Isolation and nucleotide sequence of the gene encoding human rhodopsin. *Proc. Natn. Acad. Sci.* **81**: 4851-4855 (1984)

14. Keen J, Lester D, Inglehearn CF, Curtis A, Bhattacharya SS. (1991) Rapid determination of single base pair mismatches as heteroduplexes on Hydrolink gels. *Trends Genet.* **7** (1): 5

15. Marmor MF, Holder GE, Porciatti V, Trick G, Zrenner E. (1996) Guidelines for pattern electroretinography: recommendations by the International Society for Clinical Electrophysiology of Vision. *Doc Ophthalmol.* **91**: 291-298.

16. Karnik, S.S., Sakmar, T.P., Chen, H.B. & Khorana, H.G. Cysteine residues 100 and 187 are essential for the fomration of correct structure in bovine rhodopsin. *Proc Natl Acad Sci USA* **85**, 8459-8463 (1988).

17. Doi, T., Molday, R.S. & Khorana, H.G. Role of the intradiscal domain in rhodopsin assembly and function. *Proc Natl Acad Sci USA* **87**, 4991-4995 (1990).

18. Ellgaard, L., Molinari, M. & Helenius, A. Setting the standards: quality control in the secretory pathway. *Science* **286**, 1882-1888 (1999).

19. Gething, M.J. & Sambrook, J. Protein folding in the cell. *Nature* **355**, 33-45 (1992).

20. Matouschek, A. Recognizing misfolded proteins in the endoplasmic reticulum. *Nature Struct.Biol.* **7**

21. Aridor M, Balch WE Integration of endoplasmic reticulum signaling in health and disease *Nature Med.* **7**, , 265-266 (2000).

CHRONIC ADMINISTRATION OF PHENYL N-*TERT*-BUTYLNITRONE PROTECTS THE RETINA AGAINST LIGHT DAMAGE

Isabelle Ranchon[1,3], Jeremy White[1], Sherry Chen[1,3], Kathleen Alvarez[1,3], Yashige Kotake[4] and Robert E. Anderson[1,2,3]

1. INTRODUCTION

Retinal dystrophies comprise a large number of disorders characterized by a slow and progressive retinal degeneration. Most are the results of mutations expressed in either photoreceptor or retinal pigment epithelial cells. There is currently no treatment by which the primary disorder can be modified, although researchers around the world are working on different potential therapeutic approaches to treat retinal degenerations. These are based on the possibility of transfecting the photoreceptor or retinal epithelial cells with a functioning gene, transplanting photoreceptor or retinal pigment epithelial cells into the sub-retinal space, using electronic devices to stimulate the retina, or administrating drugs to enhance the repair/defense systems of the retina. Although encouraging success has been shown in animals models, gene therapy[1], transplantation[2] or injection of growth factors[3,4,5] require invasive intraocular procedures and some have significant side effects.

Oral or systemic administration of medical drugs is currently the only non-invasive therapeutic approach. Vitamin A supplementation has been shown to slow the progression of inherited retinal degeneration, although this has been subject of controversy[6,7]. In animal models, systemic administration of antioxidants has been tested with some success[8,9,10]. In a previous study, we showed that systemic administration of phenyl-N-*tert*-butylnitrone (PBN) protects the morphological and functional properties of the retina against the damaging effects of light exposure[11]. PBN has well-defined free radical trapping capabilities[12,13] and, given the role of oxidation in retinal light damage[8,9,10], it seems logical to propose that by scavenging free radicals, PBN acts early

Departments of [1]Ophthalmology and [2]Cell Biology, University of Oklahoma Health Sciences Center and [3]Dean A. McGee Eye Institute and [4]Oklahoma Medical Research Foundation, Oklahoma City, OK, 73104.

in the degenerative cascade of events and prevents cell death. Oxidative processes have been proposed to play a causative or contributing role in inherited retinal degenerations and age-related macular degeneration[14]. Therefore, PBN and related compounds are potential candidates as therapeutic agents.

In our previous study, we showed that PBN rapidly enters into the retina after IP injection, but has a short lifetime and requires injection every 6 hours to be effective[11]. Our goal is to test this molecule on inherited retinal degenerations, which requires chronic administration of the compound. Therefore, we tested the use of an osmotic pump, which allows a continuous delivery of PBN. Albino rats were implanted with an osmotic pump containing PBN in water and exposed to bright light for 24 hours, after which their retinal function and morphology was evaluated. Our results show that chronic administration of PBN by an osmotic pump protects the retina from light damage.

2. MATERIALS AND METHODS

2.1. Materials

Phenyl-N-*tert*-butylnitrone (PBN) was synthesized as previously described . Phenacetin was from Sigma. HPLC solvents were from Fisher Scientific. Alzet osmotic pumps were from Alza Coorporation (Newark, DE).

2.1.1. Animals

Sixteen adult male albino Wistar rats (150-200 g weight) were born and kept in dim cyclic light (12 hrs ON; 12 hrs OFF; 5-10 lux). They were fed lab chow *ad libidum* and had free access to water. The animal care strictly conformed to the Association for Research in Vision and Ophthalmology (ARVO) statement for the Use of Animals in Vision and Ophthalmic Research and the University of Oklahoma Health Sciences Center Guidelines for Animals in Research. All protocols were reviewed and approved by the Institutional Animal Care and Use Committees of the OUHSC and Dean A. McGee Eye Institute.

2.1.2. Treatment

Rats were divided into 2 groups and implanted with a 2 ml osmotic pump containing either PBN (25 mg/ml) in water (PBN group) or only water (water group). The delivery flow was 10 μl/hr, which resulted in each animal receiving 6 mg of PBN per day. This is the maximum amount that could be administrated because of the limited solubility of PBN in water and the size of the pump. Rats were implanted 40 hours before exposure to the damaging light.

2.1.3 Light Damage paradigm

Light exposure was for 24 hr to fluorescent lights (cool white, 34 watt) of 1700 lux intensity at the eye level of the animals. During exposure, the rats had free access to food and water. After exposure, the animals were placed in darkness for 24 hr and then returned to the dim cyclic light for four days.

2.1.4. Electroretinography

Rats were dark-adapted overnight and prepared under dim red light for the ERG study. They were anesthetized with intramuscular injections of ketamine (120 mg/kg body weight) and xylazine (6 mg/kg body weight). One drop of 1% tropicamide was applied to the cornea to dilate the pupil and one drop of 0.5% proparacaine HCl was applied for local anesthesia. The white light stimulus used to evoke ERGs was delivered in 10 ms pulses by an integrating sphere (Labsphere, North Sutton, USA) at an interval between flashes of 60 sec. These conditions have been shown to be sufficient to prevent light adaptation to the light flashes[15]. ERGs were recorded with gold electrodes at 21 intensities presented in ascending order, starting below threshold, in order to get the b-wave sensitivity curves. The software program Origin 6.0 (Microcal, Inc, Northampton, MA, USA) was used to fit the data of each rat, giving the saturated b-wave amplitude (B_{max}) and the luminance that induces an amplitude $B_{max}/2$ (K).

ERGs were recorded before treatment and light exposure, and then four days after light exposure.

2.1.5 Histology

Rats were killed after the last ERG (4 days after the light exposure) for light microscopic evaluation of retinal structure. Immediately after death, eyes were excised, placed in Perfix fixative, and embedded in paraffin. Sections of 5 μm were cut along the vertical meridian through the optic nerve. The thickness of the outer nuclear layer (ONL) was measured at 0.5 mm distances from the optic nerve to the inferior and superior ora seratta[16]. The area under the curves was integrated using the Origin 6.0 program (Microcal Software, Inc, Northampton, MA, USA).

2.1.6. Statistical analysis

Results are plotted as mean ± SD. Significant differences across groups were assessed using an unpaired t-test for the ERG and the histology, with a level of significance set at p=0.05.

3. RESULTS

3.1. Retinal function

Electroretinograms were recorded for 8 rats from each group before any treatment or light-exposure and four days after light exposure. The b-wave sensitivity curves from the 2 groups (control and PBN) recorded before any treatment or light-exposure were not significantly different and therefore were combined, and represent the control curve in Figure 1. In the group that was implanted with the pump containing only water, 24 hours of exposure to 1700 lux-light caused a large reduction of the retinal function, evidenced by the collapse of the sensitivity curve (Figure 1A). The maximal b-wave B_{max} was reduced to 19 % of the control value (p<0.001) and the luminance K decreased by 25 %.

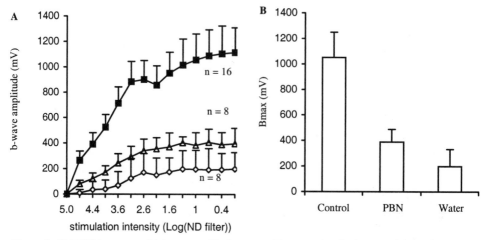

Figure 1: (A) ERG b-wave sensitivity curves. Black squares (■) are results of animals tested before treatment and light exposure. Diamonds (◇) and triangles (Δ) are results from animals tested after constant light exposure that had been implanted with osmotic pumps containing only water or PBN in water, respectively. The results are expressed as a mean voltage ± SD (n = 8 or n=16 for each point). (B) Relative area units under the sensitivity curves. The results are expressed as a mean area units ± SD (n = 8 or 16 for each group).

When the animals were implanted with an osmotic pump containing PBN and exposed to the damaging light, retinal function was better preserved. B_{max} was 37 % of the control value and significantly higher than B_{max} of the water group (Figure 1B). K was not significantly different from the control group (results not shown).

These experiments show that chronic administration of PBN significantly preserves retinal function.

3.2. Retinal Structure

Figures 2 and 3 show the morphologic and morphometric analysis of each experimental group, respectively. Exposure to a 1700 lux light for 24 hours caused extensive damage to the retinas of the animals implanted with a pump containing only water. Photomicrographs of different retinal regions (Figure 2) show that once the debris in the ONL was removed, only a few rows of nuclei remained, indicating an extensive loss of photoreceptors (a control retina has 11-12 rows). The extent of damage is evidenced by the large loss of ONL thickness (Figure 3A) over all the retina, with the superior part being more affected. When the rats were implanted with the pump containing PBN, the ONL was thicker in the inferior and central part of the retina, indicating that some photoreceptors had been preserved in these regions. Integration of the area under the ONL thickness curves gave a significantly (p=0.018) higher value for the group implanted with pump containing PBN compared to the pump containing only water.

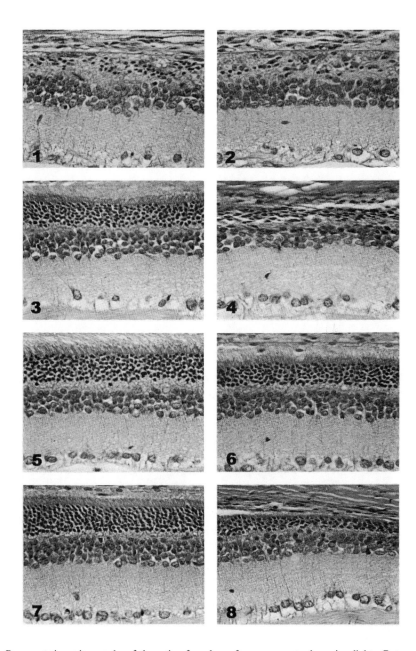

Figure 2: Representative micrographs of the retina four days after exposure to damaging light. Rats were implanted with pumps containing only water *(right column)* or PBN in water *(left column)*. (**1, 2**) 2 mm from the optic nerve in the superior retina; (**3, 4**) 0.5 mm from the optic nerve in the superior retina; (**5, 6**) 2 mm from the optic nerve in the inferior retina; (**7, 8**) 0.5 mm from the optic nerve in the inferior retina.

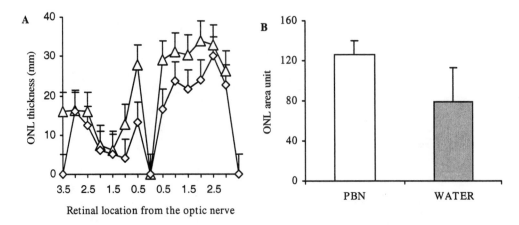

Figure 3: (A) Outer nuclear layer (ONL) thickness along the vertical meridian (superior retinal values are left and inferior values right of the minimal point represented by the optic nerve). Rats were implanted with pump containing only water (✧) or PBN in water (Δ). The results are expressed as a mean ONL thickness ± SD (n = 8 for each group). (B) Relative area units under the curves from the ONL thickness measurements. The results are expressed as a mean area units ± SD (n = 8 for each group).

4. DISCUSSION

This study shows that chronic administration of PBN can protect the morphologic and functional properties of the retina against the damaging effect of constant light exposure. When the animals were implanted with the pump containing only water and then exposed to the damaging light, we observed the characteristic retinal light-damage. The saturated amplitude reduction can be explained by the loss of photoreceptors since the ONL thickness is also reduced, and K changes suggests that four days after the end of the light-exposure, the remaining photoreceptor integrity was still altered[10, 15]. The superior part of the retina was more damaged than the inferior area[15, 16]. When animals were implanted with the pump containing PBN, the saturated amplitude and ONL thickness were significantly higher compared to the water group, implying that PBN prevented the death of some photoreceptor cells. In addition, the surviving photoreceptor cell integrity was preserved since K did not vary.

These results confirm the efficiency of PBN already observed in our previous experiment where PBN was injected every 6 hours over the 24 hours of constant light exposure at 50 mg/kg/injection[11]. It is noteworthy that the amount of PBN delivered by the pump was only 6 mg over the 24 hours of light-exposure compared 50-75 mg in the first experiment. Therefore, it is not surprising that the level of protection (37 %) was less than that observed in the earlier study (~80% protection). However, these results demonstrate that it is possible to administrate PBN chronically to animals without provoking any evident toxicity, and that the drug is efficacious in preventing light damage. Studies are currently underway to determine the effects of providing PBN in drinking water.

Since the suggestion that light induces oxidative reactions resulting in damage to the photoreceptor cells[17], evidence that lipid peroxidation occurs in the retina in response to light has gradually accumulated. An increase in the level of peroxides in ROS extracts[18] and the specific loss of docosahexaenoic acid (22:6n-3) from the rod outer segment membranes[19] have been shown after exposure to intense light. The hypothesis that oxygen free radicals mediate lipid peroxidation[20] has been supported by *in vivo* studies from several laboratories[19,21,22,23]. Since constant light damage presents an oxidative stress to the retina, it is remarkable to observe a significant protection by a so small amount of PBN.

PBN has been shown to affect several signaling pathways, such as preventing the induction of iNOS[24,25], inhibiting the activation of the transcription factor NF-κB[24], inhibiting the expression of multiple cytokine genes[26,27], activating transcription factors[26,27], inhibiting expression of multiple apoptosis-associated genes[28], and down-regulating the MAPK pathway[29]. All of these PBN-effects could be, of course, secondary to the radical scavenging properties. However, PBN alone has recently been shown to reduce basal protein phosphorylation[30] as well as to up-regulate the expression of heat shock proteins such as hsp27[29]. Therefore, the protection provided by PBN may be from a cumulative effect of its multiple pharmacologic activities.

In most cases of retinal degeneration the manifestation of the disorder is similar regardless of the specific genetic lesion, mediated by a common induction of apoptosis, which results in photoreceptor cell degeneration and blindness. Oxidative stress has been implicated in several inherited retinal degenerations as well as in age-related macular degeneration[14]. Consequently, PBN or related compounds might be good candidates as medical therapeutic agents in the treatment of neurodegenerative diseases.

5. ACKNOWLEDGEMENTS

We thank Drs. Robert Floyd, and Kenneth Hensley for helpful discussions, and Mark Dittmar for technical help. This work was supported by grants from the National Institutes of Health/National Eye Institute (EY00871, EY04149, and EY12190), Research to Prevent Blindness, Inc., New York, NY; The Foundation Fighting Blindness, Baltimore, MD; Samuel Roberts Nobel Foundation, Inc., Ardmore, OK; The IPSEN Foundation, Paris; and Presbyterian Health Foundation, Oklahoma City, OK.

REFERENCES

1. M. Cayouette, D. Behn, M. Sedtner, P. Lachapelle, C. Gravel, Intraocular gene transfer of ciliary neurotrophic factor prevents death and increases responsiveness of rod photoreceptors in the retinal degeneration slow mouse, *J. Neurosci.* **18,**9282-9293 (1998).
2. F. Ghosh, B. Ehinger, Full-thickness retinal transplants: a review. *Ophthalmologica* **214,**54-69 (2000)
3. M.M. LaVail, D. Yasumura, M.T. Matthes, C Lau-Villacorta, K Unoki, CH Sung, RH Steinberg, Protection of mouse photoreceptors by survival factors in retinal degenerations, *Invest. Ophthalmol. Vis Sci.* **39**, 592-602 (1998)

4. E.G. Faktorovich, R.H. Steinberg, D. Yasumura, M.T. Matthes, M.M. LaVail Photoreceptor degeneration in inherited retinal dystrophy delayed by fibroblast growth factor, Nature **347**,83-86 (1990)
5. M.M LaVail, K. Unok, D. Yasumura, M.T. Matthes, G.D. Yancopoulos, R.E. Steinberg, Multiple growth factors, cytokines and neurotrophins rescue photoreceptors from the damaging effects of constant light. *Proc. Natl. Acad. Sci. USA* **89**,11249-11253 (1992)
6. E.L. Berson, B. Rosner, M.A. Sandberg, K.C. Hayes, B.W. Nicholson, C. Weigel-DiFranco, W.A. Willett, Randomized trial of Vitamin A and E supplementation for retinitis pigmentosa. *Arch. Ophthalmol.* **111**,761-772 (1993).
7. R.W. Massof, D. Finkelstein, Supplemental Vitamin A retards loss of ERG amplitude in retinitis pigmentosa. Arch. Ophthalmol. **111**,751-754 (1993)
8. D.T. Organisciak, R.M. Darrow, I.R. Bicknell, Y.L. Jiang, M. Pickford, J.C. Blanks, Protection against retinal light damage by natural and synthetic antioxidants. In: *Retinal Degenerations*, edited by R.E. Anderson, J.G. Hollyfield, M.M. LaVail (Plenum Press, New York and London, 1991), pp.189-201.
9. D.T. Organisciak, R.M. Darrow, Y-L. Jiang, G.E. Marak, J.C. Blanks, Protection by dimethylthiourea against retinal light damage in rats. *Invest. Ophthalmol. Vis. Sci.* **33**,1599-1609 (1992)
10. I. Ranchon, J.M. Gorrand, J. Cluzel, M-T. Droy-Lefaix, M. Doly Functional protection of photoreceptors from light-induced damage by dimethylthiourea and Ginkgo biloba extract. *Invest. Ophthalmol. Vis. Sci.* **40**,1191-1199 (1999).
11. I. Ranchon, J. White, S. Chen, K. Alvarez, R.E. Anderson, Systemic Administration of Phenyl-N-*tert*-butylnitrone Protects the Retina from Light Damage. Submitted to Invest. Ophthalmol. Vis. Sci. (2000)
12. J.L. Poyer, R.A. Floyd, P.B. McCay, K.G. Janszen, E.R. David. Spin trapping of the trichloromethyl radical produced during enzymic NADPH oxidation in the presence of carbon tetrachloride or bromotrichloromethane. *Biochim. Biophys. Acta,* **539**,402-409 (1978).
13. J.L. Poyer, P.B. McCay, E.K. La, K.G. Janzen, E.R. David, Confirmation of assignment of the trichloromethyl radical spin adduct detected by spin trapping during 13C-carbon tetrachloride metabolism in vitro and in vivo. *Biochem. Biophys. Res. Commun.* **94**,1154-1160 (1980).
14. B.S. Winkler, M.E. Boulton, J.D. Gottsch, P. Sternberg, Oxidative damage and age related macualr degeneration, *Molecular Vision,* **5**,32 (1999).
15. I. Ranchon, J.M. Gorrand, J. Cluzel, J.C. Vennat, M. Doly. Light-induced variations of retinal sensitivity in rats. Curr. Eye Res. **17**, 14-23 (1998).
16. L.M. Rapp, M.I. Naasn, R.D. Wiegand, et al. Morphological and biochemical comparison between retinal regions having differing susceptibility to photoreceptor degeneration. In: *Retinal Degeneration: Experimental and Clinical Studies*, edited by M.M. La Vail, J.G. Hollyfield, R.E. Anderson (New York: Alan R Liss, 1985) pp. 421-437.
17. W.K. Noell, V.S. Waker, B.S. Kang, S. Berman. Retinal damage by light in rats. *Invest. Ophthalmol.,* **5(5)**, 450-473 (1966).
18. R.D. Wiegand, N.M. Giusto, L.M. Rapp, R.E. Anderson. Evidence for rod outer segment lipid peroxidation following constant light illumination of the rat retina. *Invest. Ophthalmol. Vis Sci.,* **24**,1433-1435 (1983).

19. D.T. Organisciak, H.M. Wang, Z.Y. Li, M.O.M. Tso, The protective effect of ascorbate in retinal light damage of rats. *Invest. Ophthalmol. Vis. Sci.* **26,**1580-1588 (1985).
20. L. Feeney, E.R. Berman, Oxygen toxicity: membrane damage by free radicals. *Invest. Ophthalmol. Vis. Sci.* **15,** 789-792 (1976).
21. D.T. Organisciak, I.R. Bicknell, R.M. Darrow, The effect of L and D ascorbic acid administration on retinal tissue levels and light damage in rats. *Curr. Eye. Res.* **11,**231-241 (1992).
22. W.T. Ham, H.A. Muller, J.J. Ruffolo, J.J.E. Millen, S.F. Cleary, R.K. Guerry, D Guerry, Basic mechanisms underlying the production of photochemical lesions in the mammalian retina. *Curr. Eye. Res.* **3,**165-174 (1984).
23. J. Li, D.P. Edward, T.T. Lam, M.O.M. Tso, Amelioration of retinal photic injury by a combination of flunarizine and dimethylthiourea. *Exp. Eye Res.* **56,**71-78 (1993).
24. Y. Kotake, H. Sang, T. Miyajima, G.L. Wallis, Inhibition of NF-kappaB, iNOS mRNA, COX2 mRNA, and COX catalytic activity by phenyl-N-*tert*-butylnitrone (PBN). *Biochim. Biophys. Acta,* 1448:77-84 (1998).
25. T. Tabatabaie, K.L. Graham, A.M. Vasquez, R.A. Floyd, Y. Kotake, Inhibition of the cytokine-mediated inducible nitric oxide synthase expression in rat insulinoma cells by phenyl N-*tert*-butylnitrone. *Nitric Oxide* **4,**157-67 (2000).
26. H. Pogrebniak, M. Merino, S. Hahn, J. Mitchell, H. Pass, Spin trap salvage from endotoxemia: the role of cytokine down-regulation, Surgery **112,**130-139 (1992).
27. H. Sang, G.L. Wallis, C.A. Stewar, Y. Kotake, Expression of cytokines and activation of transcription factors in lipopolysaccharide-administered rats and their inhibition by phenyl N-*tert*-butylnitrone (PBN). *Arch. Biochem. Biophys.* 363:341-348 (1999).
28. C.A. Stewart, K. Hyam, G. Wallis, et al. Phenyl-N-*tert*-butylnitrone demonstrates broad-spectrum inhibition of apoptosis-associated gene expression in endotoxin-treated rats. *Arch. Biochem. Biophys.* 363,:71-74 (1999).
29. M. Tsuji, O. Inanami, M. Kuwabara. Neuroprotective effect of alpha-phenyl-N-*tert*-butylnitrone in gerbil hippocampus is mediated by the mitogen-activated protein kinase pathway and heat shock proteins. *Neurosci. Lett.* **282,**41-44 (2000).
30. K.A. Robinson, C.A. Stewart, Q. Pye, R.A. Floyd, K. Hensley. Basal phosphorylation is decreased and phosphatase activity increased by an antioxidant and a free radical trap in Primary rat glia. *Arch. Biochem. Biophys.* **365,**211-215 (1999).

SYSTEMS FOR DELIVERY OF VITAMIN A
TO THE RETINA IN RETINITIS PIGMENTOSA

Ekaterina M. Semenova, Alan Cooper, Malcolm W. Kennedy, Clive G. Wilson, and Carolyn A. Converse[*]

1. INTRODUCTION

Vitamin A (all-trans-retinol) plays a fundamental role in the visual cycle. There have been several studies establishing a link between retinal diseases and abnormalities in the absorption, transport and delivery of vitamin A[1-4] and oral vitamin A has been reported to retard the progression of retinitis pigmentosa (RP)[1]. Large doses of vitamin A administrated orally were used in treatment of Sorsby's fundus dystrophy (SFD)[5] and of a mouse model of Leber congenital[6]. The studies were based on a hypothesis that the night blindness in these conditions is due to a chronic deprivation of vitamin A at the level of the photoreceptors caused by a thickened membrane barrier between the photoreceptor layer and its blood supply. It was found that vitamin A supplementation could lead to dramatic restoration of photoreceptor function, demonstrating that pharmacological intervention has the potential to improve vision in otherwise incurable genetic retinal degenerations.

Systemic administration of drugs into ocular tissues is often considered as first option for posterior eye diseases involving the optic nerve, retina and uveal tract. However, the systemic route has the significant disadvantage that all the organs of the body are subject to the action of the drug, when only a very small volume of tissue in the eye may need the treatment. Poor intraocular penetration is also observed due to the presence of an efficient blood-retinal barrier. Because the long-term systemic consumption of vitamin A could lead to hypervitaminosis A, with a broad spectrum of clinical abnormalities[7, 8], the direct delivery of high dosage of vitamin A to the retina may be desirable. The aim of this study has been to evaluate potential delivery of

---

[*] Ekaterina M. Semenova, Clive G. Wilson, and Carolyn A. Converse, University Of Strathclyde, Glasgow, Scotland G4 0NR. Alan Cooper, and Malcolm W. Kennedy, University of Glasgow, Glasgow, Scotland G12 8QQ.

New Insights Into Retinal Degenerative Diseases
Edited by Anderson *et al.*, Kluwer Academic/Plenum Publishers, 2001

vitamin A to the retina using cyclodextrins as the drug delivery systems. Research was also carried out to investigate the ability of interphotoreceptor retinoid-binding protein (IRBP) to bind all-trans-retinol and interference by fatty acids and some plasticisers. Continuous shuttling of retinoids between photoreceptors and pigment epithelium across the aqueous compartment that separates them, the interphotoreceptor matrix (IPM), is a critical part of the visual cycle and IRBP has been implicated in this process, although recent transgenic mouse studies cast some doubt on this mechanism[9]. IRBP, a 140 kDa retinoid and fatty acid binding glycoprotein, is the major soluble protein component of the IPM.

An aqueous transfer mechanism requires drug molecules to be sufficiently water-soluble, however retinal is poorly water-soluble and subject to chemical degradation. The pharmaceutical use of cyclodextrins (CDs) is confined mainly to the complexation with poorly soluble, unstable, irritating and difficult to formulate substances. CDs are a group of cyclic oligosaccharides with a hydrophilic outer surface consisting of six, seven or eight glucose units, named α-, β- or γ-cyclodextrin, respectively. To improve their physicochemical and biological properties the molecular structure of these natural CDs have been modified[10]. Cyclodextrins form inclusion complexes with many drugs by taking up a whole molecule, or some lipophilic part of it, into the hydrophobic central cavity. The drug displaces the water in the cavity and is held in place by hydrophobic interactions. In aqueous solution the complexes are readily dissociated. The release of the drug from the complex is achieved by using just enough cyclodextrin to solubilise the drug[11]. Thus, through CD complexation it is possible to increase aqueous solubility of some lipophilic water-insoluble drugs without affecting their intrinsic abilities to penetrate amphipathic membranes. Owing to its polyenic structure, all-trans-retinol is virtually insoluble in water and chemical labile, and it easily undergoes isomerisation and photooxidation, but these problems may be alleviated by cyclodextrin inclusion. Cyclodextrins have been widely used in analytical fluorescence techniques because the formation of inclusion complexes of hydrophobic fluorophores increases their apparent solubility and therefore improves the sensitivity of the detection[12-14]. Very often, this complex formation is associated with an enhancement of the fluorescence quantum yield as a consequence of the protection of the exited state. The fluorescence properties of a variety of all-*trans*-retinol-CD inclusion complexes have been studied in order to evaluate their potential as drug delivery systems.

2. METHODS

IRBP was prepared from the bovine eyecup and excised retinas which were incubated on the ice with PBS-P-A (phosphate-buffered saline, AEBSF, azide, pH 7.4) for 15 min and subjected to centrifugation[15]. IRBP was purified from the supernatant by Concanavalin A Sepharose affinity chromatography followed by a DEAE Sephacel column. SDS-PAGE showed a single protein band at molecular weight 140 kDa. High sensitivity fluorimetry on a SPEX FluorMax spectrofluorimeter was used to measure the binding of vitamin A to the first binding site of bovine IRBP and its displacement by oleic acid, phytanic acid and some plasticisers. Ligand binding was monitored by following either the fluorescence of retinol (excitation, 350nm; emission, 475nm) or by following the intrinsic tryptophan fluorescence of the protein (excitation, 280nm; emission, 333nm). Fluorescence data were corrected for dilution where necessary and fitted by standard nonlinear regression techniques (using Microcal ORIGIN software) to a single noncompetitive binding model to give estimates of the dissociation constant (K_d).

Spectroscopic properties of a range of CDs and vitamin A inclusion complexes have been evaluated for ease of detection in aqueous environment. Appropriate adjustment of buffer pH, amount of ethanol and the concentration of CDs necessary for stabilising and solubilizing vitamin A were studied. The fluorescence enhancement which occurs when certain CDs encapsulate all-trans-retinol was followed over a 2 hour period.

3. RESULTS

Several fatty acids are capable of displacing vitamin A from IRBP, however, using fluorescence-based studies we found a remarkably high selectivity for oleic acid ($K_d \sim 10^{-8}$M), a phenomenon which is being investigated further (Figure 1).

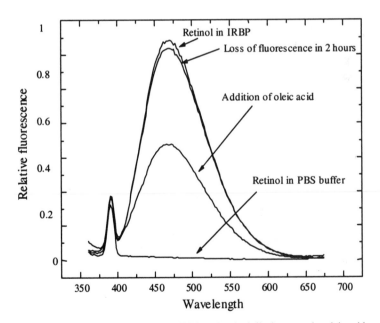

Figure 1. Retinol binding activity of IRBP and retinol displacement by oleic acid.

Competition titrations that were carried out in the presence of phytanic acid revealed that this fatty acid efficiently hindered the interactions of retinol with IRBP, suggesting that IRBP may be inhibited from normal retinoid transport in Refsums disease (Figure 2). The effects of some plasticisers on the interactions of all-trans-retinol with IRBP were investigated. The observations of the displacement of retinol by dibutyl phthalate suggest that common environmental toxins may contribute to vitamin A deficiency, a serious consideration especially where ocular retinoid transport is already impaired.

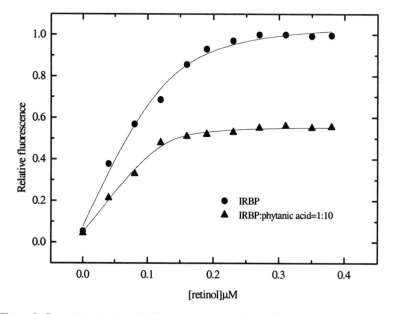

Figure 2. Competitive titration of IRBP with all-trans-retinol in the presence of phytanic acid.

A survey of nine CDs [β-CD, hydroxypropyl-β-CD, 2-hydroxypropyl-β-CD, methyl-β-CD, 2,6-dimethyl-β-CD, α-CD, γ-CD, Heptakis (2,3,6-tri-O-methyl)-β-CD ("TOM") and SBE-7-β-CD ("Captisol™"; CyDex)] showed that only TOM and Captisol were capable of stabilising and solubilising vitamin A. TOM was found to give stronger fluorescence emission intensity for all-trans-retinol compared with Captisol (Figure 3) while their degradation rates were approximately the same.

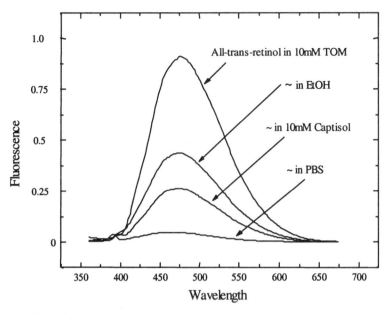

Figure 3. Fluorescence intensity of all-trans-retinol in different environments.

3. CONCLUSIONS

Fluorescence-based studies show that the protection and solubilisation of vitamin A in an aqueous environment might be accomplished by binding to a carrier retinoid-binding protein; this binding is affected by a variety of ligands. However, the immunogenicity of IRBP suggests that it would be unsafe for drug delivery unless well shielded from the immune system. The CDs surveyed are potentially safer, but stabilise vitamin A less effectively than does IRBP; additional cyclodextrins will be investigated as candidates for a drug delivery system. Due to the unique role of vitamin A in the regeneration of rhodopsin, its direct delivery to the retina may help those patients with hereditary eye diseases associated with visual cycle defects.

4. ACKNOWLEDGEMENTS

Ekaterina Semenova was supported in part by a Russian President's Studentship. We are grateful to CyDex Inc. for the gift of Captisol, to Alan Wright for illuminating discussions, and to Thomas McGrory for assistance in obtaining bovine eyes.

REFERENCES

1. E. L. Berson, B. Rosner, M. A. Sandberg, K. C. Hayes, B. W. Nicholson, C. Weigel- DiFranco, and W. Willett, A randomised trial of vitamin A and vitamin E supplementation for retinitis pigmentosa, Archives of Ophthalmology 111(11), 761-772 (1993).
2. C. D. B. Bridges, S. O'Gordon, S-L. Fong, R. A. Alvarez, and E. Berson, Vitamin A and interstitial retinol-binding protein in an eye with recessive retinitis pigmentosa, Invest. Ophthalmol. Vis. Sci. 26(5), 681-691 (1995).
3. A. F. Wright, A searchlight through the fog, Nature Genetics 17, 132-134 (1997).
4. P. Gouras, P. E. Carr, and R. D. Gunkel, Retinitis pigmentosa in abetalipoproteinemia: effects of vitamin A, Invest. Ophthalmol. Vis. Sci. 10, 1971, p. 784.
5. S. G. Jacobson, A.V. Cideciyan, G. Regunath, F. J. Rodriguez, K. Vandenburgh, V. C. Sheffield, and E.M. Stone, Night blindness in Sorsby's fundus dystrophy reversed by vitamin A, Nature Genetics 11(1), 27-32 (1995).
6. J. P. Van Hooser, T. S. Aleman, Y-G. He, A. V. Cideciyan, V. Kuksa, S. J. Pittler, E. M. Stone, S. G. Jacobson, and R. Palczewski, Rapid restoration of visual pigment and function with oral retinoid in a mouse model of childhood blindness, Proceedings of the National Academy of Sciences of the United States of America 97(15), 8623-8628 (2000).
7. N. Noy, Physical-chemical properties and action of retinoids, Handbook of Experimental Pharmacology 139, 3-29 (1999).
8. L. Sibulesky, K. C. Hayes, A. Pronczuk, C. Weigel-DiFranco, B. Rosner, and E. L. Berson, Safety of <7500 RE (<25000 IU) vitamin A daily in adults with retinitis pigmentosa, Am. J. Clin. Nutr. 69(4), 656-663 (1999).
9. K. Palczewski, J. P. Van Hooser, G. G. Garwin, J. Chen, G. I. Liou, and J. C. Saari, Kinetics of visual pigment regeneration in excised mouse eyes and in mice with a targeted disruption of the gene encoding interphotoreceptor retinoid-binding protein or arrestin, Biochemistry 38, 12012-9 (1999).
10. T. Lofssona, and T. Jarvinen, Cyclodextrins in ophthalmic drug delivery, Advanced drug delivery reviews 36(1), 59-80 (1999).
11. S. Sanvill, and R. Kolouch, Cyclodextrins: solution for insolubility, Neurotransmissions 15(1), 18-19 (1999).
12. S. M. Botella, M. A. Martin, B. del Castillo, J. C. Menendez, L. Vazquez, and D. A. Lerner, Analytical application of retinoid-cyclodextrin inclusion complexes. I Characterization of a retinal-β-cyclodextrin complex, Journal of pharmaceutical and biomedical analyses 14, 909-915 (1996).
13. S. M. Botella, D. A. Lerner, B. del Castello, and M. A. Martin, Analytical application of retinoid-cyclodextrins inclusion complexes. II Luminescence properties at room temprature, Analyst 121, 1557-1560 (1996).
14. S. M. Botella, D. A. Lerner, B. del Castillo, and M. A. Martin, Selectivity afforded by room temperature luminescence and absorption of complexes of retinoids with cyclodextrins, Biomedical chromatography 11, 91-92 (1997).
15. A. J. Adler, G. J. Chader, and B. Wiggert, Purification and assay of interphotoreceptor retinoid-binding protein from the eye, Methods Enzymol. 189, 213-223 (1990).

TIMECOURSE OF bFGF AND CNTF EXPRESSION IN LIGHT-INDUCED PHOTORECEPTOR DEGENERATION IN THE RAT RETINA

Natalie Walsh and Jonathan Stone[*]

1. INTRODUCTION

Various growth factors and cytokines are known to have survival promoting functions on neuronal populations. Within the retina, their usefulness in either preventing or delaying the progression of retinal degenerations has led to their development as potential therapeutic agents. Two such factors, basic fibroblast factor (bFGF or FGF-2) and ciliary neurotrophic factor (CNTF) are expressed by the retina during development, and are present in the normal adult retina (Chun et al., 2000; Connolly et al., 1992; Gao and Hollyfield, 1996; Ju et al., 1999; Kirsch et al., 1997; Xiao et al., 1998). bFGF and CNTF are consistently found in the macroglial cells of the retina (Mμller cells, astrocytes and RPE) with bFGF prominent in the nuclei, and CNTF in the cytoplasm (Walsh et al., 2001). Within neurones, bFGF is consistently found in ganglion cells (Connolly et al. 1992; Gao and Hollyfield, 1996), and in response to retinal stress bFGF expression becomes evident in the somas of photoreceptors (Bowers et al. 2001; Stone et al. 1999; Walsh et al. 2001). The expression of CNTF in retinal neurones has proved difficult to detect.

Both bFGF and CNTF are upregulated in response to damaging stimuli and, whether applied exogenously or expressed endogenously, bFGF and CNTF have been shown to slow the progression of photoreceptor death in both inherited (Faktorovich et al. 1990) and environmentally induced retinal degenerations (Bowers et al. 2001; Faktorovich et al. 1992; LaVail et al. 1991). This study addresses the question why, given that stress to the retina causes powerful upregulation of protective factors in and around photoreceptors, the photoreceptors nevertheless are highly vulnerable to stress. Previous studies have shown that upregulation of protective factors, induced by a submaximal stress, provides protection against subsequent stress (Liu et al. 1998; Liu et al. 1997; Nir et al. 1999).This observation suggests that the timing of expression of protective factors is critical in determining whether photoreceptors will survive the later stress or die. In the present experimental paradigm we have subjected the retina to the stress caused by switching and keeping on a bright overhead light, and documenting the time course of increases in photoreceptor death, and the upregulation of bFGF and CNTF levels.

---

[*]   Jonathan Stone, NSW Retinal Dystrophy Research Centre, Department of Anatomy and Histology, University of Sydney, NSW, 2006, Australia. jonstone@anatomy.usyd.edu.au

New Insights Into Retinal Degenerative Diseases
Edited by Anderson *et al.*, Kluwer Academic/Plenum Publishers, 2001

2. METHODS

2.1 Light Exposure

Data from three separate series (total of 24 animals) were analysed for this report. In each series 10 week old Sprague Dawley rats reared in dim (5-10 lux) cyclic light, were exposed to bright continuous light (BCL:1400-1800lux) for periods of 3, 6, 12, 24, 36 and 48h. At the end of each of these time periods animals were euthanised with an overdose of sodium pentobarbitone (60mg/kg intraperitoneal), and the superior aspect of each eye marked prior to enucleation. One animal was exposed to 48 hours of BCL then returned to dim cyclic lighting for one week. Littermates not exposed to BCL served as controls. Eyes were immersion-fixed for 1-3hrs in 4% paraformaldehyde and left overnight in 15% sucrose to provide cryoprotection. The tissues were embedded in mounting medium, snap frozen in liquid nitrogen and cyrosectioned at 20µm. All procedures had approval by the University of Sydney Animal Ethics Committee.

2.2 TUNEL Labelling and Counting

Apoptotic cells were detected *in situ* using the TdT-dUTP terminal nick-end labelling (TUNEL) technique (Gavrieli et al. 1992) according to our previous protocol (Maslim et al. 1997), except using Cy3 (Jackson Immuno Research Laboratories, West Grobe, PA) as the fluorophore. Cell counts were made in the mid-peripheral retina, using a 20x objective and an eyepiece graticule, which marked 400µm lengths along the section. Counts were averaged from at least 4 sections from each eye studied.

2.3 Immunohistochemistry

Sections were labelled with a mouse monoclonal antibody against bovine bFGF (Type 1, Upstate Biotechnology, Lake Palcid, NY) and a rabbit polyclonal antibody to CNTF (Chemicon International, Temecula,CA). An Alexa 488-conjugated goat anti-mouse IgG (H+L) and an Alexa 594-conjugated goat anti-rabbit IgG (H+L) (Molecular Probes, Portland, Oregon) were used as secondary antibodies to bFGF and CNTF respectively. Protocols for these reactions were as published previously (Valter et al. 1998; Walsh et al. 2001).

2.4 Quantitative Analysis of Immunolabelling

To allow quantitative comparison of bFGF and CNTF immunolabelling between sections from different eyes, sections were prepared in the same labelling runs, with identical incubation and processing times. When imaged by confocal microscopy photomultiplier settings were set to cover the range of labelling to be studied and then held constant. The 'analysis tool' of NIH Image allowed for examination of immunolabelling. We measured mean signal brightness over selected layers of the retina, the ONL, the INL or 'inner retina' (from the inner limiting membrane to the outer plexiform layer). We also recorded the immunofluorescence profile across the retina, averaged across a 9 pixel wide line.

3. RESULTS

3.1 bFGF and CNTF in Normal Retina

The retina of a rat reared in dim cyclic light and not exposed to BCL, shows very few TUNEL+ cells in any of the layers (2-5 per 20μm section). In mid-peripheral retinal regions, the patterns of bFGF and CNTF localisation observed (Figure 1), confirmed previous reports (Bowers et al. 2001; Kirsch et al. 1997; Xiao et al. 1998). bFGF was most prominent in Mμller cell somas in the INL, and was detectable in the somas of ganglion cells and in the occasional photoreceptor soma in the ONL (Figure 1A,B). CNTF was most prominent at the ILM, in the inner and outer plexiform layers, and was difficult to detect in the inner and outer nuclear layers (Figure 1C,D). We have previously presented (Walsh et al. 2001), evidence that CNTF is largely confined to the processes of astrocytes and Mμller cells.

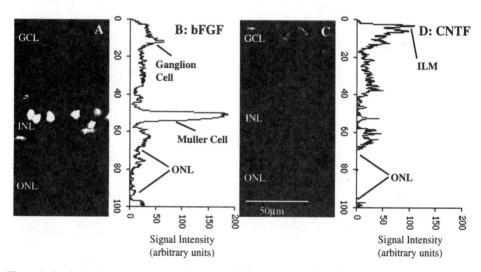

Figure 1. Confocal microscope images and immunolabelling density profiles of bFGF (A ,B) and CNTF (C,D) in mid-peripheral regions of control retina.

3.2 Timecourse of BCL induced photoreceptor death

BCL induced TUNEL-labelling was specific to the outer nuclear (ONL) and therefore to photoreceptors, confirming Bowers et al (2001). In mid-peripheral retina, TUNEL labelling was evident at 3h after the onset of BCL, the earliest time point examined (Figure 2B). At successive time-points to 48h, the number of TUNEL + cells increased (Figure 2C-G), and at 48h the entire thickness of the ONL was TUNEL+ (Figure 2G). In eyes exposed to 48h BCL, then returned to dim-cyclic light for one week, the ONL had disintegrated and TUNEL+ cells were not detected (Figure 1H). Superior-inferior differences in TUNEL+ labelling were not noted in these series.

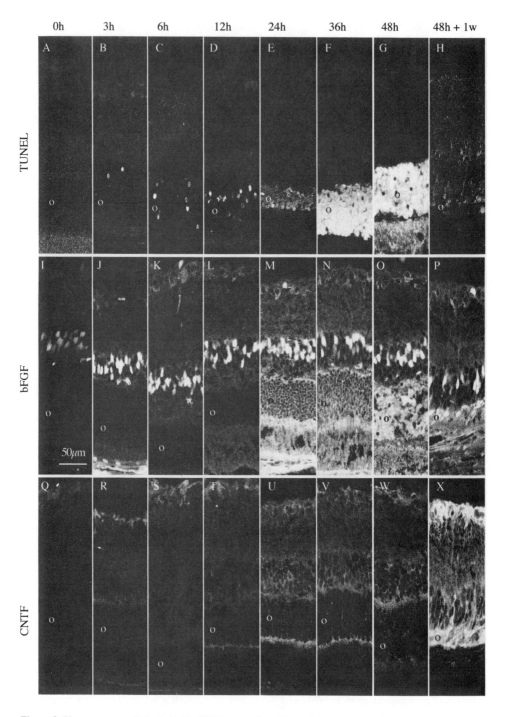

Figure 2. Photoreceptor cell death (A-H), bFGF expression (I-P) and CNTF expression (Q-X) in the superior mid-peripheral retina following the onset of bright continuous light (time in hours, at top).

3.3 Timecourse of bFGF and CNTF upregulation

Light damage induced increases in the levels of both bFGF (Figure 2I-P) and CNTF (Figure 2Q-X). By 48h (Figure 2O), bFGF levels had risen in ganglion cell somas, nuclei of astrocytes and Mμller cells, the outer plexiform layer and in the outer nuclear layer (Figure 3A). The most striking feature of these changes was the rise of bFGF levels in the cytoplasm of photoreceptors in the ONL. At 48h+1wk, bFGF levels at the site of the ONL (o, in Figure 2P), may not be photoreceptor-related, but rather in glial cells which invade the layer secondary to the photoreceptor degeneration.

The upregulation of bFGF was most spectacular in the ONL (Figure 3A) but it was striking that the ONL increase occurred with a delay. bFGF upregulation in the INL was marked at 3h, and increased only marginally thereafter (Figure 3B). By contrast, bFGF upregulation in the ONL was not prominent until 24h (Figure 2M). This difference in timecourse is shown graphically in Figure 3B, and a similar delay in ONL upregulation of bFGF was clear in the other two experimental series.

When the timecourses of TUNEL labelling and bFGF upregulation were compared, it was apparent that the upregulation of bFGF in the INL was at least as fast at the onset of photoreceptor death (Figure 3C). However, bFGF upregulation in the ONL lagged behind photoreceptor death (Figure 3D). This lag, of approximately 9h, may be a critical factor in BCL-induced photoreceptor degeneration (see Discussion).

The upregulation of CNTF during and beyond the 48h period of BCL is shown in Figure 2Q-X. At the end of the 48h period of BCL, the upregulation of CNTF was much less prominent than bFGF (compare Figure 2O and W). Systematic measurement showed nevertheless that levels of CNTF increased during the 48h, in almost all retinal layers, most markedly at the ILM, the OPL and the OLM (Figure 4A) and that these increases began early, by the 3 hour timepoint (Figure 4B). The lag in upregulation between inner and outer retina seen in the expression of bFGF was not apparent with CNTF (Figure 4B). However, it should be noted that through the 48h period of BCL, the level of CNTF in the ONL, though increasing, was approximately half the level in the inner layers of retina (Figure 4B). By 1w after BCL, CNTF levels were very high throughout the surviving layers (Figure 2X), and especially at the inner and outer surfaces of the retina.

4. DISCUSSION

Present results show that, following the onset of a light sufficiently bright to induce light damage, there is a rapid upregulation in the TUNEL labelling of photoreceptors (indicating the onset of cell death among them) and of the retina's expression of bFGF and CNTF, two trophic factors known to be highly protective of photoreceptors (Faktorovich et al. 1990; Faktorovich et al. 1992; LaVail et al. 1992; Steinberg, 1994). Comparison of the time courses of these upregulations shows that the upregulation of the growth factors is at least as fast as the onset of photoreceptor death and that the upregulation of one factor (bFGF) is most prominent in photoreceptor somas, in the ONL. A layer-by-layer comparison shows however that the upregulation of bFGF in photoreceptors somas, although spectacular by 48h, was significantly delayed after bFGF upregulation in inner retina. Conversely, although the upregulation of CNTF in the ONL is as rapid as in the inner retina, the levels of CNTF in the ONL were consistently 50% less than in inner retina.

The slow and limited upregulations of bFGF and CNTF in the ONL, compared to inner retina can be explained in terms of the cellular structure of the photoreceptor soma, and of Mμller cells. Each photoreceptor soma contains its nucleus and nuclear DNA, whose fragmentation in apoptotic death is detected by the TUNEL technique. However,

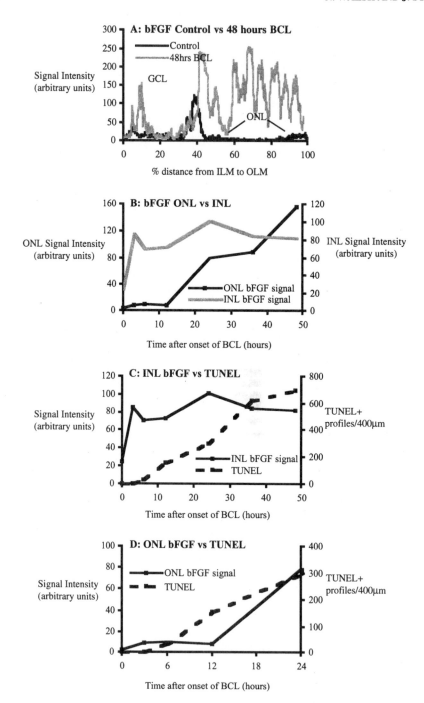

Figure 3. Effects of light damage on bFGF levels. A: bFGF immunolabel signals from control retina (black) and retina exposed to 48h BCL (grey), measured along a transect the thickness of the retina. B: Timecourse of bFGF expression in the ONL vs INL. C,D: Timecourse of bFGF expression (black) in the I NL (C) and ONL (D) vs TUNEL counts (dashed).

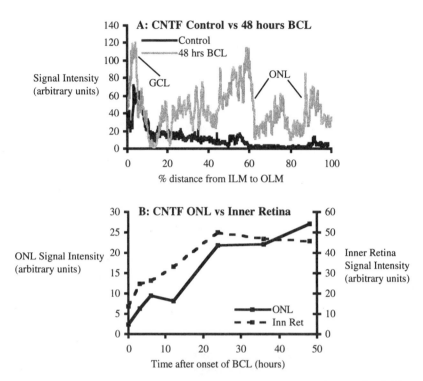

Figure 4. Light-induced changes in CNTF expression. Retinal transects across the control and 48hours BCL retina (A). Timecourse of CNTF expression in the ONL and inner retina (B).

the ribosomes of the photoreceptor, where proteins such as trophic factors are presumably translated, are segregated away from the soma, in the inner segments, up to 50um away along a thin external process. In most neurones, by contrast, the ribosomes are in the cell soma, close to the nucleus. In the photoreceptor, bFGF is produced by the photoreceptor itself; bFGF mRNA has been identified in photoreceptors, concentrating in the inner segment (Wen et al. 1998; Wen et al. 1995). However, compared to a less specialised neurone, the separation of the photoreceptor's ribosomes from the nucleus presumably adds to the time needed for mRNA to reach the ribosomes and for the bFGF protein to get back to the nucleus.

There is at least one alternative source of trophic factors for photoreceptor protection. In undamaged retina, bFGF and CNTF are prominent in Mμller cells in the INL and inner retina and could reach the photoreceptors along the Mμller cell processes, which embrace all photoreceptor somas. Again the protein must move a significant distance and, in the case of bFGF, which is protective when in the photoreceptor soma (Bowers et al. 2001; Liu et al. 1998; Nir et al. 1999; Xiao et al. 1998), must be transported from Mμller cell to photoreceptor. The delay shown in Figure 3B could thus be explained in terms of photoreceptor and Mμller cell morphology.

Previous studies have noted that, where bFGF is upregulated in photoreceptors prior to an environmental stress such as BCL, photoreceptors are highly protected. The protective effect of prior upregulation is clear for example, where the retina is pre-stressed by bright light (Gao and Hollyfield, 1996; Liu et al. 1998) or by mechanical (Cao et al. 1997) or laser (Xiao et al. 1998) damage, or at the edge of the retina where a still-unidentified stress upregulates bFGF levels in photoreceptor somas to very high levels (Bowers et al. 2001). Once trophic factors have reached the photoreceptor somas, they are well protected. Until the factors reach their somas however, photoreceptors are vulnerable and the evidence presented here of a delay in trophic factor delivery to photoreceptor somas may be a previously unrecognised factor in the vulnerability of retinal photoreceptors.

REFERENCES

Bowers, F., Valter, K., Chan, S., Walsh, N., Maslim, J. and Stone, J., 2001, Effects of oxygen and bFGF on the vulnerability of photoreceptors to light damage, *Invest Ophthalmol Vis Sci*. (in press).

Cao, W., Wen, R., Li, F., LaVail, M. and Steinberg, R., 1997, Mechanical injury increases bFGF and CNTF mRNA expression in the mouse retina, *Exp Eye Res*. **65**: 241-248.

Chun, M., Ju, W., Kim, K., Lee, M., Hofmann, H., Kirsch, M. and Oh, S., 2000, Upregulation of ciliary neurotrophic factor in reactive Müller cells in the rat retina following optic nerve transection, *Brain Research*. **868**: 358-362.

Connolly, S., Hjelmeland, L. and LaVail, M., 1992, Immunohistochemical localisation of basic fibroblast growth factor in mature and developing retinas of normal and RCS rats, *Curr Eye Res*. **11**: 1005-1017.

Faktorovich, E.G., Steinberg, R.H., Yasumura, D., Matthes, M.T. and LaVail, M.M., 1990, Photoreceptor degeneration in inherited retinal dystrophy delayed by basic fibroblast growth factor, *Nature*. **347**: 83-86.

Faktorovich, E.G., Steinberg, R.H., Yasumura, D., Matthes, M.T. and LaVail, M., 1992, Basic fibroblast growth factor and local injury protect photoreceptors from light damage in the rat, *J Neurosci*. **12**: 3554-3567.

Gao, H. and Hollyfield, J., 1996, Basic fibroblast growth factor: Increased gene expression in inherited and light-induced photoreceptor degeneration, *Exp Eye Res*. **62**: 181-189.

Gavrieli, Y., Sherman, Y. and Ben-Sasson, S.A., 1992, Identification of programmed cell death *in situ* via specific labeling of nuclear DNA fragmentation, *Journal of Cell Biology*. **119**: 493-501.

Ju, W., Lee, M., Hofmann, H., Kirsch, M. and Chun, M., 1999, Expression of CNTF in Müller cells of the rat retina, *Neuroreport*. **10**: 419-422.

Kirsch, M., Lee, M.Y., Meyer, V., Wiese, A. and Hofmann, H.D., 1997, Evidence for multiple, local functions of ciliary neurotrophic factor (CNTF) in retinal development: expression of CNTF and its receptors and in vitro effects on target cells., *J Neurochem*. **68**: 979-90.

LaVail, M.M., Faktorovich, E.G., Hepler, J.M., Pearson, K.L., Yasumura, D., Matthes, M. and Steinberg, R.H., 1991, Basic fibroblast growth factor protects photoreceptors from light-induced degeneration in albino rats, *Annals New York Academy of Sciences*. **638**: 341-347.

LaVail, M.M., Unoki, K., Yasumura, D., Matthes, M.T., Yancopoulos, G.D. and Steinberg, R.H., 1992, Multiple growth factors, cytokines, and neurotrophins rescue photoreceptors from the damaging effects of constant light, *Proc Nat Acad Sci USA*. **89**: 11249-11253.

Liu, C., Peng, M., Laties, A. and Wen, R., 1998, Preconditioning with bright light evokes a protective response against light damage in the rat retina, *J Neurosci*. **18**: 1337-1344.

Liu, C., Peng, M. and Wen, R., 1997, Pre-exposure to constant light protects photoreceptor from subsequent light damage in albino rats, *Invest Ophthalmol Vis Sci*. **38**: S718.

Maslim, J., Valter, K., Egensperger, R., Hollander, H. and Stone, J., 1997, Tissue oxygen during a critical developmental period controls the death and survival of photoreceptors, *Invest. Ophthal. Vis. Sci.* **38**: 1667-1677.

Nir, I., Liu, C. and Wen, R., 1999, Light treatment enhances photoreceptor survival in dystrophic retinas of Royal College of Surgeons rats., *Invest Ophthalmol Vis Sci*. **40**: 2383-90.

Steinberg, R.H., 1994, Survival factors in retinal degenerations, *Curr Opin Neurobiol*. **4**: 515-524.

Stone, J., Maslim, J., Valter-Kocsi, K., Mervin, K., Bowers, F., Chu, Y., Barnett, N., Provis, J., Lewis, G., Fisher, S., Bisti, S., Gargini, C., Cervetto, L., Merin, S. and Pe'er, J., 1999, Mechanisms of photoreceptor death and survival in mammalian retina, *Prog Ret Eye Res*. **18**: 689-735.

Valter, K., Maslim, J., Bowers, F. and Stone, J., 1998, Photoreceptor dystrophy in the RCS rat: Roles of oxygen, debris and bFGF, *Invest Ophthalmol Vis Sci*. **39**: 2427-2442.

Walsh, N., Valter, K. and Stone, J., 2001, Cellular and subcellular patterns of expression of bFGF and CNTF in the normal and light stressed adult rat retina, *Exp Eye Res*. (in press).

Wen, R., Cheng, T., Song, Y., Matthes, M., Yasumura, D., La Vail, M. and Steinberg, R., 1998, Continuous exposure to bright light upregulates bFGF and CNTF expression in the rat retina, *Curr Eye Res*. **17**: 494-500.

Wen, R., Song, Y., Cheng, T., Matthes, M., Yasamura, D., LaVail, M. and Steinberg, R., 1995, Injury-induced upregulation of bFGF and CNTF mRNAS in the rat retina, *J Neurosci*. **15**: 7377-7385.

Xiao, M., Sastry, S., Li, Z.-Y., Possin, D., Chang, J., Klock, I. and Milam, A., 1998, Effects of retinal laser photocoagulation on photoreceptor basic fibroblast growth factor and survival., *Invest. Ophthal. Vis. Sci.* **39**: 618-630.

ROLE OF PIGMENT EPITHELIUM-DERIVED FACTOR (PEDF) IN PHOTORECEPTOR CELL PROTECTION

Wei Cao*, Joyce Tombran-Tink, Rajesh Elias, Steven Sezate and James F. McGinnis

1. INTRODUCTION

Pigment epithelium-derived factor (PEDF), a 50-kDa glycoprotein, was first isolated from medium conditioned by human fetal retinal pigment epithelial (RPE) cells and was shown to be made *in vivo* by both fetal and adult RPE cells. After release from the RPE, PEDF binds to the glycosaminoglycans of the interphotoreceptor matrix,[1,2] placing it in a prime physical location to affect the underlying neural retina. The PEDF gene has been cloned, sequenced and shown to have a tight linkage with a retinitis pigmentosa locus (RP13) on chromosome 17p13.3 making it a candidate gene for this form of retinal degeneration.[3,4] It has been reported that PEDF supported normal development of photoreceptor neurons and opsin expression after retinal pigment epithium removal.[5] PEDF acts as a survival factor for cultured cerebellar granule cells,[6-8] spinal motor neurons,[9,10] hippocampal neurons[11] and recently we demonstrated the survival-promoting activity of PEDF on retinal neurons *in vitro*[12]. In addition to its effects on neurons, PEDF is a potent anti-angiogenesis factor,[13] and is considered as a key coordinator of retinal neuronal and vascular functions.[14] However, the role of PEDF as a neuroprotective factor in the retina and its association with specific retinal degenerative diseases are still being elucidated. The aim of the present research is to examine whether PEDF can prevent or delay photoreceptor degeneration. Using H_2O_2-induced cell death in cultured retinal neurons as *in vitro* model and the light-dependent degeneration of rat photoreceptor cells as *in vivo* model, our data demonstrate that PEDF results in significant protection of photoreceptor cells.

* Wei Cao, Department of Ophthalmology, University of Oklahoma Health Sciences Center, Dean McGee Eye Institute, 608 Stanton L. Young Blvd, Oklahoma City, OK 73104, (405) 271-3370 (voice), (405) 271-3721 (fax), E-mail: wei-cao@ouhsc.edu.

New Insights Into Retinal Degenerative Diseases
Edited by Anderson *et al.*, Kluwer Academic/Plenum Publishers, 2001

2. MATERIALS AND METHODS

2.1. Primary Cultures of Postnatal Retinal Neurons

Animals used in these studies were cared for and handled according to guidelines of the Association for Research in Vision and Ophthalmology (ARVO) for the Use of Animals in Vision and Ophthalmic Research and the University of Oklahoma Faculty of Medicine Guidelines for Animals in Research. Preparation of primary retinal neuronal culture was as previously described.[15] Briefly, retinas of 10-15 rat pups, 0-2 days old, were removed, suspended and mechanically dissociated. The concentration of cells was determined with a cell counter or hemocytometer and the suspension diluted with medium to 1×10^5 cells per ml. The cells (1 ml) were plated in 24 well tissue culture plates on 12 mm coverslips that had been pre-treated overnight with poly-D-lysine (10 μg/ml). The cultures at 14-20 days after plating were used for experiments.

2.2. TUNEL Assay

TUNEL (TdT-mediated digoxigenin-dUTP nick-end labeling) method was carried out using a commercially available *in situ* apoptosis detection kit. Positive controls were carried out by treating coverslips with DNAse I (1 μg/ml) prior to the assay. Negative controls were carried out by omitting TdT from the protocol. Cells were fixed in 2% paraformaldehyde for 10 min at room temperature and washed twice with phosphate-buffered saline (PBS) for 5 min each at room temperature. Staining for the TUNEL technique was according to the manufacturer's protocol. TUNEL positive cells were visualized using diaminobenzidine as substrate for the kit's horseradish peroxidase. The results were viewed with a Nikon Eclipse 800 microscope. The pictures were transferred by a digital camera and stored in the computer. The percentage of apoptotic cells was calculated by counting TUNEL positive cells from the digitized image stored in the computer, and by dividing TUNEL positive cells by total cells visualized by nomarski optics in the same field. Three digitized images of similar total cell numbers were selected from each coverslip for counting and averaging, and were considered as one independent experiment. Three independent experiments were then averaged.

2.3. MTT Assay

The MTT assay is useful for the measurement of cytotoxicity.[16] MTT was dissolved at a concentration of 5 mg/ml in PBS. Lysing buffer was prepared as follows: sodium dodecyl solfate (20% w/v) was dissolved at 37°C in a 50% solution of DMF in deionized water. The pH was adjusted to 4.7. 25 μl of the stock solution of MTT was added to each well and incubated for 2 hours at 37°C, and then 100 μl of the lysing buffer was added. After an overnight incubation at 37°C, absorbance of the samples was read at 562 nm using a microtiter plate ELISA reader.

2.4. Intravitreal Injection and Constant Light Exposure

Sprague-Dawley albino rats were obtained at 2-5 months of age and maintained in our cyclic light environment [12 hr on, 12 hr off at an in-cage illuminance of less than 250 lux] for 10 or more days before use. Rats were anesthetized with a ketamine (80 mg/kg)-xylazine (6 mg/kg) mixture, and then administered intravitreally 2 μg PEDF or 1 μg bFGF or a mixture of 1 μg PEDF/1 μg bFGF in a volume of 1 μl. Human recombinant bFGF was purchased from R&D System (Minneapolis). The injections were made using a 30-gauge needle inserted through the sclera, choroid, and retina approximately midway between the ora serrata and equator of the eyeball. To simplify manipulations and to avoid errors, it was arbitrarily decided that the left eye would be used as a sham-operated control (PBS-injected) and the right eye subjected to either PEDF, bFGF, or a PEDF/bFGF mixture. Animals were exposed to constant light for 7 days at an illuminance level of 1200 - 1500 lux provided by two 40 W white fluorescent light bulbs.

2.5. Functional Evaluation of Photoreceptor Cell Rescue by Electroretinogram

Animals were kept in total darkness for a minimum of 60 minutes before recording electroretinograms (ERG).[17] Pupils were dilated with 1% atropine and 2.5% phenylephrine HCL. Animals were anesthetized intramuscularly with a ketamine-xylazine mixture. ERG responses were recorded with a silver chloride needle electrode placed in the cornea with 1% tetracaine topical anesthesia. A reference electrode was positioned at the nasal fornix, and a ground electrode on the foot. The duration of light stimulation was 10 milliseconds. The band pass was set at 0.3-500 Hz. Fourteen responses were averaged with flash intervals of 20 seconds. For quantitative analysis, the B-wave amplitude was measured between A- and B-wave peaks.

2.6. Morphological Evaluation of Photoreceptors Rescue by Quantitative Histology[18,19]

Animals were sacrificed by an overdose of carbon dioxide after electroretinographic testings. The eyes were enucleated, fixed, embedded in paraffin, and 5-μm thick sections were taken along the vertical meridian to allow comparison of all regions of the eye. In each of the superior and inferior hemispheres, outer nuclear layer (ONL) thickness was measured at nine defined points. Each point was centered on adjacent 450 μm lengths of retina. The first point of measurement was taken at approximately 450 μm from the optic nerve head, and subsequent points were located more peripherally. In addition to mean ONL thickness for the entire retinal section, ONL thickness of the region of retina most sensitive to the damaging effects of light was compared among different groups of rats. In each of the experiments where ONL thickness was quantified, a single section from the retinas of at least 6 (usually 10 or more) eyes was measured.

2.7. Statistical Analysis

Results are expressed as mean ± S.D. Data were analyzed by means of analysis of variance (ANOVA), and further assessed by Dunnett tests. Accepting significance was $p < 0.05$ and $p < 0.01$.

3. RESULTS

3.1. PEDF Prevents Photoreceptor Cell Apoptosis Induced by H_2O_2

A few positive staining cells were noted in control cultures (Figure 1A), whereas cultures treated with 60 μM H_2O_2 for 24 hrs contained large numbers of cells undergoing apoptotic cell death (Figure 1B). We have also noted that at lower concentrations of H_2O_2 (60-100 μM), the TUNEL positive cells exhibited classical apoptotic morphology. Double staining with a rod photoreceptor cell marker showed that most apoptotic cells were photoreceptor cells (Fig. E). Parallel cultures treated with 100 ng/ml PEDF 1 hr before 60 μM H_2O_2 exposure showed a dramatic decrease in the numbers of apoptotic cells (Figure 1C).

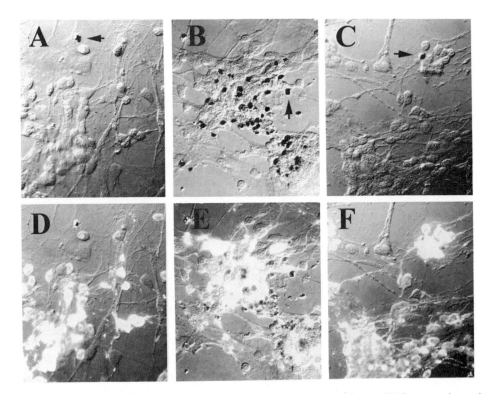

Figure 1. PEDF inhibits H_2O_2–induced photoreceptor cell apoptosis. A: Nomarski image of H_2O_2-untreated control; arrows in A, B and C indicate an apoptotic cell. D: Superimposition of fluorescein stained cells (rhodopsin-positive rods) on the Nomarski image of the same field shown in A. B: Nomarski image of H_2O_2-treated cells (60 μM, 24 hr). E: Superimposition of fluorescein stained cells on the Nomarski image of the same field shown in B shows that the most apoptotic cells are rhodopsin-positive rod cells. C: Nomarski image of cultured retinal neurons pretreated with PEDF (100 ng/ml) 1 hr before 60 μM H_2O_2 exposure. F: Superimposition of fluorescein stained cells (rhodopsin-positive rods) on the Nomarski image of the same field shown in C.

3.2. Effects of PEDF and Multi-Survival-Factors on H₂O₂-Induced Cytotoxicity

Effects of multi-survival factors on H_2O_2–induced cytotoxicity were evaluated by MTT assay as percentage of cell viability. Two hours before 100 μM H_2O_2 exposure for 24 hrs, the cultures were treated with PEDF (100 ng/ml) or bFGF (100 ng/ml) or BDNF (50 ng/ml) or CNTF (100 ng/ml) or PEDF+bFGF or PEDF+BDNF or PEDF+CNTF. The results show that pretreated with either PEDF or bFGF or BDNF or CNTF alone increased cell viability from 45% (100 μM H_2O_2 only) to 70%, 80%, 56% and 78% respectively. When PEDF was combined with bFGF or BDNF or CNTF, cell viability increased from 45% to 98%, 86% and 93% respectively. Compared to PEDF pretreatment alone, the combination sets significantly improved the cell viability, and the effect was additive. These data indicate that combination of PEDF with other growth factors known to protect retinal neurons could improve the protection

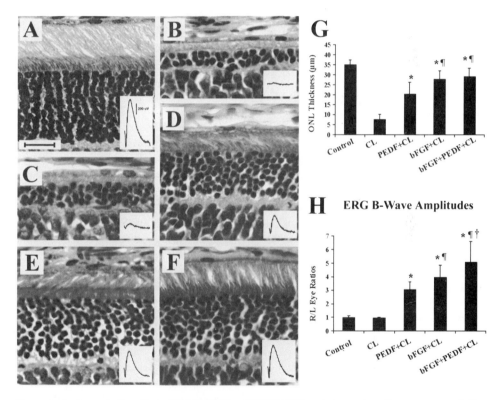

Figure 2. *In vivo* protective effects of PEDF, bFGF and PEDF/bFGF on photoreceptor cell rescue measured after a 14-day recovery time. (A) Normal retina from an uninjected rat reared in cyclic light. The insert shows a typical ERG waveform. (B) Retina from an uninjected rat. (C) Retina from a PBS-injected rat. (D) Retina from a PEDF-injected rat. (E) Retina from a rat injected with bFGF. (F) Retina from a rat injected with a mixture of PEDF/bFGF. (G) Mean ONL thickness of retinas. * p<0.05 versus the uninjected group, & p<0.05 versus PEDF-injected group (mean ± SD, n=10). (H) Mean right/left eye ratios of ERG B-wave amplitudes. * p<0.05 versus the uninjected light damaged or control group, & p<0.05 versus PEDF-injected group, H p<0.05 versus bFGF-injected group (mean ± SD, n=10).

3.3. *In vivo* Protection of Photoreceptors from Light Damage by PEDF

As shown in normal retina (Fig.2A), thickness of the outer nuclear layer (ONL) of photoreceptor cell nuclei is 35-40 μm (9-11 rows) with a ERG B-wave amplitude at 800-1000 μV (see insert). One week of exposure to flourescent light (1200-1500 lux) reduced thickness of the ONL to 1-3 rows (5-10 μm) in the most degenerated region of the uninjected eye with almost distinguished ERG waveform (Fig. 2B) or the PBS-injected sham control with ERG B-wave amplitude less than 100 μV (Fig.2C). However, in PEDF injected eyes, there was significant rescue of photoreceptors with the ONL having 6-7 rows (25-30 μm) of nuclei with ERG B-wave amplitude of 400 μV or more (Fig. 2D). Injection of bFGF alone (Fig. 2E) does provide better protection than PEDF alone (Fig. 2D) but together (Fig. 2F) the preservation of rod outer segment morphology is more impressive than with bFGF by itself. A quantitation of the thickness of the ONL in the retinas from all of the eyes (Fig. 2G) supports these conclusions. As with morphology, functional rescue (ERGs) improves over time with either PEDF or bFGF alone and together they act synergistically. A graphic summary of the change in ratios of the B-wave amplitudes of experimental and control eyes is presented in Figure 2H and these data corroborate the synergistic benefits of using both factors simultaneously.

4. DISCUSSION

Hydrogen peroxide, a membrane permeable oxidant, is a precursor of highly oxidizing radicals such as hydroxyl radicals and has been shown to induce cell injury in a number of neuronal cell culture models.[20] It is known that the H_2O_2 level was increased approximately 15 fold in the vitreous of rabbit eyes after exposure to constant light. The estimated concentration of H_2O_2 in the vitreous of the eyes after 3-7 day exposure of constant light was 110 μM.[21] Since in the retina, H_2O_2 is released by illuminated photoreceptors, one could expect that in the light damage model, the level of H_2O_2 could be even higher in the retina than that in the vitreous. We used H_2O_2-induced cell death in cultured retinal neurons as an *in vitro* model and the light-dependent degeneration of rat photoreceptor cells as an *in vivo* model. Our data demonstrate that PEDF results in significant protection of photoreceptor cells.

As shown in our *in vitro* study, we observed that exposure to H_2O_2 caused the death of retinal neurons via an apoptotic pathway. This is in agreement with the observation in other neuronal culture systems.[20] We have also noted that at lower concentrations of H_2O_2, most TUNEL positive cells exhibited an apoptotic morphology. At higher concentrations of H_2O_2, however, most cells showed signs of necrotic cell death. This dose-dependent induction of apoptosis in our culture may represent a typical feature of neuronal death caused by oxidative radical stress. PEDF clearly inhibits H_2O_2-induced photoreceptor cell apoptosis. The data also showed a high degree of retinal neuron protection with bFGF and CNTF, while lesser protection with BDNF. It is not surprising that bFGF was the most effective factor against H_2O_2-induced retinal neuron death in this study because bFGF has been shown to be very effective in rescuing retinal neurons in several *in vivo* models of retinal degeneration.[18,19] CNTF has also been shown to delay photoreceptor cell loss in the retinal degeneration. The peak protective effect of BDNF occurred at 50 ng/ml. Therefore, a less protective effect with

BDNF is not due to insufficient supply of BDNF. One possible explanation is that BDNF is more effective for ganglion cells and photoreceptor cells are more susceptible to oxidative damage than ganglion cells. The benefits of various combinations of neurotrophic factors in retinal degeneration have been reported.[19] We have shown in the present study that combinations of PEDF with bFGF, BDNF, or CNTF increase the protection, and the effect is additive. This suggests that a cocktail of growth factors may provide optimum rescue in conditions of retinal degeneration.

PEDF is synthesized and secreted by the RPE cells in early embryogenesis and in the adult stages of the retina and is found in abundance in the IPM which surrounds retinal neurons.[2] It is therefore in a prime location to affect both the differentiation and survival of the retina, especially the photoreceptor cells which lie directly adjacent to the RPE cells and surrounded by the PEDF rich IPM. Light damage is thought to result from the generation of oxygen free radicals and the ensuing peroxidation of lipids. The present data on the *in vivo* protective effects of PEDF support the results and conclusions obtained in cell culture and extend them to include the demonstration that PEDF protects retinal neurons from light damage. Basic FGF has been shown to be the most potent survival factor in preventing photoreceptor cell death in several rat models of retinal degeneration including the light damage model.[22] However, the mitogenic and angiogenic properties of bFGF substantially limit its usefulness in the treatment of retinal degeneration. In a recent study, PEDF was shown to have anti-angiogenic activity and to inhibit the mitogenic and angiogenic properties of bFGF.[13] These observations, in combination with our data showing a protective effect of PEDF on photoreceptor cells, suggest that these two factors together could be beneficial to photoreceptor protection. Even a modest reduction in the rate of photoreceptor cell death might lead to a significant prolongation of useful vision.

5. Acknowledgments

Supported by NIH grants (EY06085, EY13050), and by a Jules and Doris Stein Professorship from Research to Prevent Blindness (RPB) to JFM; by a core facilities grant from NEI (EY12190), and by an unrestricted grant from RPB to the Department of Ophthalmology. The authors thank Paula Pierce, Na Wei and Mark Dittmar for technical support.

REFERENCES

1. J. Tombran-Tink, G. J. Chader, and L. V. Johnson, PEDF: a pigment epithelium-derived factor with potent neuronal differentiative activity [letter], *Exp. Eye Res.* 53, 411-414 (1991).
2. J. Tombran-Tink, S. M. Shivaram, G. J. Chader, L. V. Johnson, and D. Bok D, Expression, secretion, and age-related downregulation of pigment epithelium-derived factor, a serpin with neurotrophic activity, *J. Neurosci.* 15, 4992-5003 (1995).
3. R. Goliath, J. Tombran-Tink, I. R. Rodriquez, G. J. Chader, R. Ramesar, and J. Greenberg, The gene for PEDF, a retinal growth factor is a prime candidate for retinitis pigmentosa and is tightly linked to the RP13 locus on chromosome 17p13.3, *Mol Vis.* 2, 5-8 (1996).
4. J. Tombran-Tink, H. Pawar, A. Swaroop, I. Rodriguez, and G. J. Chader, Localization of the gene for pigment epithelium-derived factor (PEDF) to chromosome 17p13.1 and expression in cultured human retinoblastoma cells, *Genomics* 19, 266-272 (1994).

5.   M. M. Jablonski, J. Tombran-Tink, D. A. Mrazek, and A. Iannaccone, Pigment epithelium-derived factor supports normal development of photoreceptor neurons and opsin expression after retinal pigment epithium removal, *J Neurosci.* **20**, 7149-7157 (2000).

6.   T. Taniwaki, S. P. Becerra, G. J. Chader, and J. P. Schwartz, Pigment epithelium-derived factor is a survival factor for cerebellar granule cells in culture, *J Neurochem.* **64**, 2509-2517 (1995).

7.   T. Taniwaki, N. Hirashima, S. P. Becerra, G. J. Chader, R. Etcheberrigaray, and J. P. Schwartz, Pigment epithelium-derived factor protects cultured cerebellar granule cells against glutamate-induced neurotoxicity, *J Neurochem.* **68**, 26-32 (1997).

8.   T. Araki, T. Taniwaki, S. P. Becerra, G. J. Chader, and J. P. Schwartz, Pigment epithelium-derived factor (PEDF) differentially protects immature but not mature cerebellar granule cells against apoptotic cell death, *J Neurosci Res.* **53**, 7-15 (1998).

9.   L. J. Houenou, A. P. D'Costa, L. Li, V. L. Turgeon, C. Enyadike, E. Alberdi, and S. P. Becerra, Pigment epithelium-derived factor promotes the survival and differentiation of developing spinal motor neurons, *J Comp Neurol.* **412**, 506-514 (1999).

10. M. M. Bilak, A. M. Corse, S. R. Bilak, M. Lehar, J. Tombran-Tink, and R. W. Kuncl, Pigment epithelium-derived factor (PEDF) protects motor neurons from chronic glutamate-mediated neurodegeneration, *J Neuropathol Exp Neurol.* **58**, 719-728 (1999).

11. M. A. DeCoster, E. Schabelman, J. Tombran-Tink, and N. G. Bazan, Neuroprotection by pigment epithelial-derived factor against glutamate toxicity in developing primary hippocampal neurons, *J. Neurosci. Res.* **56**, 604-610 (1999).

12. W. Cao, J. Tombran-Tink, W. Chen, D. Mrazek, R. Elias, and J.F. McGinnis, Pigment epithelium-derived factor protects cultured retinal neurons against hydrogen peroxide-induced cell death, *J Neurosci Res.* **57**, 789-800 (1999).

13. D. W. Dawson, O. V. Volpert, P. Gillis, S. E. Crawford, H. Xu, W. Benedict, and N. P. Bouck, Pigment epithelium-derived factor: a potent inhibitor of angiogenesis, *Science* **285**, 245-248 (1999).

14. G. L. King, and K. Suzuma, Pigment-epithium-derived factor — a key coordinator of retinal neuronal and vascular functions, *N Engl J Med.* **342**, 349-351 (2000).

15. W. Cao, W. Chen, R. Elias, and J. F. McGinnis, Recoverin Negative photoreceptor cells, *J Neuosci Res.* **60**, 195-201 (2000).

16. J. Y. Chang, and L. Z. Liu, Toxicity of cholesterol oxides on cultured neuroretinal cells, *Curr. Eye Res.* **17**, 95-103 (1998).

17. W. Cao, M. Zaharia, A. Drumheller, C. Casanova, G. Lafond, J. R. Brunette, and F. B. Jolicoeur, Effects of dextromethorphan on ischemia induced electroretinogram changes in rabbit, *Curr Eye Res.* **13**, 97-102 (1994).

18. E. G. Faktorovich, R. H. Steinberg, D. Yasumura, M. T. Matthes, and M. M. LaVail, Basic fibroblast growth factor and local injury protect photoreceptors from light damage in the rat, *J Neurosci.* **12**, 3554-3567 (1992).

19. M. M. LaVail, K. Unoki, D. Yasumura, M. T. Matthes, G. D. Yancopoulos, and R. H. Steinberg, Multiple growth factors, cytokines, and neurotrophins rescue photoreceptors from the damaging effects of constant light, *Proc Natl Acad Sci U.S.A.* **89**, 11249-11253 (1992).

20. H. Taguchi, Y. Ogura, T. Takanashi, M. Hashizoe, and Y Honda, In vivo quantitation of peroxides in the vitreous humor by fluorophotometry, *Invest. Ophthalmol. Vis. Sci.* **37**, 1444-1450 (1996).

21. B. Hivert, C. Cerruti, and W. Camu, Hydrogen peroxide-induced motoneuron apoptosis is prevented by poly ADP ribosyl synthetase inhibitors, *Neuroreport* **9**, 1835-1838 (1998).

22. R. H. Steinberg, Survival factors in retinal degenerations, *Curr Opin Neurobiol.* **4**, 515-424 (1994).

CHARACTERIZATION AND LOCALIZATION OF PIGMENT EPITHELIUM-DERIVED FACTOR BINDING SITES IN THE BOVINE RETINA

Maria S. Aymerich, Alfredo Martinez, and S. Patricia Becerra[*]

1. INTRODUCTION

Several lines of evidence support a neurotrophic role for pigment epithelium-derived factor (PEDF) in the retina and central nervous system. PEDF induces neuronal differentiation in retinoblastoma cells,[1] and can protect cultured retinal neurons from induced-apoptosis.[2] After RPE removal, PEDF mimics the supportive role of the RPE on photoreceptors during the final stages of retinal morphogenesis[3]. In animal models of retinitis pigmentosa, the *retinal degeneration* (*rd*) and *retinal degeneration slow* (*rds*) mice, PEDF transiently delays photoreceptor loss.[4] In addition, the PEDF gene shows a tight linkage to a retinitis pigmentosa locus (RP13) on chromosome 17p13.3, suggesting its involvement in this disease[5]. Besides its effects on retina cells, PEDF also has neuronal differentiating and survival effects on neurons from the cerebellum,[6, 7, 8] hippocampus[9] and spinal cord,[10, 11] providing further evidence for its neurotrophic role.

The interactions between the RPE and the retina play an important role in the visual system. It has been demonstrated that the RPE releases factors, which promote survival, proliferation and differentiation on cells of retinal origin.[12, 13, 14] The RPE is the major source for the neurotrophic factor PEDF. Whereas, high levels of PEDF mRNA can be detected in the RPE, they remain at a low level and even undetectable in the neural retina from several species.[15, 16] RPE cells produce and secrete the PEDF protein to the extracellular interphotoreceptor matrix, where most of it is found in association to glycosaminoglycans/polyanions and next to the photoreceptors of neural retina[1, 17, 18].

The first step in the mechanism of action of PEDF is its interaction to cell-surface receptors. The presence of PEDF receptors has been studied and characterized in cultured cells[19]. PEDF binds to a single class of receptors on the surface of human retinoblastoma Y-79 cells and rat cerebellar granule cells neurons. Receptors on both types of cells have

---

[*] Maria S. Aymerich and S. Patricia Becerra, Laboratory of Retinal Cell and Molecular Biology, National Eye Institute, NIH, Bethesda, Maryland, 20892. Alfredo Martinez, CBD, DCS, National Cancer Institute, NIH, Bethesda, Maryland, 20892.

New Insights Into Retinal Degenerative Diseases
Edited by Anderson *et al.*, Kluwer Academic/Plenum Publishers, 2001

similar dissociation constant (K_D = 3 nM) and molecular weight (80,000). It is not known yet whether the retina has cell-surface receptors for PEDF that account for its neurotrophic activity. Here we have investigated the presence and distribution of receptors for this neurotrophic factor in the retina.

2. BINDING OF PEDF TO BOVINE RETINA

2.1. *In Situ* Binding of ^{125}I-PEDF to Bovine Retina

First, we investigated the presence of binding sites for PEDF in bovine retina pieces. The ligand was recombinant human PEDF radiolabeled with ^{125}I and obtained in a biologically active form[19]. Bovine eyes were cut below the iris, then the vitreous was removed from the inner retinal surface and the retina was gently separated from the RPE with forceps. The retina was placed extended in a Petri dish, on a grid, and pieces of 1 cm² were cut. Figure 1 shows the time curve for binding to freshly extracted pieces of neural retina. ^{125}I-PEDF bound directly to the retina as early as 30 minutes incubation (shorter incubation times were not technically possible). After 3 hours, the amount of bound ligand decreased significantly. These results indicate that the bovine retina contains sites for ^{125}I-PEDF binding.

2.2. Characterization of ^{125}I-PEDF Binding To Bovine Retina Membranes

By sequence homology, PEDF belongs to the superfamily of serine protease inhibitors (serpin)[20]. However, it does not have inhibitory activity against serine proteases and lacks the serpin stressed-to-relaxed (S→R) conformational change upon cleavage of its serpin-exposed loop, similarly to other non-inhibitory serpins, such as ovalbumin,

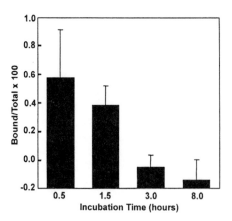

Figure 1. Time curve for *in situ* binding of ^{125}I-PEDF to bovine retina. Pieces of 1 cm² of bovine retina were cut, incubated with 1 nM ^{125}I-PEDF in binding buffer (1%BSA/PBS) at 4°C and collected at different time points. Pieces were washed with binding buffer to separate free and bound ligand. They were transferred to a scintillation vial for radioactivity determination using a β-counter. Background for each time point was determined from reactions with 1000-fold excess of unlabeled PEDF.

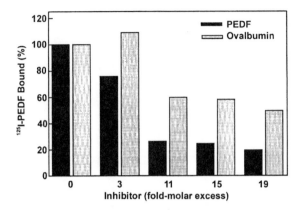

Figure 2. Effect of unlabeled PEDF and ovalbumin on ^{125}I-PEDF binding to bovine retina membranes. Membrane extracts were mixed with 0.25 nM ^{125}I-PEDF in the absence or presence of increasing concentrations of unlabeled PEDF (solid bars) or ovalbumin (open bars) and incubated at 4°C for 16 h. Free and bound ligand was separated by filtration through glass fiber filters under vacuum. Bound PEDF was calculated from radioactivity in filters using a β-counter. Binding in the absence of inhibitor corresponds to 100% and binding in the presence of 40-fold molar excess of unlabeled rhPEDF to 0%.

angiotensinogen and maspin[21, 22]. To characterize the binding of ^{125}I-PEDF to the retina we used unlabeled PEDF and ovalbumin as competitors of the binding. The binding reactions were performed with plasma membrane extracts obtained from bovine neural retinas by differential centrifugation. Figure 2 shows that, in contrast to ovalbumin, increasing concentrations of PEDF displaced efficiently the binding of the radioligand to retina plasma membranes. These results reveal that the binding is specific for PEDF and that other serpins do not bind efficiently to PEDF cell-surface receptors in the retina.

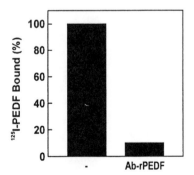

Figure 3. Effect of Ab-rPEDF on ^{125}I-PEDF binding to bovine retina membrane extracts. ^{125}I-PEDF was preincubated with or without Ab-rPEDF at 4°C for 16 h, before adding to membrane extracts to a final radioligand concentration of 1 nM. After incubation at 4°C for 90 min, the reaction mixture was treated as in figure 2. Binding in the absence of Ab-rPEDF corresponds to 100% and binding in the presence of 50-fold molar excess of unlabeled PEDF to 0%.

The polyclonal antiserum against PEDF (Ab-rPEDF) blocks the PEDF's neuronal differentiating and survival activities on Y-79 cells[17] and cerebellar granule cells[7], respectively, by inhibiting the binding of PEDF to receptors[19]. To evaluate the effect of Ab-rPEDF on PEDF binding to the retina, the radioligand was preincubated with the antiserum and then added to the retina membrane extracts. The Ab-rPEDF prevented 90% of ^{125}I-PEDF binding to bovine retina membranes (Figure 3).

Together, these characteristics of ^{125}I-PEDF binding to the retina are similar to those for PEDF receptors on Y-79 cells[19] and imply that the PEDF binding sites in both retinoblastoma and retina are similar.

3. LOCALIZATION OF PEDF BINDING SITES IN THE RETINA

To determine the distribution of PEDF binding sites in the retina, we chose a PEDF protein conjugated with fluorescein as ligand and frozen sections of bovine retinas as a samples with assumed receptors.

3.1. Characterization of Fluoresceinated PEDF

Fluorescein was coupled to primary amines in lysine residues of recombinant human PEDF. The chemically modified PEDF (Fl-PEDF) retained its biological activity on retinoblastoma cells and cerebellar granule cell neurons. Figure 4 shows the effect of Fl-PEDF on retinoblastoma Y-79 cells. Although, Fl-PEDF retained its binding affinity for cell-surface receptors in retinoblastoma cells, it lost its binding affinity for glycosaminoglycans/polyanions, as expected from a chemical modification of positively charged lysine residues[18]. These are important characteristics for Fl-PEDF because they provide the ligand with capabilities for detecting interactions with PEDF receptors excluding those with glycosaminoglycan-rich areas of cells and tissues.

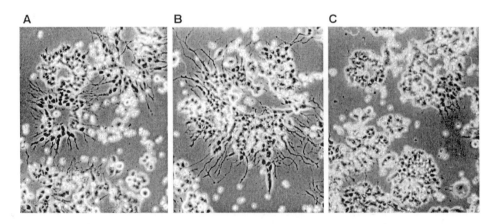

Figure 4. Fl-PEDF induces morphological differentiation in human retinoblastoma Y-79 cells. *A*, 2 nM Fl-PEDF. *B*, 0.5 nM bacterially-derived recombinant PEDF (positive control). *C*, no addition (negative control). Photographs of representative fields are shown.

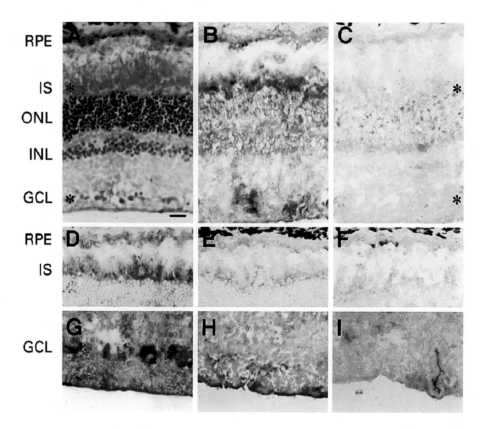

Figure 5. Fl-PEDF binding sites in bovine retina. Cryosections of bovine retina were incubated with 20 nM Fl-PEDF at 4°C for 30 min. Fluorescein was detected with an anti-fluorescein antibody bound to alkaline phosphatase and colorimetric development of its enzymatic activity (purple). *A*, bovine retina stained with Hematoxilin and Eosin. *B*, *D* and *G*, incubations with Fl-PEDF. *C*, *F* and *I*, incubations in the absence of Fl-PEDF. *E* and *H*, incubations with Fl-PEDF and 200-fold molar excess of rhPEDF. Each row corresponds to data obtained from three different eyes. RPE, retinal pigment epithelium; IS, inner segments of photoreceptors; ONL, outer nuclear layer; INL, inner nuclear layer; and GCL, ganglion cell layer. The bar corresponds to 30µm for *A* to *F* and 10 µm for *G* to *I*. Asterisks align with the IS and GCL.

3.2. Distribution of PEDF Binding Sites in the Retina

To detect Fl-PEDF binding sites on cryosections of bovine retina, we used a primary antibody against fluorescein followed by colorimetric detection and visualization of the fluorescein signal by light microscopy. Cryosections incubated with Fl-PEDF showed dense anti-fluorescein staining in the region of the inner segments (IS) of photoreceptor cells (Figs. 5B and D) and in the ganglion cell layer (GCL) (Figs. 5B and G). The anti-fluorescein signal decreased significantly in these regions when cryosections of retinas were incubated with Fl-PEDF plus excess of unmodified protein (Figs. 6E and H), or in the absence of ligand (Figs. 5C, F and I). These observations indicate that the distribution of Fl-PEDF binding, which predominated in the IS of photoreceptor cells and ganglion cells, was specific and competitive. The binding sites in the IS represent the likely cell-surface receptors, which may interact directly with the PEDF ligand residing naturally in the IPM. These observations are also in agreement with the PEDF protective effect on photoreceptors of the *rd* and *rds* mice[4]. On the other hand, the Fl-PEDF binding in ganglion cells suggests for the first time that these cells might be targets for a yet unknown PEDF activity. Our data provide a basis for both *in vitro* and *in vivo* studies to assess potential action for this neurotrophic factor.

4. SUMMARY

We have demonstrated the presence of PEDF binding sites in the neural retina from bovine eyes. Furthermore, these PEDF binding sites have characteristics similar to the ones described in human retinoblastoma Y-79 cells. Our data correlates with PEDF effects on the survival of photoreceptor cells *in vivo* and retinal cells in culture[2, 3, 4], and suggest that PEDF might also have an effect on ganglion cells. The demonstration of ^{125}I-PEDF and Fl-PEDF binding in the bovine retina supports a role for this factor in the adult retina and provides an anatomical basis for investigations of retina functions for PEDF.

REFERENCES

1. J. Tombran-Tink, G. J. Chader, and L. V. Johnson. PEDF: a pigment epithelium-derived factor with potent neuronal differentiative activity, *Exp. Eye Res.* **53**, 411-414 (1991).
2. W. Cao, J. Tombran-Tink, W. Chen, D. Mrazek, R. Elias, and J. F. McGinnis. Pigment epithelium-derived factor protects cultured retinal neurons against hydrogen peroxide-induced cell death, *J. Neurosci. Res.* **57**, 789-800 (1999).
3. M. M. Jablonski, J. Tombran-Tink, D. A. Mrazek, and A. Iannaccone. Pigment epithelium-derived factor supports normal development of photoreceptor neurons and opsin expression after retinal pigment epithelium detachment. *J. Neurosci.* **20**, 7149-7157 (2000).
4. M. Cayouette, S. B. Smith, S. P. Becerra, and C Gravel. Pigment epithelium-derived factor delays the death of photoreceptors in mouse models of inherited retinal degenerations. *Neurobiol. Dis.* **6**, 523-532 (1999).
5. R. Goliath, J. Tombran-Tink, I. R. Rodriquez, G. Chader, R. Ramesar, and J. Greenberg. The gene for PEDF, a retinal growth factor is a prime candidate for retinitis pigmentosa and is tightly linked to the RP13 locus on chromosome 17p13.3. *Mol. Vis.* **2**, 5-8 (1996).
6. T. Taniwaki, S. P. Becerra, G. J. Chader, and J. P. Schwartz. Pigment epithelium-derived factor protects cultured cerebellar granule cells against glutamate-induced neurotoxicity *J. Neurochem.* **64**, 2509-2517 (1995).
7. T. Taniwaki, N. Hirashima, S. P. Becerra, G. J. Chader, R. Etcheberrigaray, and J. P. Schwartz. Pigment epithelium-derived factor protects cultured cerebellar granule cells against glutamate-induced neurotoxicity. *J. Neurochem.* **68**, 26-32 (1997).
8. T. Araki, T. Taniwaki, S. P. Becerra, G. J. Chader, and J. P. Schwartz. Pigment epithelium-derived factor (PEDF) differentially protects immature but not mature cerebellar granule cells against apoptotic cell death. *J. Neurosci. Res.* **53**, 7-15. (1998).
9. M. A. DeCoster, E. Schabelman, J. Tombran-Tink, N. G. Bazan. Neuroprotection by pigment epithelial-derived factor against glutamate toxicity in developing primary hippocampal neurons. *J. Neurosci. Res.* **56**, 604-10 (1999).
10. L. J. Houenou, A. P. D'Costa,, L. Li, V. L. Turgeon, C. Enyadike, E. Alberdi, and S. P. Becerra. Pigment epithelium-derived factor promotes the survival and differentiation of developing spinal motor neurons. *J. Comp. Neurol.* **412**, 506-514 (1999).
11. M. M. Bilak, A. M. Corse, S. R. Bilak, M. Lehar, J. Tombran-Tink, and R. W. Kuncl. Pigment epithelium-derived factor (PEDF) protects motor neurons from chronic glutamate-mediated neurodegeneration. *J. Neuropathol. Exp. Neurol.* **58**, 719-728 (1999).
12. V. P. Gaur, Y. Liu, and J. E. Turner. RPE conditioned medium stimulates photoreceptor cell survival, neurite outgrowth and differentiation *in vitro. Exp. Eye. Res.* **54**, 645-659 (1992).
13. H. J. Scheedlo, L. Linxi, and J. E. Turner. Effects of RPE cell factors secreted from permselective fibers on retinal cells in vitro. *Brain Res.* **587**, 327-337 (1992).
14. C. D. Jaynes, and J. E. Turner. Muller cell survival and proliferation in response to medium conditioned by the retinal pigment epithelium. *Brain Res.* **678**, 55-64 (1995).
15. L. A. Perez-Mediavilla, C. Chew, P. A. Campochiaro, R. W. Nickells, V. Notario, D. J. Zack, and S. P. Becerra. Sequence and expression analysis of bovine pigment epithelium-derived factor. *Biochim. Biophys. Acta* **1398**, 203-214 (1998).
16. J. Ortego, J. Escribano, S. P. Becerra, and M. Coca-Prados. Gene expression of the neurotrophic pigment epithelium-derived factor in the human ciliary epithelium. *IOVS* **37**, 2759-2767 (1996).

17. Y. Q. Wu, V. Notario, G. J. Chader, and S. P. Becerra. Identification of pigment epithelium-derived factor in the interphotoreceptor matrix of bovine eyes. *Protein Expr. Purif.* **6**, 447-456 (1995).
18. E. Alberdi, C. C. Hyde, and S. P. Becerra. Pigment epithelium-derived factor (PEDF) binds to glycosaminoglycans: analysis of the binding site. *Biochemistry* **37**, 10643-10652 (1998).
19. E. Alberdi, M. S. Aymerich, and S. P. Becerra. Binding of pigment epithelium-derived factor (PEDF) to retinoblastoma cells and cerebellar granule neurons. Evidence for a PEDF receptor. *J. Biol. Chem.* **274**, 31605-31612 (1999).
20. F. R. Steele, G. J. Chader, L. V. Johnson, and J. Tombran-Tink. Pigment epithelium-derived factor: neurotrophic activity and identification as a member of the serine protease inhibitor gene family. *Proc. Natl. Acad. Sci. USA* **90**, 1526-1530 (1993).
21. S. P. Becerra, I. Palmer, A. Kumar, F. Steele, J. Shiloach, V. Notario, and G. J. Chader. Overexpression of fetal human pigment epithelium-derived factor in Escherichia coli. A functionally active neurotrophic factor. *J. Biol. Chem.* **268**, 23148-23156 (1993).
22. S. P. Becerra, A. Sagasti, P. Spinella, and V. Notario. Pigment epithelium-derived factor behaves like a noninhibitory serpin. Neurotrophic activity does not require the serpin reactive loop. *J. Biol. Chem.* **270**, 25992-25999 (1995).

LEUKEMIA INHIBITORY FACTOR PREVENTS PHOTORECEPTOR CELL DEATH IN rd$^{-/-}$ MICE BY BLOCKING FUNCTIONAL DIFFERENTIATION

LIF alters gene expression in photoreceptors

John D. Ash

SUMMARY

Stress cytokines including leukemia inhibitory factor (LIF) and ciliary neurotrophic factor (CNTF) have been shown to be upregulated in the retina following injury, stress, or ischemia. It is believed that LIF and/or CNTF are expressed by Muller glial cells during periods of stress to protect the retinal neurons from cell death. Injection of LIF or CNTF either in the vitreous or sub-retinal space has been shown to be effective in blocking or delaying photoreceptor degeneration in experimental light-damage models and inherited retinal degenerative diseases. The mechanisms by which these cytokines protect photoreceptors are unknown. Recently, we have generated transgenic mice that express LIF in the lens. We have found that LIF inhibits photoreceptor maturation, and thereby protects photoreceptors from apoptosis in rd homozygous mice. LIF did not appear to alter early events in retinal differentiation such as cell fate determination, or the establishment of the laminar organization of the retina. However, LIF inhibits the expression of function-related genes in rod photoreceptors and bipolar cells. The data suggest that LIF can act as a neuroprotective agent to photoreceptors by reducing the expression of genes that are involved in the perception and response to light.

1. INTRODUCTION

Animal models of retinitis pigmentosa (RP) provide insights into mechanisms of photoreceptor degeneration as well as models to test potential therapeutic strategies. The

Address correspondence to: John D. Ash, Department of Ophthalmology, University of Oklahoma Health Sciences Center, Oklahoma City, Oklahoma, 73104. Tel: (405) 271-3642, Fax: (405) 271 3721, email: john-ash@ouhsc.edu.

New Insights Into Retinal Degenerative Diseases
Edited by Anderson *et al.*, Kluwer Academic/Plenum Publishers, 2001

rd$^{-/-}$ mouse is a well-characcterized model which has a defect in the β-subunit of the cyclic guanosine mono-phosphate-phosphodiesterase gene.[1, 2] Mice that are homozygous for this recessive mutation undergo rapid rod photoreceptor cell death between postnatal days 10 and 20.

LIF and CNTF have been shown to be effective at protecting retinal neurons from a variety of toxic stimuli in vivo, including prolonged exposure to constant light, inherited retinal degenerative mutations, and transient ischemia.[3-8]. The changes that take place within photoreceptors to affect this protection have not been identified. In order to study the in vivo effects of LIF on photoreceptors we have generated transgenic mice that express human LIF in the lens. This strategy has allowed us to analyze the effects of LIF from embryonic day 11 (E11) to adults. We found that transgenic expression of LIF in the lens can protect rd$^{-/-}$ photoreceptors from cell death in vivo. To characterize the mechanism for this neuro-protection we mated the transgenic mice to wildtype C57BL/6 mice so that we could analyze the effects of LIF in the absence of retinal degeneration. We analyzed retinas for changes in organization and morphology of neuronal cell types, and for changes in expression of differentiation markers. Our observations suggest that LIF can inhibit functional maturation of rod photoreceptors. LIF inhibited the assembly of photoreceptor outer segments and the expression of genes necessary for the function of photoreceptors including NRL and opsin. The data suggest that photoreceptors that do not complete functional differentiation are not sensitive to cell death due to the rd mutation. Therefore the neuroprotective effects of LIF and CNTF may be mediated by suppressing functional differentiation in photoreceptors.

2. MATERIALS AND METHODS

2.1. Generation and screening of transgenic mice

All procedures were in accordance with the ARVO Statement for the Use of Animals in Ophthalmic and Vision Research. We generated transgenic mice by microinjection of an αA-crystallin-LIF minigene. A 600pb cDNA encoding the 180 amino acid secreted human LIF protein[9] was cloned into BamHI and Hind III sites downstream of a minimal 360 bp αA-crystallin promoter.[10] SV40 sequences were included to provide an intron and polyadenylation signals. The minigene was released from the plasmid vector by Not I digestion and purified by agarose gel electrophoresis using a quick spin gel extraction kit (Qiagen). The DNA fragment was eluted in 10 mM Tris-HCl pH 7.4, 0.1 mM EDTA, and diluted to 2 ng/μl prior to microinjection into pronuclei of FVB/N embryos.[11, 12] To identify transgenic mice, PCR amplifications were carried out on 5 μl of proteinase K digested, phenol extracted, tail DNA samples using a sense primer (SV40A, GTGAAGGAACCTTACTTCTGTGGTG) and an antisense primer (SV40-B, GTCCTTGGGGTCTTCTACCTTTCTC) which amplify the SV40 sequences. PCR reactions (50 μl) consisted of 10 mM Tris-HCl (pH 8.3), 50 mM KCl, 1.5 mM MgCl$_2$, 0.1% gelatin, 25 μM of each dNTP, 4 μM of each primer, and 2.5 units of *Taq* DNA polymerase (Perkin-Elmer Cetus). Reactions were denatured at 94°C for 2 minutes, then subjected to 30 cycles of 94°C for 30 seconds, 58°C for 30 seconds, and

72°C for 1 minute. A 15 μl aliquot from each reaction was analyzed by gel electrophoresis in a 1% agarose gel for the presence of a 300bp band. Founder mice were mated to FVB/N mice in order to establish transgenic lines. The FVB mouse strain is homozygous for the retinal degeneration mutation ($rd^{-/-}$), which is caused by a mutation in the gene encoding rod-specific cGMP phosphodiesterase.[1, 2] This autosomal recessive defect results in rapid rod cell death followed by a slow loss of cones. To analyze the effects of LIF on normal retinas, LIF transgenic mice were mated to C57BL/6$^{rd+/+}$ mice. The F1 progeny of this mating are heterozygous for rd ($rd^{+/-}$) and do not have photoreceptor degeneration. Therefore, the changes in retinal phenotype and function described for mice in the C57BL/6 background are caused by the expression of LIF in the lens and not a complication of the rd mutation.

2.2. Tissue preparation and routine histology

Eyes were enucleated, fixed in 10% neutral buffered formalin for 24 hours, dehydrated, and embedded in paraffin. All histological procedures were carried out on 5 μm paraffin sections, which had been deparaffinized in xylene and rehydrated in an ethanol series. Sections were stained with hematoxylin and eosin for routine histology.

2.3. Tunel Assay

For in situ analysis of cell death, paraffin sections were dewaxed and rehydrated to PBS, then incubated with 50 U of terminal deoxynucleotidyl transferase (Pharmacia) and 1 nmol of biotin-16-dUTP (Boehringer Mannheim), as described previously.[13] Biotin incorporation was detected with the avidin-biotin peroxidase complex reagent (Vector Laboratories). Sections were counterstained with hematoxylin. To quantify cell death the TUNEL stained cells (brown stain) were counted in the tissue sections.

2.4. In situ hybridizations

In situ hybridizations were carried out using ^{35}S riboprobes as described previously.[10] To detect Nrl expression, a plasmid containing 450 bp of the mouse Nrl cDNA (obtained from Anand Swaroop) was linearized with HindIII and transcribed with T3 RNA polymerase. The silver grains were visualized using darkfield illumination on a Nikon Eclipse microscope. The darkfield images were superimposed upon Nomarski images in PhotoShop.

2.5. Immunofluorescence

To detect opsin expression, paraffin embedded tissue sections on glass slides were rehydrated in an ethanol series then blocked in 10% horse serum in PBS for 30 minutes at room temperature prior to incubation with primary antibody. The mouse anti-opsin antibody (obtained from Colin Barnstable) was diluted 1:200 in 10% horse serum/PBS.

Sections were incubated overnight at 4°C, washed three time in PBS, then incubated for one hour with Texas Red anti-mouse secondary antibody diluted 1:200 in 10% horse serum/PBS. Slides were coverslipped with 50% glycerol in PBS, and fluorescent complexes were detected using a Nikon Eclipse fluorescent microscope. The fluorescent images were superimposed upon Nomarski images in PhotoShop.

3. RESULTS

3.1. Transgenic expression of LIF rescues photoreceptors in adult rd$^{-/-}$ mice

Description of founders: The αA-LIF minigene utilizes the αA-crystallin promoter to drive the lens-specific expression of a 600pb cDNA which encodes the secreted human LIF protein.[9] Eight αA-LIF transgenic lines were generated, and all but two of the founders developed cataracts as adults. Transgenic eyes either lacked an iris or had an asymmetrical iris located on the dorsal side of the eye that did not close in response to light. Eyes from all of the founders were analyzed by histology. Two lines did not develop any identifiable phenotype and were not analyzed further. The transgenic mice were generated in the FVB/N strain which is an albino strain which are homozygous for the *rd* mutation causing them to loose their photoreceptors by three weeks of age. Surprisingly, the histology demonstrated that six of the founders retained their

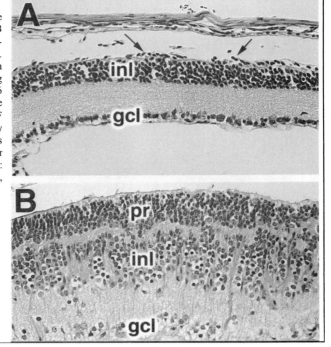

Figure 1. Representative histology images for FVB (rd$^{-/-}$) mice (A) and for αA-LIF transgenic mice (B). The arrows in A point to a single row of remaining photoreceptor nuclei in a 6 week old FVB mouse. The 6 week old αA-LIF transgenic mice have nearly normal numbers of cells located in the photoreceptor layer (pr). Abbreviations: inl, inner nuclear layer; gcl, ganglion cell layer.

photoreceptors to a varying degree as adult mice (fig. 1). We picked one of the lines, OVE 736, for further analysis because it was the line with the most surviving photoreceptors. A second line OVE 774 had few surviving photoreceptors as adults, but did show a delay in photoreceptor loss. To determine if the difference in photoreceptor survival correlated with levels of transgene expression we analyzed the abundance of transgenic mRNA in tissue sections by in situ hybridization. We found that photoreceptor survival correlated with increased LIF expression (not shown).

3.2. Photoreceptors in the αA-LIF transgenic mice have reduced cell death

Photoreceptor loss in rd$^{-/-}$ mice is due to cell death which can be detected in situ using TUNEL staining. At Postnatal day 10 (P10) photoreceptors in FVB mice have a high rate of cell death as demonstrated by TUNEL positive cells (fig. 2A). However, there were significantly fewer dying cells in the αA-LIF transgenic retina (fig. 2B). To quantify the effect we counted positive cells in the photoreceptor layer in sections that pass through the center of the eye. At P10 there were 272 TUNEL positive photoreceptors in the

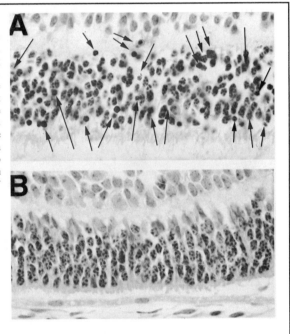

Figure 2. Analysis of cell-death. Representative images of TUNEL stained sections. Cells undergoing apoptosis are stained brown in the Tunel assay. The brown staining is not apparent in the grayscale image. Twenty brown cells were observed in the image of the P10 FVB retina (Arrows in A). However, no brown cells were found in the section from the αA-LIF transgenic mouse (B).

sections of FVB eyes. This age represents the most active period of cell death in rd$^{-/-}$ mice (fig. 2). In contrast, the αA-LIF transgenic retina section contained only 19 TUNEL positive cells at P10. Therefore, the LIF transgenic mice have a dramatically slower rate of photoreceptor death compared to FVB. At 6 weeks of age the αA-LIF transgenic mice in the FVB background had outer nuclear layers containing of 5 to 6 rows of nuclei.

3.2. Histological analysis

In order to study the effect of LIF on a normal retina, transgenic mice were mated to C57BL/6 mice which are rd$^{+/+}$. The resulting F_1 progeny are all rd $^{+/-}$ and the non-transgenic mice are phenotypically normal. Eyes were collected and analyzed at various ages including new-born (P1), postnatal day 5 (P5), P6, P7, P10, P14, P21, P28, and at

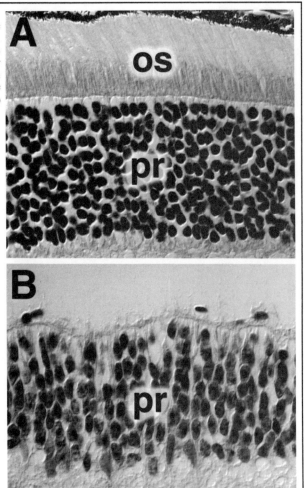

Figure 3. Representative histology of rd$^{+/-}$ retinas at P21. Non-transgenic photoreceptors (A) have elaborate outer segments (os) at this age. In contrast photoreceptors in the αA-LIF transgenic mice (B) have only short inner segments.

six and eight weeks. Histological examination of hemotoxylin and eosin stained sections suggested that LIF blocked aspects of retinal development. The morphogenesis of the three nuclear layers, which occurs in the normal retina during the first postnatal week, occurred at the appropriate time in the transgenic retinas (not shown). However, the appearance and arrangement of nuclei within each layer was less organized than in the normal retina (not shown), and the separation between layers was less obvious in transgenic retinas than in normal retinas. The nuclei of photoreceptors in the outer nuclear layer were less organized than in the normal retinas and were less condensed (fig 3). In addition, the photoreceptors failed to develop outer segments (fig. 3).

3.3. Expression of differentiation-specific genes is blocked. Photoreceptors in the αA-LIF transgenic mice failed to synthesize the outer segments suggesting the expression of differentiation specific genes may be inhibited. We analyzed the expression of genes that are important for the function of photoreceptors including rod opsin, ABCR transporter, inter-photoreceptor retinal binding protein (IRBP), and the expression of NRL and CRX, which are transcription factors that are involved in photoreceptor gene expression. We found that IRBP, CRX and ABCR were expressed in the αA-LIF transgenic mice, although at reduced levels (not shown). This demonstrates that the expression of some photoreceptor genes is not blocked in the LIF transgenic mice. However, the expression of NRL and rod opsin was inhibited (fig. 4C and D). The results demonstrate that LIF can block the expression of genes that are necessary for photoreceptor function.

4. DISCUSSION

The present study demonstrates that LIF can block the functional maturation of rod photoreceptors and that this block can protect photoreceptors from cell death in rd[-/-] mice. The block of functional maturation might be mediated by inhibition of expression of the transcription factor NRL. The absence of outer segment synthesis suggests that NRL is essential for the expression of genes involved in outer segment synthesis, including the expression of opsin.

4.1. Relevance to stress-induced neuroprotection.

CNTF and LIF are members of the interleukine 6 (IL-6) family of cytokines. Members of this family do not share sequence homology but are grouped together based on activation of a common tyrosine kinase receptor, gp130.[14] The high-affinity receptor for LIF is a heterodimer between a low affinity LIF receptor and the tyrosine kinase gp130.[15] CNTF also utilizes the LIF receptor complex, but requires an additional subunit, CNTFRα, for efficient binding.[16] Because of the receptor similarities, cells that can respond to CNTF will also respond to LIF. The expression of CNTF and LIF has been shown to be increased in the retina following injury, stress, or induced ischemia.[3, 17-21] Previous studies have suggested that LIF and/or CNTF are expressed during periods of stress to protect the retinal neurons from cell death.[3, 4, 22, 23] In support of this hypothesis other studies have shown that intraocular injection of LIF or CNTF is effective at

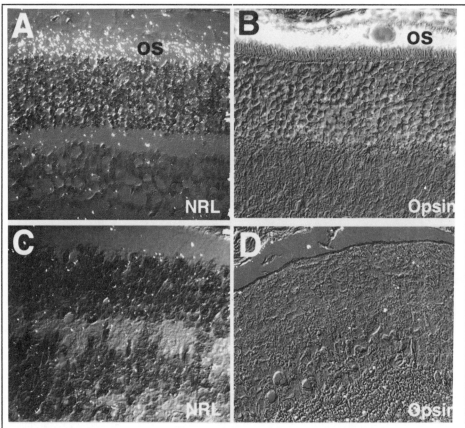

Figure 4.   Inhibited of NRL and Opsin expression.   The expression of differentiation-specific genes
was analyzed in the retinas from non-transgenic (A, and B) and αA-LIF transgenic mice (C, and D)
in the C57BL/6 background.   The transcription factor NRL was analyzed by in situ hybridization.
NRL expression was detected in the non-transgenic retina (indicated by the silver grains located in the
outer segments (os) in the darkfield image shown in panel A), but was not detected in the αA-LIF
retina (C).   Opsin expression was analyzed by immunofluorescence, and was abundantly expressed in
the outer segments in the non-transgenic retina (B), but not in the αA-LIF transgenic retina (D).

protecting retinal neurons from a variety of toxic stimuli, including prolonged exposure
to constant light, inherited retinal degenerative mutations, and transient ischemia.[3-8].
Recent studies have shown that following CNTF stimulation, the MAP kinase pathway is
activated in Muller glial cells, while the Stat3 pathway is activated in astrocytes, ganglion
cells and Muller glial cells.[23, 24] These studies indicate that only astrocytes, ganglion cells,
or Muller glial cells directly respond to CNTF and (by extrapolation) to LIF. Therefore
the responses in photoreceptors and bipolar cells may be mediated by a second signal
originating from the inner retina, possibly from Muller cells. We did observe GFAP
expression in Muller cells in the LIF transgenic families (data not sown), which is
consistent with this mechanism. While the mechanisms for neuroprotection are not well
understood, in vitro studies have shown that LIF and CNTF can suppress the

development of opsin-positive cells in primary retinal cultures from rodents.[25, 26] One consequence of stimulation by LIF appears to be decreased expression of the cell type specific transcription factor NRL.[27] Taken together, these studies suggest a model where LIF or CNTF can protect photoreceptors through an indirect mechanism involving activation of glial cells. Activated glial cells in an unknown mechanism induce photoreceptors to turn off or reduce expression of function-related genes. Part of the mechanism for down regulating gene expression in photoreceptors is likely to involve regulation of NRL expression. To further develop this hypothesis it will be necessary to identify the second signal from the glial cells and to identify the mechanism of NRL inhibition.

5. ACKNOWLEDGEMENTS

This work was partially accomplished in the laboratory of Paul Overbeek (Baylor College of Medicine). This work was supported by National Institute of Health grants EY06708 (JDA), EY10448 (PAO), EY10803 (PAO), EY11348(PAO), and by a National Eye Institute core grant to OUHSC. Additional support was provided by Research to Prevent Blindness (New York, New York), and the Retina Research Foundation (Houston, TX.).

REFERENCES

1. Bowes C, Li T, Danciger M, et al. Retinal degeneration in the rd mouse is caused by a defect in the beta subunit of rod cgmp-phosphodiesterase. *Nature.* 1990;347:677-680.
2. Pittler SJ, Baehr W. Identification of a nonsense mutation in the rod photoreceptor cgmp phosphodiesterase beta-subunit gene of the rd mouse. *Proc Natl Acad Sci U S A.* 1991;88:8322-8326.
3. LaVail MM, Yasumura D, Matthes MT, et al. Protection of mouse photoreceptors by survival factors in retinal degenerations. *Invest Ophthalmol Vis Sci.* 1998;39:592-602.
4. Unoki K, LaVail MM. Protection of the rat retina from ischemic injury by brain-derived neurotrophic factor, ciliary neurotrophic factor, and basic fibroblast growth factor. *Invest Ophthalmol Vis Sci.* 1994;35:907-915.
5. Chong NH, Alexander RA, Waters L, et al. Repeated injections of a ciliary neurotrophic factor analogue leading to long-term photoreceptor survival in hereditary retinal degeneration. *Invest Ophthalmol Vis Sci.* 1999;40:1298-1305.
6. Cayouette M, Gravel C. Adenovirus-mediated gene transfer of ciliary neurotrophic factor can prevent photoreceptor degeneration in the retinal degeneration (rd) mouse. *Hum Gene Ther.* 1997;8:423-430.
7. Cui Q, Lu Q, So KF, Yip HK. CNTF, not other trophic factors, promotes axonal regeneration of axotomized retinal ganglion cells in adult hamsters. *Invest Ophthalmol Vis Sci.* 1999;40:760-766.
8. Ogilvie JM, Speck JD, Lett JM. Growth factors in combination, but not individually, rescue rd mouse photoreceptors in organ culture. *Exp Neurol.* 2000;161:676-685.
9. Gough NM, Gearing DP, King JA, et al. Molecular cloning and expression of the human homologue of the murine gene encoding myeloid leukemia-inhibitory factor. *Proc Natl Acad Sci U S A.* 1988;85:2623-2627.

10. Robinson ML, Overbeek PA, Verran DJ, et al. Extracellular FGF-1 acts as a lens differentiation factor in transgenic mice. *Development*. 1995;121:505-514.

11. Taketo M, Schroeder AC, Mobraaten LE, et al. Fvb/n: An inbred mouse strain preferable for transgenic analyses. *Proc Natl Acad Sci U S A*. 1991;88:2065-2069.

12. Hogan B, Beddington R, Costantini F, Lacy E. Manipulating the mouse embryo : A laboratory manual. 2nd. Plainview, N.Y.: Cold Spring Harbor Laboratory Press; 1994:xvii, 497

13. Gavrieli Y, Sherman Y, Ben-Sasson SA. Identification of programmed cell death in situ via specific labeling of nuclear DNA fragmentation. *J Cell Biol*. 1992;119:493-501.

14. Ip NY, Nye SH, Boulton TG, et al. Cntf and lif act on neuronal cells via shared signaling pathways that involve the il-6 signal transducing receptor component gp130. *Cell*. 1992;69:1121-1132.

15. Gearing DP, Comeau MR, Friend DJ, et al. The IL-6 signal transducer, gp130: An oncostatin m receptor and affinity converter for the lif receptor. *Science*. 1992;255:1434-1437.

16. Ip NY, McClain J, Barrezueta NX, et al. The alpha component of the CNTF receptor is required for signaling and defines potential cntf targets in the adult and during development. *Neuron*. 1993;10:89-102.

17. Cao W, Wen R, Li F, Lavail MM, Steinberg RH. Mechanical injury increases bFGF and CNTF mRNA expression in the mouse retina. *Exp Eye Res*. 1997;65:241-248.

18. Wen R, Cheng T, Song Y, et al. Continuous exposure to bright light upregulates bFGF and CNTF expression in the rat retina. *Curr Eye Res*. 1998;17:494-500.

19. Wen R, Song Y, Cheng T, et al. Injury-induced upregulation of bFGF and CNTF mRNAs in the rat retina. *J Neurosci*. 1995;15:7377-7385.

20. Ju WK, Lee MY, Hofmann HD, Kirsch M, Chun MH. Expression of CNTF in Muller cells of the rat retina after pressure- induced ischemia. *Neuroreport*. 1999;10:419-422.

21. Chun M, Ju W, Kim K, et al. Upregulation of ciliary neurotrophic factor in reactive Muller cells in the rat retina following optic nerve transection. *Brain Res*. 2000;868:358-362.

22. LaVail MM, Unoki K, Yasumura D, et al. Multiple growth factors, cytokines, and neurotrophins rescue photoreceptors from the damaging effects of constant light. *Proc Natl Acad Sci U S A*. 1992;89:11249-11253.

23. Peterson WM, Wang Q, Tzekova R, Wiegand SJ. Ciliary neurotrophic factor and stress stimuli activate the JAK-STAT pathway in retinal neurons and glia. *J Neurosci*. 2000;20:4081-4090.

24. Wahlin KJ, Campochiaro PA, Zack DJ, Adler R. Neurotrophic factors cause activation of intracellular signaling pathways in Muller cells and other cells of the inner retina, but not photoreceptors. *Invest Ophthalmol Vis Sci*. 2000;41:927-936.

25. Neophytou C, Vernallis AB, Smith A, Raff MC. Muller-cell-derived leukaemia inhibitory factor arrests rod photoreceptor differentiation at a postmitotic pre-rod stage of development. *Development*. 1997;124:2345-2354.

26. Kirsch M, Schulz-Key S, Wiese A, Fuhrmann S, Hofmann H. Ciliary neurotrophic factor blocks rod photoreceptor differentiation from postmitotic precursor cells in vitro. *Cell Tissue Res*. 1998;291:207-216.

27. Kumar R, Chen S, Scheurer D, et al. The bzip transcription factor NRL stimulates rhodopsin promoter activity in primary retinal cell cultures. *J Biol Chem*. 1996;271:29612-29618.

GLYCOBIOLOGICAL APPROACH TO CONTROL OF RETINAL DEGENERATION

Fumiyuki Uehara[*], Norio Ohba, and Masayuki Ozawa

1. INTRODUCTION

Medical treatment of degenerative disorders of the central nervous system has been attempted using a number of neurotrophic agents. In the rat retina, direct injection of growth factors or neurotrophic factors into the eye delays photoreceptor degeneration in inherited retinal dystrophy (Faktorovich et al., 1990) and light damage (Faktorovich et al., 1992). Although a combination of different factors enhances the degree of photoreceptor survival (LaVail et al., 1992), the effects of other types of agents should be examined to determine the perfect combination.

We have previously examined the relation between retinal degeneration and glycoconjugates, especially paying attention to the terminal sialyl residues on the glycans of the interphotoreceptor matrix (IPM) and photoreceptor cells (Uehara et al., 1988, 1989, 1996, 1997). Based on these glycobiological findings, we are now trying to control retinal degeneration. Different mutations and constant light cause photoreceptor cell death via the common pathway of apoptosis (Chang et al., 1993). In this process, we have shown that β-galactoside residues on photoreceptor cells are exposed by intrinsic increased neuraminidase (Uehara et al., 1997). Galectin is an animal lectin which recognizes β-galactoside residues (Barondes et al., 1994). We expect that exogenous galectin-3 may bind to photoreceptor cells with β-galactoside residues and hence inhibit their apoptosis since galectin-3 has not only β-galactoside-binding ability (Barondes et al., 1994) but also an antiapoptotic property (Yang et al., 1996; Akahani et al., 1997). Light-damage model involving rats is an excellent one for studying the molecular mechanisms

* Fumiyuki Uehara[1], Norio Ohba[1], and Masayuki Ozawa[2], Departments of Ophthalmology[1] and Biochemistry[2], Kagoshima University Faculty of Medicine, 8-35-1 Sakuragaoka, Kagoshima-shi, Japan 890-8520

of therapeutic agents on the degeneration of photoreceptor cells (LaVail et al., 1987). We report here that galectin-3 delays photoreceptor cell-degeneration due to constant light in the rat retina, showing potential promise for the glycobiological control of retinal degeneration.

2. METHODS

2.1. Expression of Galectin-3 as a Fusion Protein

A cDNA clone for rat retinal galectin-3 was isolated by means of the polymerase chain reaction (PCR) from the rat retinal cDNA in the pAP3neo vector (Takara Shuzo, Shiga, Japan) using a pair of oligonucleotides, 5'CGGGATCCAGGAAAATGGCAGA CGGCTTC3' and 5'GGGGTACCTCATAACACACAGGGCAGTTC3', as PCR primers according to the manufacturer's instructions for an Advantage cDNA PCR kit (CLONTECH Laboratories, Palo Alto, CA). A PCR-product of 0.85 kbp was isolated, cloned into Bluescript II SK (Stratagene, La Jolla, CA), and then sequenced according to a protocol for ABI Prism BigDye terminator cycle sequencing (PE Applied Biosystems, Foster City, CA). A search of the GenBank revealed that the nucleotide sequence encoded by the cDNA is identical to that of galectin-3 from rat basophilic leukemia cells (Albrand et al., 1987; J02962) except for one nucleotide (nt-59: G→A). The 20th amino acid of glutamine, deduced from the cDNA sequence of nt-58-60 determined in the present study, is the same as that in human galectin-3 (Oda et al., 1991; AB006780, M36682) and mouse galectin-3 (Jia et al., 1988; J03723, X16074). To express rat galectin-3 as a fusion protein with maltose-binding protein (MBP) or glutathione S-transferase (GST) in *Escherichia coli* (*E. coli*) BL21 cells, cDNA encoding the lectin was cloned into an MBP fusion vector (pMALc2; New England Biolabs, Beverly, MA) or a GST fusion vector (pGEX-4T1; Amersham Pharmacia, Piscataway, NJ). Bacterial cells containing a fusion plasmid were cultured, and then the fusion protein was induced and affinity-purified on columns of amylose-resin (New England Biolabs; for the MBP fusion protein) or glutathione-sepharose (Amersham Pharmacia; for the GST fusion protein). The pMALc2 or pGEX-4T-1 vector without the cDNA-insert was also introduced into *E. coli* BL21 cells, and then the protein, MBP or GST, was affinity-purified by the method previously described (Uehara et al., 2000).

2.2. Purification of an Anti-Galectin-3 Antibody

Rabbit antisera were raised against the MBP/galectin-3-fusion protein in a New Zealand white rabbit, and the antibody against galectin-3 was affinity-purified on a column of the GST/galectin-3-fusion protein coupled to Affi-Gel 10 (BIO-RAD, Hercules, CA) by the method previously described (Uehara et al., 1985; Ozawa et al., 1995).

2.3. Animals, Injection, Histological Procedures and Quantification

Wistar albino rats of 2 to 3 months of age (Kyudo, Kumamoto, Japan) were maintained with a 12-hour light (in-cage illumination, less than 15 foot-candles)/12-hour dark cycle for more than 1 week before use. All animal procedures conformed to the Guidelines of the Kagoshima University Faculty of Medicine for Animal Experiments, and the ARVO Statement for the Use of Animals in Research. Two days before constant light exposure, we anesthetized the rats with an intramuscular injection of a ketamine-xylazine mixture and injected 1 µl of a solution containing MBP-galectin-3 (1 mg/ml PBS), GST-galectin-3 (1 mg/ml PBS), or anti-galectin-3 antibodies (0.1 mg/ml PBS) into the vitreous cavity of the inferior hemisphere of one eye with a 32-gauge beveled needle. The other eye of each rat was injected with the same volume of MBP (1 mg/ml PBS), GST (1 mg/ml PBS), or PBS, respectively, as a control. One group of rats (at least 6 rats with the same injection) was maintained under cyclic lighting (12-hour on/12-hour off) conditions for 9 days following the injection. On the other hand, two days after the injection, the other group of rats (at least 6 rats with the same injection) was placed in constant light, at an illumination of 130 fc to 150 fc, for 1 week. After this exposure, all the rats were killed with an overdose of carbon dioxide, and then perfused intravascularly with a fixative containing 2.5% glutaraldehyde and 2% paraformaldehyde. The eyes were enucleated, bisected along the vertical meridian, rinsed in PBS, dehydrated and then embedded in paraffin. The eyes were sectioned at 3 µm thickness and then stained with hematoxylin-eosin. The degree of light-induced retinal degeneration was determined by measuring the thickness of the outer nuclear layer (ONL), which is used as an index of photoreceptor cell loss (LaVail et al., 1987). The mean ONL thickness was obtained for single sections of at least 6 eyes in each experiment. For each of the superior hemispheres, two sets of color slides were taken and the ONL thickness was determined by the method previously reported (Unoki et al., 1994). Since we injected the agents through the inferior hemispheres in the present study, only the superior ones were used for the quantification. The difference in the ONL thickness between two groups was analyzed using a paired t-test.

3. RESULTS

When the rats were maintained with a 12-hour light/12-hour dark cycle after the injection, almost no morphological change was observed throughout the retina whatever the injected agents were (Fig. 1a, b). However, when the rats were kept under constant light after the injection, degenerative changes were detected in the photoreceptor layer and ONL to varying extents, depending on the injected materials and ocular regions (Fig. 1c-h). In general, degeneration of photoreceptor cells after 1 week of constant light exposure was most severe in the posterior to equatorial region of the superior hemisphere, as expected from results of previous studies (LaVail et al., 1987; LaVail et al., 1992; Unoki et al., 1994). Although the photoreceptor layer and ONL were reduced in thickness, the inner nuclear layer (INL), inner plexiform layer (IPL), ganglion cell layer (GC) and nerve fiber layer (NF) showed an almost intact morphology with constant light. The thickness of the ONL in the PBS-injected eyes from the rats maintained under constant light was approximately one-fourth of that in the case of those kept with the usual light/dark cycle (Fig. 1a, c). The injection of an antibody against galectin-3 by itself did not affect the ONL-thickness in the rats maintained under the normal light conditions in comparison with in the case of PBS (Fig. 1a, b; $p < 0.05$). However, the antibody accelerated the photoreceptor cell-loss in those kept under constant light, the thickness being reduced more than with PBS-injection (Fig. 1c, d; $p < 0.05$). The thickness of the ONL in either MBP- or GST-injected eyes from the rats exposed to constant light was similar to that in the PBS-injected ones from those under the same light conditions (Fig. 1c, e, g; $p < 0.05$). On the other hand, intravitreal injection of either MBP-galectin-3 or GST-galectin-3 significantly protected the photoreceptors and ONL in comparison with in the case of either PBS, MBP or GST (Fig. 1c, e, f, g, h; $p < 0.05$). In these injected eyes

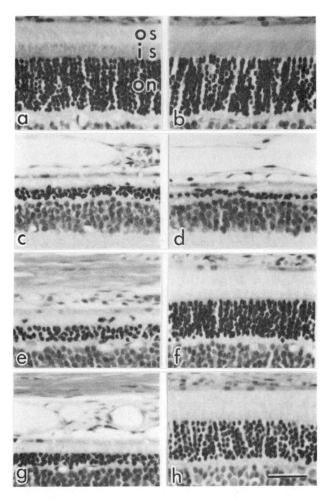

Figure 1.   Hematoxylin and eosin staining of a rat retina taken from the posterior superior region of the eye, which is most sensitive to the damaging effects of constant light.   The retinas of rats, maintained in cyclic light after the injection, show almost no morphological change (a, PBS-injected; b, anti-galectin-3 antibody-injected). Those kept under constant light after the injection (c-h) exhibit degenerative changes of the photoreceptor cells (os, outer segments; is, inner segments; on, outer nuclear layer) to varying extents (c-h) depending on the injected materials [c, PBS; d, anti-galectin-3 antibody; e, maltose binding protein (MBP); f, MBP-galectin-3; g, glutathione S-transferase (GST); h, GST-galectin-3]. Scale bar=50 μm.

after constant light exposure, the photoreceptors (inner and outer segments) were similar in thickness to those without constant light although they appeared a little disorganized (Fig. 1a, b, f, h). The thickness of the ONL in either MBP-galectin-3 or GST-galectin-3 injected eyes from the rats maintained under constant light was approximately 60% of that in the case of those kept in the usual light/dark cycle (Fig. 1a, b, f, h). The mean ONL thickness of the retinas in individual experiments are summarized in Fig. 2.

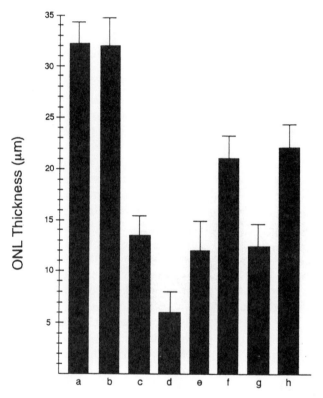

Figure 2. Measurement of the outer nuclear layer (ONL) thickness of rat retinas either maintained in cyclic light (a, b), or after 1 week of constant light (c-h) with intravitreous injection of various agents as in Fig. 1 (a, c, PBS; b, d, anti-galectin-3 antibody; e, MBP; f, MBP-galectin-3; g, GST; h, GST-galectin-3). Each data point represents the mean ($\pm$S.D.) for 6 or more eyes. A paired t-test analyzing differences in ONL thickness revealed that a, b>f, h>c, e, g>d (p<0.05), implying that intravitreous injection of galectin-3 significantly prevents photoreceptor cell loss due to constant light, while that of anti-galectin-3 antibodies accelerates it.

4. DISCUSSION

Our previous study showed that sialic acids at the termini of sugar chains on glycoconjugates of photoreceptor cells decrease on constant light exposure, the β-

galactoside residues on them becoming exposed (Uehara et al., 1997). The carbohydrate-binding domain of the injected galectin-3 fused with MBP or GST may have bound to these cells with exposed β-galactoside residues, since we have confirmed that both MBP-galectin-3 (7.4 kDa) and GST-galectin-3 (5.7 kDa) bind to a biotinylated β-galactose probe by means of Western blotting. Then, apoptosis-inhibitory signals to the photoreceptor cells may have been generated. The NWGR motif, which has only been detected in galectin-3 among galectin family members (Yang et al., 1996; Akahani et al., 1997), may be one of the candidates responsible for this inhibition. Although this motif can inhibit photoreceptor cell death through a *bcl*-2 (Yang et al., 1996) or the cysteine protease pathway (Akahani et al., 1997), intracellular ingestion of galectin-3 should have occurred. However, galectin-3 at least binds to the surfaces of the photoreceptor cells with β-galactoside residues and may pull the trigger of intracellular transmission of anti-apoptotic signal. Then, the photoreceptor cells to which galectin-3 binds may become resistant to constant light-induced apoptosis.

On the other hand, since the injection of an antibody against galectin-3 accelerated the photoreceptor degeneration due to constant light, it was thought that the expression of endogenous galectin-3 may also occur as an anti-apoptotic system in the rat retina on constant light exposure. In fact, we have recently found that the endogenous expression of galectin-3 increased in the constant light-exposed rat retina by means of immunohistochemistry. The antibody reaction products appeared as strands extending from the inner limiting membrane (ILM) to the outer limiting membrane (OLM) across the neural retina with side branches through the outer plexiform layer (OPL). This staining pattern closely matched the distribution of Mueller cells and their processes (Uga and Smelser, 1973). These findings suggest that endogenous expression of galectin-3 increases in Mueller cells in response to constant light exposure. In connection with this observation, a previous report describing the labeling of the radial fibers of Mueller cells with glial fibrillary acidic protein (GFAP) after constant light (Eisenfeldt et al., 1984) is noticeable. In this report (Eisenfeldt et al., 1984), increased labeling with GFAP in radial fibers of the inner retina was shown. Unlike this observation, the staining of galectin-3 most increased in some parts of Mueller cells of the ONL, OPL and INL, including both radial and horizontal fibers. These different distributional patterns of GFAP and galectin-3 imply that they are involved in different functions. The increase of galectin-3 in the ONL may affect photoreceptor cells unlike GFAP, which is not observed in the ONL (Eisenfeldt et al., 1984).

The injection of the antibody against galectin-3 by itself did not cause photoreceptor cell loss with the usual cyclic lighting without constant light exposure. This injection accelerated the light damage on constant light, implying that endogenous galectin-3, which is increased in Muller cells, may be secreted out and at least partially protect photoreceptor cells from light damage. Thus, the increased expression of galectin-3 in Mueller cells in response to constant light exposure may also be a part of the endogenous neuroprotective system in the retina. The amount of endogenous galectin-3 expressed in Mueller cells was not enough for inhibition of photoreceptor cell-apoptosis under such severe conditions as the present constant light exposure, although it was enough for Mueller cells to survive. Exogenous galectin-3 also did not completely suppress the photoreceptor degeneration under the present experimental light conditions. However, a combination of other factors may enhance the degree of photoreceptor survival. Based on glycobiological viewpoints, we already have several other reagents that are effective to inhibit photoreceptor degeneration. Our preliminary studies have shown that an inhibitor of neuraminidase delays the degenerative process of photoreceptor cells. We also have observed that the subretinal application of jacalin stabilizes the structure of the IPM and alters photoreceptors to be resistant to light damage. Since galectin-3 binds to the target cells, it can be used to deliver another anti-apoptotic agent to the photoreceptor cells with β-galactoside residues. A drug delivery system or gene therapy involving galectin-3 as a targeting carrier may be applied for other cells with β-galactoside residues as well. The effects of various anti-apoptotic agents coupled or fused with galectin-3 on apoptosis should be examined in future studies to control retinal degeneration.

Acknowledgments
The authors thank Yoshiko Maeda for her technical assistance. This work was supported by a Grant-in-Aid for Scientific Research (C) from the Japanese Ministry of Education, Science, Sports and Culture (11671747).

REFERENCES

Akahani, S., Nangia-Makker, P., Inohara, H., Kim, H.R., and Raz, A., 1997, Galectin-3: a novel antiapoptotic molecule with a functional BH1 (NWGR) domain of Bcl-2 family, *Cancer Res.* **57**: 5272-5276.

Albrand, K.A., Orida, N.K., and Liu, F.T., 1987, An IgE-binding protein with a distinctive repetitive sequence and homology with an IgG receptor, *Proc. Natl. Acad. Sci. USA.* **84**: 6859-6863.

Barondes, S.H., Cooper, D.N.W., Gitt, M.A., and Leffler, H., 1994, Galectins, structure and function of a large family of animal lectins, *J. Biol.Chem.* **269**: 20807-20810.

Chang, G.Q., Hao, Y., and Wong, F., 1993, Apoptosis: Final common pathway of photoreceptor death in rd, rds, and rhodopsin mutant mice, *Neuron* **11**: 595-605.

Eisenfeld, A.J., Bunt-Milam, A.H., and Sarthy, P.V., 1984, Mueller cell expression of glial fibrillary acidic protein after genetic and experimental photoreceptor degeneration in the rat retina, *Invest. Ophthalmol. Vis. Sci.* **25**: 1321-1328.

Faktorovich, E.G., Steinberg, R.H., Yasumura, D., Matthes, M.T., and LaVail, M.M., 1990, Photoreceptor degeneration in inherited retinal dystrophy delayed by basic fibroblast growth factor, *Nature* **347**: 83-86.

Faktorovich, E.G., Steinberg, R.H., Yasumura, D., Matthes, M.T., and LaVail, M.M., 1992, Basic fibroblast growth factor and local injury protect photoreceptors from constant light damage in the rat, *J. Neurosci.* **12**: 3554-3567.

Jia, S., and Wang, J.L., 1988, Carbohydrate binding protein 35: Complementary DNA sequence reveals homology with proteins of the heterogeneous nuclear RNP, *J. Biol. Chem.* **263**: 6009-6011.

LaVail, M.M., Gorrin, G.M., Repaci, M.A., Thomas, L.A., and Ginsberg, H.M., 1987, Genetic regulation of light damage to photoreceptors, *Invest. Ophthalmol. Vis. Sci.* **28**: 1043-1048.

LaVail, M.M., Unoki, K., Yasumaura, D., Matthes, M.T., et al., 1992, Multiple growth factors, cytokines and neurotrophins rescue photoreceptors from the damaging effets of constant light, *Proc. Natl. Acad. Sci. USA.* **89**: 11249-11253.

Oda, Y., Leffler, H., Sakakura, Y., Kasai, K., and Barondes, S.H., 1991, Human breast carcinoma cDNA encoding a galactose-binding lectin homologous to mouse Mac-2 antigen, *Gene.* **99**: 279-283.

Ozawa, M., Terada, H., and Pedraza, C., 1995, The fourth armadillo repeat of plakoglobin (-catenin) is required for its high affinity binding to the cytoplasmic domains of E-cadherin and desmosomal cadherin Dsg2, and the tumor suppressor APC protein, *J. Biochem.* **118**: 1077-1082.

Uehara, F., Muramatsu, T., Ozawa, M., Koide, H., et al., 1986, Purification of antibody against peanut agglutinin-receptors of bovine interphotoreceptor matrix, *Jpn. J. Ophthalmol.* **30**: 56-62.

Uehara, F., Sameshima, M, Takumi, K., Unoki, K., et al., 1988, Sialic acid in the retina and its significance in the retinal degeneration, in: *Glycoconjugates in Medicine*, M. Ohyama, ed., Professional Postgraduate Services, Tokyo, pp. 310-315.

Uehara, F., Ohba, N., Sameshima, M., Takumi, K., et al., 1989, Nucleotide induced retinal changes, in: *Inherited and Environmentally Induced Retinal Degenerations,* M.M. LaVail, R.E. Anderson, and J.G. Hollyfield, ed., Alan R. Liss, New York, pp. 577-583.

Uehara, F., Ohba, N., Sameshima, M., Yanagita, T., et al., 1996, Sialoglycoconjugates and retinal degeneration, in: *Retinal Degeneration and Regeneration,* S. Kato, N.N. Osborne and M. Tamai, ed., Kugler Publications, Amsterdam, pp. 73-80.

Uehara, F., Ohba, N., Yanagita, T., Sameshima, M., et al., 1997, Glycohistochemical study of light-induced retinal degeneration, in: *Degenerative Retinal Diseases,* M.M. LaVail, J.G. Hollyfield and R.E. Anderson, ed., Plenum Press, New York, pp.181-191.

Uehara, F., Ohba, N., and Ozawa, M., 2000, Isolation and characterization of mucinlike glycoprotein associated with photoreceptor cells, *Invest. Ophthalmol. Vis. Sci.* **41**: 2759-2765.

Uga, S., and Smelser, G.K., 1973, Comparative study of the fine structure of retinal Mueller cells in various vertebrates, *Invest. Ophthalmol.* **12**: 434-448.

Unoki, K., Ohba, N., Arimura, H., Muramatsu, H., and Muramatsu, T., 1994, Rescue of photoreceptors from the damaging effects of constant light by midkine, a retinoic acid-responsive gene product, *Invest. Ophthalmol. Vis. Sci.* **35**: 4063-4068.

Yang, R.Y., Hsu, D.K., and Liu, F.T., 1996, Expression of galectin-3 modulates T-cell growth and apoptosis, *Proc. Natl. Acad. Sci. USA.* **25**: 6737-6742.

A MULTIDISCIPLINARY APPROACH TO INVESTIGATING OPTIC NERVE REGENERATION IN THE GOLDFISH

Malini Devadas, Toru Matsukawa, Zhongwu Liu, Kayo Sugitani, Kiyoshi Sugawara, Manabu Kaneda, and Satoru Kato[*]

1. SUMMARY

Unlike the mammalian central nervous system, the goldfish optic nerve can regenerate after transection. We have used this regenerative ability to examine possible factors involved in this process. By monitoring goldfish behavior, we found two phases of recovery. Simple behavioral patterns such as tilting and turning return one month after optic nerve transection (fast recovery phase), whereas more complex behaviors such as schooling behavior require up to 4 months or more to recover (slow recovery phase). We also demonstrated that severe hypertrophy occurs in retinal ganglion cells after optic nerve transection, taking about 4 months to recover. Our most recent investigations have looked at various aspects of the fast recovery phase using histochemical and molecular techniques. We have observed that there is an increase in the amount of the nitric oxide (NO) synthesizing enzyme (NOS) in retinal ganglion cells during the first month after optic nerve section. During optic nerve regeneration NO may play a role in strengthening regenerating connections between the retina and tectum. Finally, we have investigated changes in gene expression in the retina during optic nerve regeneration. Of the molecules that we have studied, transglutaminase (TG) and Na-K ATPase appear to be the most interesting. There is an increase in tissue TG gene expression in the retina at about one week after optic nerve section. Na-K ATPase alpha-3 subunit activity in the retina also peaked around this time. Using this multidisciplinary approach, we hope to gain a better understanding of many aspects of regeneration, from genetic control to behavioral modification.

[*] Correspondence to: Satoru Kato, Dept Molecular Neurobiology, Graduate School of Medicine, Kanazawa University, 13-1 Takara-machi Kanazawa, Ishikawa Japan 920-8640

2. INTRODUCTION

The teleost visual system has long been used as a model of neuronal regeneration. Unlike neurons in the mammalian central nervous system, which usually die after damage such as axon ligation, neurons in fish and amphibians can survive such trauma and can eventually function again. The axons of retinal ganglion cells, which form the optic nerve, terminate in the optic tectum of the teleost brain. After cutting the optic nerve, there is regeneration of the axons of the retinal ganglion cells and they can re-innervate the tectum, eventually forming the correct retino-topic pattern (Attardi and Sperry, 1963). In earlier investigations, various behavioral tests have supported this regenerative process by demonstrating a return to normal reflexes to sudden illumination (Splinger et al, 1977) and the ability to discriminate color (Arora and Sperry, 1963) which occurs by 20-30 days after optic nerve transection. Morphological studies investigating the retina and tectum have shown that while changes in the retinal ganglion cells seem to disappear within 2 months after optic nerve section (Murray and Grafstein, 1969), changes in the tectum last up to 6 months or longer (Murray and Edwards, 1982) Therefore, we are interested in examining both short- and long-term changes which occur in the retina and tectum after optic nerve section.

In the present study, we have investigated various aspects of goldfish behavior in addition to looking at morphological changes which occur in the retinal ganglion cells, and we have performed quantitative analysis of both of these parameters. We have also started to investigate the activity and mRNA expression of enzymes which may be involved in the behavioral and morphological changes that we have observed. In addition to this we are also searching for new factors which may be involved in the process of neuronal regeneration both in the retina and in the optic tectum.

3. METHODS

3.1. Animals

Common goldfish (*Carassius auratus*; body length about 5cm) were used throughout this study. Goldfish were anesthetized with 0.05% FA 100 (Tanabe Pharmaceuticals, Osaka, Japan) and tissue surrounding the eye was cut in order to access the optic nerve. The optic nerve of one eye was sectioned 1 mm away from the posterior of the eyeball with scissors. In experiments using pairs of goldfish, both the optic nerves of the two goldfish were sectioned as described above. At specific times after optic nerve section, various parameters were measured.

3.2. Behavioral measurements

The computer set-up used in the following experiments have been described in detail elsewhere (Kato et al, 1996). In brief, fish were tracked as they swam in a tank (450mm x 300mm x 300mm) with a water depth of 170 mm. The computer image processing system was composed of 2 CCD cameras set in the top and side of the tank, 2 video interfaces, and a personal computer. From the views through the two cameras the 2D and 3D positional coordinates of freely moving fish could be captured at a speed of 4.0 frames/s. Programs were written in Borland C++ (Version 3.0) to collect and analyze data. The algorithms for determining movement direction and tilting have been previously described (Kato et al, 1996). Using these programs we could determine the number of right and left turns the fish made and the tilting angle that they swam at after optic nerve transection. In experiments where the interaction of two fish was measured, the distance between the two fish (two point distance) was calculated from the positional coordinates of both fish in every frame (4 frames/s).

3.3. Horseradish peroxidase tracing into the optic tectum

To confirm the regeneration of the ganglion cell axon terminals into the tectum we used horseradish peroxidase tracing into the tectum. In brief, 4% wheat germ agglutinin conjugated HRP (WGA-HRP; Toyobo) in phosphate-buffered saline was injected into the eye 1-30 days after optic nerve section. Two days after injection the fish were anesthetized and the tectum was systemically fixed in 1% formaldehyde and 2.5% glutaraldehyde in 0.2 M phosphate buffer pH 7.4 (PB). Frozen sections (30μm) were incubated with tetramethylbenzidine (TMB; Merck) and 3% hydrogen peroxide. Neutral red was used for background staining and the TMB products were observed under a polarizing microscope (Nikon HFM).

3.4. Ganglion cell morphology

At various times after optic nerve transection ganglion cells were labeled by applying 4,6-diamidino-2-phenylindole (DAPI, WAKO) applied to the optic nerve using the tip of a needle. Two to 3 days after DAPI labeling the eye was removed, the retina was isolated and the pigment epithelium was removed. The retina were briefly fixed in 4% paraformaldehyde in 0.1 M PB and then placed ganglion cell side up in a tissue chamber on the stage of a fluorescent microscope (Nikon M-plan, x40). Using an ultra-violet filter, the nuclei of DAPI-labeled ganglion cells could be observed. The tip of a glass micropipet filled with 5% LY (Sigma) in distilled water was directed toward the blue fluorescent nucleus by microscopy. When the electrode tip had penetrated a cell, the dye was injected with a 4Hz sinusoidal current of $\pm$ 10 nA for 3-4 minutes. The LY injected ganglion cells were observed using a blue-violet filter. The photomicrographs of LY injected ganglion cells were scanned into a computer at a resolution of 1360 dpi. The area of each soma was calculated by comparing the number of pixels with that of a standard.

3.5. Localization of nitric oxide synthase

The posterior half of the eye was isolated and fixed in 4% paraformaldehyde in 5% sucrose in PB for 1 hour. It was then washed in 5% PB and then put into 20% sucrose in PB overnight. Frozen sections (15μm) on coverslips were pre-incubated in 2 ml of 0.3% Triton X-100 in PB (PBT) at room temperature for 2-3 hours. They were then incubated in 2ml of PBT containing NADPH (2mg/ml) and NBT (0.2mg/ml) at 37°C for 2 hours. The coverslips were rinsed in water, air dried and then mounted.

3.6. Isolation of goldfish retinal transglutaminase

A goldfish retinal cDNA library was constructed in the lambda ZAP II vector (Stratagene) and was screened using a fish liver TG cDNA probe (kindly given to us by Dr Yasueda, ,Ajinomoto Co. Ltd.). Positive clones were sequenced and analyzed using NCBI's BLAST algorithm.

3.7. Differential hybridization

In brief, total RNA from control retinas and from retinas 5 days after optic nerve transection was extracted from goldfish retina using standard methods (Sambrook et al, 1989). Poly(A)+ RNA was obtained with an oligo(dT)-cellulose column. Radiolabeled single stranded cDNA probes were made using the mRNA from each group of fish as a template. A cDNA library from retinas 5 days after optic nerve transection was constructed in the lambda ZAP II vector (Stratagene) and then screened using the 2 probes. Colonies which were more strongly hybridized using the 5 day after optic nerve transection probe were isolated and sequenced. The sequences were analyzed using NCBI's BLAST algorithm.

3.8. Northern blot analysis

To quantitatively assess the mRNA expression of various genes after optic nerve transection, Northern blot analysis was performed. Messenger RNA was extracted as described above. Northern blotting was done according to standard protocols (Sambrook et al, 1989) using an appropriate radiolabeled cDNA probe which was made as described above. Positive bands were quantified using an image analyzer (Bio-Rad).

4. RESULTS

The results are summarized in Table 1.

4.1. Behavioral changes after optic nerve transection

Using the algorithms discussed in the methods section, we could created a histogram of directional movement. In control goldfish, going straight ahead was the most frequent event (59%) of all events with an equal number of left and right turns (20% each) recorded. However, within a day of cutting of the optic nerve, statistically significant preferential turning to the intact eye side was observed, which lasted for about 10 days and disappeared by 30 days after optic nerve transection.

As in the turning behavior patterns, fish which had one optic nerve transected swam at an angle, tilting towards the intact eye side. We quantified the angle of body tilting over time after optic nerve transection. The ratio of goldfish image sizes (tilting:upright) increased significantly 1 day after cutting the optic nerve. This abnormal tilting behavior was statistically significant until about 15 days and had almost disappeared by one month.

Fish with both optic nerves cut showed normal turning and tilting behavior. However, when pairs of these fish were put in a tank together they swam independently of each other, unlike control pairs of fish who show schooling behavior. Although other behavior

Table 1. Summary of events which occur after optic nerve transection

Phase	Time after optic nerve transection	Stage	Peak of enzyme activity or mRNA expression	Morphological changes	Behavior
FAST	1 day				Abnormal turning, tilting and schooling behaviors begin
	1-2 weeks	Neurite extension	TG NaK-ATPase		
	3-4 weeks	Arrival at tectum	NOS	Ganglion cell hypertrophy begins	Normal turning and tiling behaviors return
				HRP tracing into tectum	
	2 months			Peak of ganglion cell hypertrophy	
SLOW	3-4 months	Synaptic refinement	?	Recovery from ganglion cell hypertrophy	Normal schooling behavior returns

returns after about one month after optic nerve section and responses to light stimuli have also returned by this time, the pairs of fish with bilateral optic nerve section continued to swim independently for up to 4 months. Control pairs of fish swim with an average 2-point distance of about 88mm. A long 2-point distance was observed soon after optic nerve section (180mm), which slowly decreased over a period of 4 months. Based on these data, normal chasing behavior took more than 4 months to return.

4.2. Morphological changes after optic nerve section

Using the HRP tracing technique we could monitor the time course of retinal ganglion cell axon terminals re-entering the optic tectum after optic nerve transection. Only one eye in each fish was injected with WGA-HRP. Therefore, only the contralateral tectum showed changes with the other tectum remaining as an internal control. No HRP-positive signals were seen in the first 1-2 weeks after optic nerve section. However, by 3 weeks, small particles of the regenerating optic nerve terminals were observed in the laminae of the tectum. By one month after optic nerve transection the HRP products were similar to those in control tecta.

In total, the morphology of 529 ganglion cells was studied. At each time point (0, 30, 60, 90 and 120 days after optic nerve section) a total of about 100-120 cells were examined from 10-20 goldfish. Because we used whole-mounted retinas we could measure the distance of the injected cell from the optic disc. In normal goldfish retinas ganglion cells in the central region of the retina (1mm from the optic disc) had diameters of 7-8μm. In the intermediate part of the retina (1-2 mm from the optic disc) cells were a little bit larger, being 8-9μm in diameter. Ganglion cells in the outermost area that we measured (2-3 mm from the optic disc) were 9-10μm in diameter. At 30 days after optic nerve section all ganglion cells studied from all parts of the retina were hypertrophic with a soma diameter of 12-14μm. Hypertrophy peaked at 60 days after optic nerve section with diameters of ganglion cells from all regions being 13-16μm. By 90 days after optic nerve section, hypertrophy had decreased slightly, but by 120 days the diameter of ganglion cell had returned to that control retinas with soma sizes being 8-10μm.

4.3. Changes in enzyme activity and mRNA expression after optic nerve section

NOS enzyme activity was monitored using NADPH diaphorase histochemistry. In normal goldfish retinas, NADPH diaphorase was strongly localized to the inner plexiform layer (IPL) with weak staining of the ganglion cells. However, by about 20 days after optic nerve transection the hypertrophied ganglion cells were strongly positive for NADPH diaphorase. This intense staining remained until about 45 days after optic nerve transection and had disappeared by 60 days, although, as described above, ganglion cells are still hypertrophic at that time

The mRNA expression of TG in the retina was investigated after optic nerve transection using Northern hybridization. Expression peaked between 5-10 days and did not return to control levels until more than one month after optic nerve section. Similarly, the peak of NaK-ATPase mRNA expression was also between 5-10 days and returned to control levels at around 20 days after optic nerve transection.

5. DISCUSSION

5.1. Quantitative analysis after optic nerve section reveals 2 phases of recovery

Earlier studies have focused on the return of simple behavioral reflexes (such as color discrimination and the startle reflex to light) after optic nerve section (Arora and Sperry, 1963; Splinger et al, 1977). However, these studies were qualitative. In the present study we were able to collect data and analyze them quantitatively. While normal goldfish make an equal number of left and right turns and swim in an upright position, we found that after unilateral optic nerve transection, fish preferred to turn toward the side of the intact eye when swimming in a tank and that they had an oblique stance, tilting towards the intact eye side. These abnormal behaviors were not seen in fish where both optic nerves were transected suggesting that they were due to the visual unbalance caused by unilateral optic nerve transection. As these behaviors return to normal at about one month after optic nerve transection, we have called this the fast phase of recovery. On the other hand, normal chasing behavior between two fish with both optic nerves cut did not return until more than 4 months after optic nerve section, which we have called the slow phase of recovery. Quantitative analysis of ganglion cell morphology also showed an approximate 4 month period of recovery from hypertrophy after optic nerve section. This is in contrast to a previous work which found an earlier recovery of ganglion cell morphology (Murray and Grafstein, 1963). However, they did not continue their investigation beyond 40 days after optic nerve section so they could have missed our later peak in hypertrophy. Therefore, we have revealed both a fast phase and a slow phase of recovery after optic nerve section and we have started to look for factors involved in both of these phases.

5.2. Identifying factors involved in both phases of regeneration and their roles

Previous studies have suggested that TG may play a role in optic nerve regeneration in fish, as fish TG could promote regeneration in the mammalian visual system after optic nerve injury (Eitan et al, 1994). However, TG is postulated to have a variety of roles (for review see Lesort et al, 2000) including a role in normal development, with an increase in TG activity reported during mouse brain development (Maccioni and Seeds, 1986). In the present study we have shown an increase in mRNA expression in the retina within a week after optic nerve section, suggesting a role during the period when new neurites are being extended. In situ hybridization will confirm the localization of this mRNA and further work is need to elucidate TG's role goldfish optic nerve regeneration.

NO is also thought to have varying roles in the central nervous system, in addition to other systems. In the rat retina, NADPH-diaphorase staining was most prominent during the period of synaptogenesis (Mitrofanis, 1989). The targets of retinal ganglion cell axons in rats (Gonzalez-Hernandez et al, 1992) and chickens (Williams et al, 1994) have also been shown to express NOS during embryonic development, again suggesting a role in synaptogenesis. In the present study NADPH diaphorase staining in the retinal ganglion cells peaked around 1 month after optic nerve section, which is when synaptogenesis is occurring, which seems to agree with the previous findings. We are currently investigating NADPH diaphorase staining in the optic tectum during regeneration.

So far, using a retinal cDNA library from goldfish 5 days after optic nerve transection, we have identified NaK-ATPase as a potential enzyme involved in the regeneration process. Of course, it is possible that this enzyme's activity increased as a result of other processes which occur after optic nerve transection, rather than being a cause or a direct consequence of it. These questions will have to be addressed in future work by investigating the pathways involved. Nevertheless, this model of optic nerve regeneration in the goldfish will allow us to search for genes which may be related to neuronal regeneration. By screening retinal and tectal cDNA libraries at various times after optic nerve section, we will be able to identify genes which increase their expression at certain stages of axonal regeneration. In the retina, this may help us find genes which are involved in axon outgrowth and the establishment of a rough retino-tectal map. These processes correspond to the time that simple behaviors return. Therefore, it is possible that the factors which increase during the first month after optic nerve transection and the morphological changes that occur may be involved in controlling such simple behaviors. On the other hand, complex behaviors take many months to return and this may be related to synaptic refinement within the tectum which occurs during that time (Murray and Edwards, 1982). Therefore, molecules which increase between 2-4 months after optic nerve transection may be involved in providing signals for the ingrowing axons and for other events during axonal regeneration such as synaptogenesis. Again, these processes might lead to the re-establishment of complex behaviors.

6. CONCLUSIONS

We have demonstrated two phases of recovery after transection of the optic nerve in goldfish by monitoring changes in goldfish behavior. The fast phase lasts for about 1 month whereas the slow phase lasts for at least 4 months. During the fast phase, tilting and turning behaviors return to normal. There is an increase in the activity of NOS. Also, in the early part of this phase, both TG and NaK-ATPase increase their mRNA expression in the retina. During the slow phase, retinal ganglion cells recover from their hypertrophy and complex chasing behaviors return to normal. In the future we hope to identify enzymes and other factors which are involved in this slow phase of recovery after optic nerve section in the goldfish.

7. ACKNOWLEDGMENTS

We thank Ms T. Urano and Ms T. Kano for secretarial and technical assistance. This study was supported in part by research grants from the Ministry of Education, Science, Culture and Sport, Japan (No. 12680750 to SK, No. 12878134 to TM) and a Japan Society for the Promotion of Science Postdoctoral Fellowship and Special Grant for Foreign Researchers (No. 99186) to MD.

REFERENCES

Arora, H. L., and Sperry, R. W., 1963, Color discrimination after optic nerve regeneration in the fish, *Astronatus ocellatus, Devl Biol.* 7:234.
Attardi, D. G., and Sperry, R. W., 1963, Preferential selection of central pathways by regenerating optic fibers, *Expl Neurol.* 7:46
Eitan., S., Solomon, A., Lavie, V., Yoles, E., Hirschberg, D. L., Belkin, M., and Schwartz, M., 1994, Recovery of visual response of injured adult rat optic nerves treated with transglutaminase, *Science.* 264:1764
Gonzalez-Hernandez, T., Conde-Sendin, M., Conazlez-Gonzalez, B., Mantolan-Satmiento, B., Perez-Gonzalez,

H., and Meyer, G., 1993, Postnatal development of NADPH-diaphorase activity in the super colliculus and the ventral lateral geniculate nucleus of the rat, *Dev. Brain Res.* **76**:141

Kato, S., Tamada, K., Shimada, Y., and Chujo, T., 1996, A quantification of goldfish behavior by an image processing system, *Behav. Brain Res.* **80**:51

Lesort, M., Tucholski, J., Miller, M., and Johnson, G. V. W., 2000, Tissue transglutaminase: a possible role in neurodegenerative diseases, *Prog. in Neurobiol.* **61**:439

Maccioni, R. B., and Seeds, N. W., 1986, Transglutaminase and neuronal differentiation, *Mol. Cell. Biochem.*

Mitrofanis, J., 1989, Development of NADPH-diaphorase cells in the rat's retina, *Neurosci. Lett.*, **102**:165-172

Murrary, M., and Edwards, 1982, A quantitative study of the reinnervation of the goldfish optic tectum following optic nerve crush. *J. Comp. Neurol.* **209**:363

Murray, M., and Grafstein, B., 1969, Changes in the morphology and amino acid incorporation of regenerating goldfish optic neurons, *Exp. Neurol.* **23**:544

Sambrook, J., Fritsch, E. F., and Maniatis, T., 1989, *Molecular Cloning; A Laboratory Manual*, Cold Spring Harbor Laboratory Press, New York, Chapter 7

Splinger, A. D., Easter, S. S., and Agranoff, B. W., 1977, The role of the optic tectum in various visually mediated behaviors of goldfish, *Brain Res.* **128**:393

Williams, C. V., Nordquist, D., and McLoon, S. C., 1994, Correlation of Nitric oxide synthase expression with changing patterns of axonal projections in the developing visual system, *J. Neurosci.* **14**:1746

INTACT-SHEET FETAL RETINAL TRANSPLANTS CAN REPAIR DEGENERATED RETINAS

Magdalene J. Seiler, Robert B. Aramant [*]

1. SUMMARY

We have established a rat model to reconstruct the retina morphologically after destruction of photoreceptors. In blindness due to degeneration of photoreceptors and/or RPE, the remaining retina that connects to the brain can still be functional. Our hypothesis is that transplantation of intact sheets of fetal retina alone or together with its RPE has the potential to restore a damaged retina. The donor tissue is dissected from rat fetuses (embryonic day = E 18 – 20). Retinal pieces, gel-embedded for protection, are stored in cold medium until use. Using a custom-made implantation tool, the tissue is placed into the subretinal space of different rat models of retinal degeneration.

This chapter describes results after transplantation of fetal retina into one model, transgenic rats with a mutated human rhodopsin gene. Such transplants can form parallel layers resembling a normal retina. The graft photoreceptors develop inner and outer segments when they are in contact with healthy RPE, and express phototransduction proteins like in normal retina. Transplants develop rosettes if the host RPE has been damaged, if trauma to the donor tissue occurs during preparation or transplantation, or if the donor tissue is misplaced. If the surgery is done when retinal degeneration of the host retina has progressed too far, causing secondary RPE damage, transplants do not develop parallel layers of photoreceptor outer segments. Transplants can integrate well with the host retina, indicating a potential for synaptic connections with the host retina. In summary, transplantation of intact sheets of fetal retina alone or together with its RPE is technically feasible and results in organized transplants.

---

[*] Departments of Ophthalmology & Visual Sciences and Anatomical Sciences and Neurobiology, University of Louisville, School of Medicine, KY 40202

2. INTRODUCTION

Retinal diseases that affect the photoreceptors and/or RPE are among the leading causes of blindness.   In many inherited retinal degenerations, the neural retina that connects to the brain remains functional.[1-3]   If the diseased photoreceptors and/or RPE can be replaced, and the new cells can connect to the functional part of the retina, it is possible that further retinal degeneration might be prevented and eyesight restored. Clinical trials of retinal transplantation include transplantation of RPE,[4] adult photoreceptor sheets,[5] dissociated fetal retinal cells[6] and fetal retinal sheets.[7]

In animal experiments, donor tissue can be derived from various sources: mature[8, 9] or fetal,[10-14] freshly harvested[8, 12, 15, 16] or cultured[17]. Donor tissue has been transplanted as cell aggregates[10-12, 18-20] or cell suspensions.[15, 21]   Few groups have transplanted donor tissue as sheets.   Intact sheets of eight-day postnatal or mature retina or photoreceptors, but not of fetal retina, have been transplanted to rat retinas with photoreceptor degeneration[8, 9]. Transplantation of intact sheets of fetal retina was first reported in 1995[22] and later.[23, 24]   However, so far, only our group has transplanted intact sheets of fetal retina to animal models of retinal degeneration.   Photoreceptors in intact-sheet transplants showed a normal shift in the expression of phototransduction proteins in response to changes in light, and, in this way, function like normal photoreceptors.[25]

Our laboratory has consistently used fresh fetal retinal donor tissue. Fetal neuronal donor tissue has exceptional potential for transplantation.   It has been established that fetal neural cells transplanted to the brain grow and make connections, whereas older cells do not survive (review in[26]).   This has culminated in successful clinical trials with Parkinson patients where fetal dopaminergic transplants have been able to reduce the symptoms of this incurable disease (review:[27, 28]). Allografts of fetal neuronal tissue have been shown to be immunologically tolerated in an adult host.[29] Host immune responses against allogeneic fetal neural brain transplants were not sufficient to cause transplant rejection,[30] and allogeneic fetal transplants survived better when a chemical inflammation was induced prior to implantation.[31]

Our research has evolved from transplantation of retinal cell aggregates into lesioned adult retinas.[32]   Such aggregate transplants have provided a good model for study of the developmental capacity of fetal retinal cells after transplantation (review[33]). However, they appear to have reduced potential for functional replacement of lost photoreceptors.

Our hypothesis is that organized retinal transplants with parallel layers and with outer segments in contact with the host RPE will have the greatest potential to improve function in a damaged retina. It would be especially important that transplant photoreceptors or interneurons establish functional synaptic contacts with such host retinal cells, as ganglion, bipolar and horizontal cells. Fetal cells have a high capacity to re-establish neuronal connections.   We therefore sought to transplant fetal retina in intact sheets in the hope that such transplants would develop parallel layers resembling normal retina. Because of the extraordinary difficulty of handling fragile fetal tissue, it needs to be protected during transplantation. This can be done by embedding it in a gel. We have recently shown that fetal retinal sheets can "repair" light-damaged retinas.[24] In addition, fetal retina can also be cografted together with its RPE and restore retinas of RCS rats with dysfunctional RPE.[34]

Recently, new rat models of retinitis pigmentosa have become available; transgenic rats with the mutated human rhodopsin gene S334ter.[35] This chapter presents unpublished data of transplants to these rats.

3. METHODS

3.1. Experimental animals

Timed-pregnant Long-Evans rats (gestational day E18 – E20) were used as donors for fetal tissue. Recipients were transgenic rats carrying a mutated rhodopsin gene (S334ter, line 5). Founder transgenic rats were produced by Chrysalis DNX Transgenic Sciences, Princeton, NJ, and kindly by supplied Dr. M.M. LaVail, UCSF, San Francisco. The original rats of this strain are derived from albino Sprague-Dawley rats. Heterozygous rats were mated with normal Long-Evans rats to produce pigmented transgenic rats. The presence of the transgene was determined by PCR. All animals were treated according to the regulations in the ARVO and NIH guidelines.

An overview of the experiments is shown in Table 1.

3.2. Preparation of donor tissue

Donor retinal tissue was obtained from embryonic day (E) 18-20 pigmented Long-Evans rat fetuses. The retina was dissected free of surrounding tissues and embedded in 0.4% MVG alginate (Pronova, Oslo, Norway). Tissue pieces (0.8 x 1.5 mm) were taken up in a custom-made implantation instrument prior to implantation.

3.3. Transplantation procedure

The procedure has been described in detail previously[24, 36]. A small (0.8 – 1.2 mm) incision was cut in the peripheral eyeball, through sclera, choroid and retina. The curved nozzle tip of the instrument was inserted into the subretinal space with extreme care to leave the host RPE as undisturbed as possible while delivering the embedded donor retina. The tissue was usually placed into the dorso-nasal retinal quadrant. Each animal received a transplant in one eye, to have the other eye as a control.

3.4. Tissue processing

Most animals were perfusion-fixed with 4% paraformaldehyde with 0.18% picric acid in 0.1 N Na-phosphate buffer, followed by embedding of the eye cups in low-melting paraffin or in histocryl plastic (Electron Microscopy Sciences, Fort Washington, PA). Eight-micron paraffin or 2 μm plastic sections were cut and stained with hematoxylin/eosin. Selected sections were processed for immunocytochemistry (not shown).

Selected rats were perfused with 2% paraformaldehyde, 2.5% glutaraldehyde in 0.1 N phosphate buffer (PB) for electron microscopy. After dissection, eyecups were postfixed overnight, and, after several washes with PB, were postfixed in 2% OsO_4 in PB.

Tissue pieces were cut through the transplant center to face the knife edge, dehydrated through graded ethanols and embedded in Epon. Semithin sections (0.5 µm) were stained with toluidine blue. Ultrathin sections (60-80 nm), stained with uranyl acetate and lead citrate, were viewed in a Philips SM20 electron microscope.

4. RESULTS

The retinal degeneration in the pigmented S334ter line 5 rat was several months delayed compared to the albino S334ter line 5 rat (data not shown). The albino strain's photoreceptors are reduced to one row at the age of 2 months, whereas this is not seen until the age of 4 –6 month in the pigmented strain.

Well-laminated transplants, with parallel layers and photoreceptor outer segments towards the host RPE, were seen in 4 of 14 transplants (see Table 1). Four more transplants showed a small area of lamination, the rest of the transplants were organized in rosettes (spherical structures with photoreceptor nuclei around an outer limiting membrane, inner and outer segments toward the lumen and with the inner retinal layers in the periphery of the rosette).

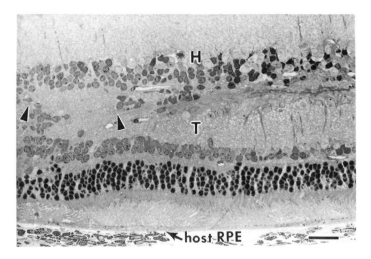

Figure 1.   Note the area of integration between laminated transplant (T) and host (H) retina (indicated by arrowheads). E18 retinal transplant to pigmented transgenic rat (S334ter line 5), 5.7 months after surgery. Host age is 7.6 months.   Bar = 20 µm.

Table 1: Overview of experiments

# of animals	host age at surgery	donor age	survival time (mo.)	# of transplants	transplants w. well-laminated area		rosettes only	graft in choroid
					large	small		
3	55 - 56 d	E19 – E20	1.9 - 3.8	2*	1	1		
6	59 d	E18	2.9 – 5.7	6	2	2	2	
3	65 d	E18	5.0 – 5.2	2*		1	1	
5	75-78 d	E18	4.0 – 4.5	4*	1		1	2
1	112 d	E20	n.a.	0*				
Total								
18	55-112d	E18-E20	1.9 – 5.7	14	4	4	4	2

* tissue came out in one animal

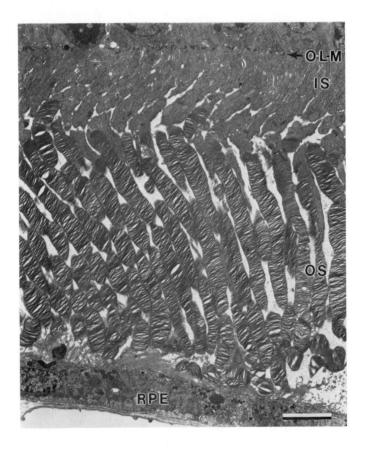

Figure 2. Transplant outer segments (OS) in contact with host RPE. Same transplant as in Figure 1. Bar = 5 μm. OLM = outer limiting membrane, IS = inner segments.

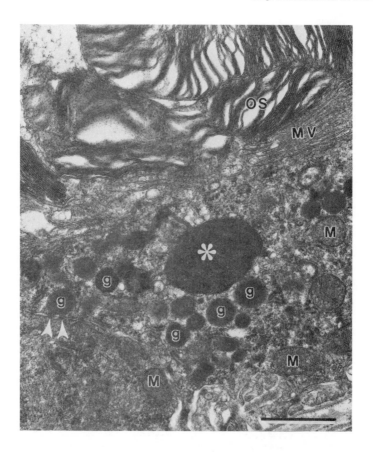

Figure 3. Active phagocytosis of shed discs of transplant outer segments by host RPE cell. Ingested discs are indicated by an asterisk, white arrowheads indicate junctions between adjacent RPE cells. Bar = 1 μm. OS = outer segments, MV = RPE microvilli, g = melanin granules, m = mitochondria

There appeared to be some age dependence in the transplant success. With a host age of 55-58 days, 3 of 8 transplants contained a large area and four more a small area of good lamination. In contrast, with a host age of 65 days or older, only 1 of 6 transplants was well-laminated; and it was impossible to transplant something into a rat of 112 days of age because of the reduced subretinal space due to the ongoing retinal degeneration and an increased ocular pressure.

A typical example of these well-laminated transplants is shown in Figures 1-5.

Transplant photoreceptors stained with antibodies for S-antigen, α-transducin and rhodopsin (data not shown). Electron microscopical evaluation of such a well-laminated transplant (shown in Figure 1) revealed well organized outer segments in contact with host RPE (Figures 2 and 3). Phagocytosis of transplant outer segments by host RPE could be observed (Figure 3). The transplant showed well-organized synaptic layers (an example of the inner plexiform layer is shown in Figure 4).

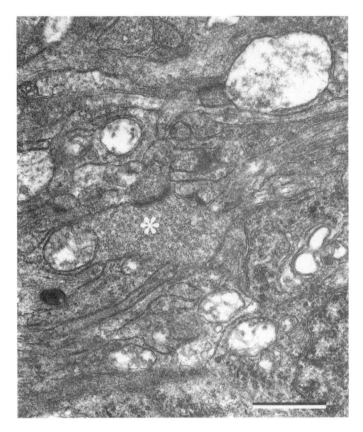

Figure 4. Inner plexiform layer of transplant. The asterisk indicates a bipolar cell terminal. Bar = 1 μm.

The interface between the transplants and the host retina varied from integration (used to describe the apposition of the transplant towards the host retina in the absence of glial barriers) to the presence of a clear glial barrier (both can be seen in Figure 1). In a well-integrated area, it was impossible to discern the exact border between transplant and host retina electron microscopically (Figure 5).

5. DISCUSSION

Transgenic rats with a mutant rhodopsin[35] are a new model of retinal degeneration, in addition to the previously established rat models, the RCS rat[37, 38] or light damage.[39] We have now also shown in this model of retinal degeneration that intact sheets of fetal retina can be transplanted to a degenerating rat retina, and develop parallel layers resembling a normal retina, with photoreceptor inner and outer segments in contact with the host RPE.

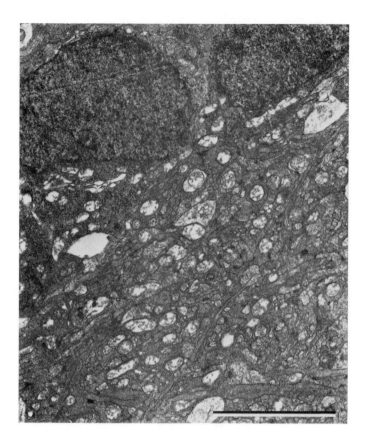

Figure 5. Fusion area between transplant and host. Cells in host inner nuclear layer facing plexiform area shared with transplant. It is impossible to say where the limit between transplant and host is. Bar = 5μm.

Not all intact-sheet transplants develop parallel layers, however. Development of rosettes instead of parallel layers can be due either to problems with dissection of the fragile donor tissue or the surgery itself. Since the surgeon cannot see how the transplant is placed after entering the subretinal space, the surgery of the small rat eye is very difficult. Transplants can survive well after transplantation into the choroid, but as they do not have contact with host RPE they will form rosettes. Likewise, damage to host RPE before or during surgery (by the implantation instrument) .will hinder the development of photoreceptor outer segments and lead to rosette formation.

Due to the internal pressure of the eye, retinal transplants tend to be displaced from the initial placement area when the insertion tool is withdrawn. This can be eliminated by letting the instrument nozzle stay in place for one minute to equilibrate the pressure before retracting. Another reason that the transplants tend to be displaced, is that part of the donor tissue may stick to the tool and may be dragged out. This can be eliminated to a large part by using a non-stick silicon nozzle.

The time point of surgery appears to be important for transplant success. Transplantation into recipients 65 days or older was not as successful as transplantation to younger recipients, due to the increase in intraocular pressure that occurs in these rats with aging, or because the RPE cells start to degenerate and neovascularization occurs after all photoreceptors are lost. Healthy RPE appear to be necessary for the full development of transplant photoreceptors.

Several other groups have reported the development of transplant outer segments in contact with host RPE after transplantation to animals with retinal degeneration.[8, 9, 12, 21, 40] The donor ages in these studies ranged from early postnatal to adult. However, photoreceptor transplants from mature tissue only maintained their outer segments when part of the inner retina was included with the transplant.[41]

Our approach is unique in that we use sheets of fetal retina to restore a damaged retina. In contrast, another group has transplanted fetal sheets to a *normal* rabbit retina which will cause damage to the host (local photoreceptor degeneration).[23, 42, 43] However, after longer survival times, connectivity between transplant and host retina can be observed.[43]

In many of our experiments, it was encouraging to see a good degree of integration (sharing of plexiform layers). Integration was usually better after long-term survival.

Do transplant cells form synapses with the host retina? This hypothesis is supported by our observation that injections of a retrogradely and transsynaptically transported pseudorabies virus[44] into the superior colliculus of transplanted rats can label transplant cells in the eye (manuscript submitted).[45] Some physical connections between long-term fetal retinal transplants and normal host retinas have been demonstrated in a rabbit model.[43] In addition, we have recently shown that transplants can restore visual responses in RCS rats (manuscript submitted)[46] and transgenic rats.[47]

In sum, the photoreceptor organization of intact-sheet fetal retinal transplants and their integration with the host retina makes it plausible that such transplants might restore light perception in recipients with retinal degeneration.

6. ACKNOWLEDGEMENTS

The authors want to thank Kelly Volk, Barbara Kalinowska, Kenneth Herman, and Lyndsay Tucker for their technical assistance. The authors thank Mary Gail Engel, University of Kentucky, Lexington, KY, for cutting ultrathin sections. The authors thank Dr. Norman D. Radtke, Audubon Hospital, Louisville, for his expert advice and support.

This work was supported by NIH grant EY08519; the Foundation Fighting Blindness; the Murray Foundation Inc.; the Vitreoretinal Research Foundation, Louisville; the Jewish Hospital Foundation, Louisville; the Kentucky Lions Eye Foundation; an unrestricted grant from the Research to Prevent Blindness; the German Humboldt Foundation; and funds from an anonymous sponsor.

Robert Aramant and Magdalene Seiler have a proprietary interest in the implantation instrument and method.

REFERENCES

1. D. Papermaster, J. Windle. Death at an early age. Apoptosis in inherited retinal degenerations. *Investigative Ophthalmology & Visual Science* **36**(6):977-83 (1995).
2. A. H. Milam, Z. Y. Li, R. N. Fariss. Histopathology of the human retina in retinitis pigmentosa. *Prog Retin Eye Res* **17**(2):175-205 (1998).
3. M. S. Humayun, M. Prince, E. de Juan, Jr., et al. Morphometric analysis of the extramacular retina from postmortem eyes with retinitis pigmentosa. *Invest Ophthalmol Vis Sci* **40**(1):143-8 (1999).
4. P. V. Algvere, P. Gouras, E. Dafgard Kopp. Long-term outcome of RPE allografts in non-immunosuppressed patients with AMD. *Eur J Ophthalmol* **9**(3):217-30 (1999).
5. H. J. Kaplan, T. H. Tezel, A. S. Berger, M. L. Wolf, L. V. Del Priore. Human photoreceptor transplantation in retinitis pigmentosa. A safety study. *Archives of Ophthalmology* **115**(9):1168-72 (1997).
6. T. Das, M. del Cerro, S. Jalali, et al. The transplantation of human fetal neuroretinal cells in advanced retinitis pigmentosa patients: results of a long-term safety study. *Exp Neurol* **157**(1):58-68 (1999).
7. N. D. Radtke, R. B. Aramant, M. J. Seiler, H. M. Petry. Preliminary report: indications of improved visual function following retina sheet transplantation to retinitis pigmentosa patients. *Am. J. Ophthalmol.* **128**:384-387 (1999).
8. M. S. Silverman, S. E. Hughes, T. Valentino, Y. Liu. Photoreceptor transplantation: Anatomic, electrophysiologic, and behavioral evidence for the functional reconstruction of retinas lacking photoreceptors. *Exp. Neurol.* **115**:87-94 (1992).
9. M. S. Silverman, S. E. Hughes. Transplantation of photoreceptors to light-damaged retina. *Invest. Ophthalmol. Vis. Sci.* **30**:1684-1690 (1989).
10. R. Aramant, M. Seiler, B. Ehinger, A. Bergström, A. R. Adolph, J. E. Turner. Neuronal markers in rat retinal grafts. *Dev. Brain Res.* **53**:47-61 (1990).
11. M. Del Cerro, J. R. Ison, G. P. Bowen, E. Lazar, C. Del Cerro. Intraretinal grafting restores function in light-blinded rats. *NeuroReport* **2**:529-532 (1991).
12. P. Gouras, J. Du, H. Kjeldbye, S. Yamamoto, D. J. Zack. Long-term photoreceptor transplants in dystrophic and normal mouse retina. *Investigative Ophthalmology & Visual Science* **35**(8):3145-53 (1994).
13. M. Seiler, R. Aramant. Transplantation of embryonic retinal donor cells labelled with BrdU or carrying a genetic marker to adult retina. *Exp. Brain Res.* **105**:59-66 (1995).
14. M. Seiler, J. E. Turner. The activities of host and graft glial cells following retinal transplantation into the lesioned adult rat eye: developmental expression of glial markers. *Dev. Brain Res.* **43**:111-122 (1988).
15. L. Li, J. E. Turner. Inherited retinal dystrophy in the RCS rat: prevention of photoreceptor degeneration by pigment epithelial cell transplantation. *Exp. Eye Res.* **47**:911-917 (1988).
16. M. J. Seiler, R. B. Aramant. Photoreceptor and glial markers in human embryonic retina and in human embryonic retinal transplants to rat retina. *Brain Research. Developmental Brain Research* **80**(1-2):81-95 (1994).
17. R. Lopez, P. Gouras, H. Kjeldbye, et al. Transplanted retinal pigment epithelium modifies the retinal degeneration in the RCS rat. *Investigative Ophthalmology & Visual Science* **30**:586-588 (1989).
18. R. Aramant, M. Seiler, J. E. Turner. Donor age influences on the success of retinal transplants to adult rat retina. *Invest. Ophthalmol. Vis. Sci.* **29**:498-503 (1988).
19. R. Aramant, M. Seiler. Cryopreservation and transplantation of rat embryonic retina into adult rat retina. *Dev. Brain Res.* **61**:151-159 (1991).
20. M. J. Seiler, R. B. Aramant, A. Bergström. Co-transplantation of embryonic retina and retinal pigment epithelial cells to rabbit retina. *Current Eye Research* **14**(3):199-207 (1995).
21. A. S. Kwan, S. Wang, R. D. Lund. Photoreceptor layer reconstruction in a rodent model of retinal degeneration. *Exp Neurol* **159**(1):21-33 (1999).
22. R. B. Aramant, M. J. Seiler. Organized embryonic retinal transplants to normal or light-damaged rats. *Soc. Neurosci. Abstr.* **21**:1308 (1995).
23. F. Ghosh, K. Arnér, B. Ehinger. Transplant of full-thickness embryonic rabbit retina using pars plana vitrectomy. *Retina* **18**:136-142 (1998).
24. M. J. Seiler, R. B. Aramant . Intact sheets of fetal retina transplanted to restore damaged rat retinas. *Invest. Ophthalmol. Vis. Sci.* **39**:2121-31 (1998).
25. M. J. Seiler, R. B. Aramant, S. L. Ball. Photoreceptor function of retinal transplants implicated by light-dark shift of S-antigen and rod transducin. *Vision Res* **39**:2589-2596 (1999).
26. S. B. Dunnett. Repair of the damaged brain. The Alfred Meyer Memorial Lecture 1998. *Neuropathol Appl Neurobiol* **25**(5):351-62 (1999).

27. G. K. Wenning, P. Odin, P. Morrish, et al. Short- and long-term survival and function of unilateral intrastriatal dopaminergic grafts in Parkinson's disease. *Ann Neurol* **42**(1):95-107 (1997).
28. E. D. Clarkson, C. R. Freed. Development of fetal neural transplantation as a treatment for Parkinson's disease. *Life Sci* **65**(23):2427-37 (1999).
29. H. Widner, P. Brundin. Immunological aspects of grafting in the mammalian central nervous system. A review and speculative synthesis. *Brain Res. Rev.* **13**:287-324 (1988).
30. W. M. Duan, H. Widner, P. Brundin. Temporal pattern of host responses against intrastriatal grafts of syngeneic, allogeneic or xenogeneic embryonic neuronal tissue in rats. *Exp Brain Res* **104**(2):227-42 (1995).
31. W. Duan, R. Cameron, H. Widner, P. Brundin. Quinolinic acid-induced inflammation in the striatum does not impair the survival of neural allografts in the rat. *Eur J. Neurosci* **10**(8):2595-606 (1998).
32. J. E. Turner, J. R. Blair. Newborn rat retinal cells transplanted into a retinal lesion site in adult host eyes. *Dev. Brain. Res.* **26**:91-104 (1986).
33. R. B. Aramant, M. J. Seiler. Retinal cell transplantation. In: *1996/1997 Yearbook of cell and tissue transplantation*, edited by R. P. Lanza, W. L. Chick, (Kluwer Academic Publishers, Dordrecht, NL 1996) pp. 193-201.
34. R. B. Aramant, M. J. Seiler, S. L. Ball. Successful cotransplantation of intact sheets of fetal retina with retinal pigment epithelium. *Invest Ophthalmol Vis Sci* **40**(7):1557-64 (1999).
35. S. Nishikawa, W. Cao, D. Yasumura, et al. Comparing the ERG to retinal morphology in transgenic rats with inherited degenerations caused by mutants opsin genes [ARVO abstract]. *Invest Ophthalmol Vis Sci* **38** (**ARVO Suppl.**):S33 (1997).
36. R. B. Aramant, M. J. Seiler, S. L. Ball. Successful cotransplantation of intact sheets of fetal retinal pigment epithelium with retina. *Invest Ophthalmol Vis Sci* **40**:1557-1564 (1999).
37. R. J. Mullen, M. M. LaVail. Inherited retinal dystrophy: primary defect in pigment epithelium determined with experimental rat chimeras. *Science* **192**:799-801 (1976).
38. P. M. D'Cruz, D. Yasumura, J. Weir, et al. Mutation of the receptor tyrosine kinase gene Mertk in the retinal dystrophic RCS rat. *Hum Mol Genet* **9**(4):645-51 (2000).
39. M. J. Seiler, O. L. Liu, N. G. Cooper, T. L. Callahan, H. M. Petry, R. B. Aramant. Selective photoreceptor damage in albino rats using continuous blue light. A protocol useful for retinal degeneration and transplantation research. *Graefes Arch Clin Exp Ophthalmol* **238**(7):599-607 (2000).
40. P. Gouras, J. Du, M. Gelanze, et al. Survival and synapse formation of transplantated rat rods. *Journal of Neural Transplantation & Plasticity* **2**(2):91-100 (1991).
41. M. S. Silverman, S. E. Hughes, T. L. Valentino, Y. Liu. Photoreceptor transplantation to dystrophic retina. In: *Retinal Degenerations*, edited by R. E. Anderson, J. G. Hollyfield, M. M. LaVail, (CRC Press, Boca Raton, Fl 1991) pp. 321-335.
42. F. Ghosh, K. Johansson, B. Ehinger. Long-term full-thickness embryonic rabbit retinal transplants. *Invest Ophthalmol Vis Sci* **40**(1):133-42 (1999).
43. F. Ghosh, A. Bruun, B. Ehinger. Graft-host connections in long-term full-thickness embryonic rabbit retinal transplants. *Invest Ophthalmol Vis Sci* **40**(1):126-32 (1999).
44. J. P. Card. Practical considerations for the use of pseudorabies virus in transneuronal studies of neural circuitry. *Neurosci Biobehav Rev* **22**(6):685-94 (1998).
45. R. B. Aramant, M. J. Seiler, G. Woch. Retinal transplants form synaptic connections with host retinas with photoreceptor degeneration - demonstrated by transsynaptic tracing from host brain [ARVO abstract]. *Invest. Ophthalmol. Vis Sci.* **41** (**Suppl.**):S101; program # 528 (2000).
46. M. A. McCall, G. Woch, R. B. Aramant, M. J. Seiler. Transplanted fetal retina can restore visual responses in a rat model of retinal degeneration [ARVO abstract]. *Invest. Ophthalmol. Vis Sci.* **41** (**Suppl.**):S855; program # 4542 (2000).
47. B. T. Sagdullaev, R. B. Aramant, M. J. Seiler, G. Woch, M. A. McCall. Retinal transplantation restores visual responses in a transgenic rat model of retinal degeneration [Abstract]. *Soc Neurosci Abstr.* **26**:1099; program # 415.4. (2000).

EVALUATION OF AN ARTIFICIAL RETINA IN RODENT MODELS OF PHOTORECEPTOR DEGENERATION

Sherry L. Ball, Machelle T. Pardue, Alan Y. Chow, Vincent Y. Chow and Neal S. Peachey*

1. INTRODUCTION

In many retinal disorders, the photoreceptor layer degenerates while inner retinal layers are spared (Eisenfeld et al., 1984; Flannery et al., 1989; Stone et al., 1992; Santos et al., 1997). Based on evaluation of patients with photoreceptor degeneration, the inner layers retain some capacity to transmit and process visual information (Humayun et al., 1995, 1996). Consequently, replacing the photoreceptor layer with healthy retina or an "artificial retina" could restore vision in affected individuals. This objective has been approached by: transplantation of adult photoreceptors (Gouras et al., 1994; Silverman and Hughes, 1989; Huang et al., 1998), embryonic retina (Humayun et al. 2000; Juliusson et al. 1993; Seiler and Aramant, 1998; Seiler et al. 1999), or full thickness retina (Aramant et al., 1999; Ghosh et al. 1998; Seiler et al. 1995) or by implantation of electrodes onto the retinal surface (Eckmiller, 1999; Grumet, et al. 2000; Humayun et al., 1999) or an electronic device into the subretinal space (Peachey and Chow, 1999; Zrenner et al., 1999; Chow et al., 2001a,b).

Initial studies involving implantation of a subretinal microphotodiode array (MPA) utilized normal adult cats and focused on determining the biocompatibility and feasibility of this approach (Chow et al., 2001a,b). However, to determine whether an MPA can activate inner portions of a degenerated retina and transmit meaningful visual signals to cortical areas requires the use of an animal with photoreceptor degeneration. A major

* Sherry L. Ball, and Neal S. Peachey, Cleveland VA Medical Center, Cleveland, OH 44106 and Cole Eye Institute, Cleveland Clinic Foundation, Cleveland, OH 44195 Machelle T. Pardue, Atlanta VA Medical Center, Decatur, GA 30033. Alan Chow and Vincent Chow, Optobionics Corporation, Wheaton, IL 60187.

New Insights Into Retinal Degenerative Diseases
Edited by Anderson *et al.*, Kluwer Academic/Plenum Publishers, 2001

drawback in the use a feline model is the known existence of only one strain with retinal degeneration, the Abyssinian cat, which has a slow and incomplete degeneration (Narfström and Nilsson, 1986). Additionally, recent evidence has demonstrated that the intact native cat and rat retina shows sensitivity to infrared light complicating the interpretation of results (Pardue et al., 2001).

The project described here utilizes two rodent models of retinal degeneration: the RCS rat (Dowling and Sidman, 1962) and a transgenic rat expressing a truncated form of rhodopsin (S334ter). In each case, by 7 months of age the photoreceptor layer is dramatically reduced and no appreciable electrical response can be measured from the corneal surface of the retina to either white or infrared light (data not shown). Although the smaller rat eye size presents a challenge in terms of accurate placement of the implant, this study demonstrates the feasibility of the surgery and serves as an impetus for more studies involving the testing of subretinal implants in rodent models of retinal degeneration.

2. METHODS

2.1. Surgery

Animals were anesthetized with ketamine (60 mg/kg) and xylazine (7.5 mg/kg); pupils were dilated with 1% tropicamide, and 2.5% phenylephrine HCl and anesthetic eye drops were applied. The posterior portion of the eye was exposed by cutting the conjunctiva and rotating the eye globe ventrally as shown in figure 1A. After clearing the connective tissue to expose the scleral surface, an incision was made with a 1.0 mm sapphire blade in an area relatively devoid of blood vessels. Figure 1A shows an incision made through sclera and completely penetrating the retina through to the vitreous at a point approximately 3 mm anterior to the optic nerve. Within 15 minutes after the incision was made, the neural retina naturally detaches from the retinal pigment epithelium (RPE) allowing the implant to be inserted into the final position. The implant was gently guided into the subretinal space using blunt tipped probes constructed from 27 gauge needles. Proper placement of the implant was confirmed by rotating the eye back to normal position and viewing the fundus through a flat pediatric lens under the surgical microscope. Placement of the implant was evaluated based on the ability to view the entire MPA with blood vessels overlying the implant. Any portion of an implant lying superficial to the retina was easily identified by the obvious direct reflection of light from the exposed mirror-like surface. Once correct placement was verified, the eye was rotated to its normal position and antibiotic ointment was applied. Following recovery from surgery, animals were returned to the rat colony where they were housed under normal 12:12 hour light/dark cycles.

2.2. Implant Design

The silicon MPA implants were fabricated using standard thin-film technique and ion implantation with iridium oxide and platinum or platinum as the electrode layer. The devices measured 0.5 or 1.0 mm in diameter and 25- 50 µm in thickness (for more details cf. Chow et al. 2001). Surgical procedures and follow-up evaluations were video taped.

2.3. Post Implantation Status of MPA

During the post-operative period, the electrical response of the implant was monitored using a standard IR LED stimulus (100 msec at 2 Hz) driven by a SRS function generator. A small subgroup (n=3) was measured at bimonthly intervals, while other animals were monitored at later time points. These responses were measured from the corneal surface of implanted animals following anesthesia and pupil dilation. A wire loop electrode placed on the surface of the cornea served as the recording electrode with reference and ground needle electrodes placed in the cheek and tail, respectively. Responses were differentially amplified (1 to 1,500 Hz), averaged (n = 400) and stored using an LKC UTAS E-2000 signal averaging system.

2.4. Histological Examination of Retinal Tissue Following Implantation

To determine the impact of the MPA on the health of the retina, eyes were collected and prepared for immunohistochemical examination. Following enucleation, eyes were immersion fixed in 4% paraformaldehyde for 10 minutes or 2.5% glutaraldehyde/2% paraformaldehyde phosphate buffered solution and fixed overnight. For immunohistochemical studies, eyes were cryoprotected overnight in 10% sucrose, were embedded in OCT, and sectioned at 7 µm on a cryostat. Eye cups embedded in epoxy-resin, sectioned at 0.5 µm, and stained with toluidine blue were used for gross morphological examination. In order to fully evaluate cell and tissue types surrounding the implant, the implant was left in place throughout the histological processing and sectioning. Although this method compromised the morphological quality of retinal sections, it allows for a more complete examination of the retina to be undertaken.

For glial fibrillary acidic protein (GFAP) labeling, retinal sections were blocked with 1% goat serum with 0.1% Triton-X for 30 minutes at room temperature in a humid chamber before being incubated in primary antibody, GFAP, (1:2,000; Sigma) overnight at 4° in a humid chamber.   Sections incubated in GFAP were then exposed to the secondary antibody for 30 minutes and reacted with an ABC kit (Vector) using DAB. Sections treated with no primary were included as controls.

3.   RESULTS AND DISCUSSION

3.1. Surgery

Implantation surgery was attempted on 17 RCS and 7 transgenic S334ter rats between 3-7 months of age. Although the disease process is not complete within this age range, clear signs of photoreceptor degeneration are present. Based on placement of the MPA within 2 mm of the optic nerve head, as assessed by repeated fundus examination, the surgeries on 11 RCS and 2 S344ter rats were successful. Of these 13 successful surgeries, 85% of the implants were 1.0 mm in diameter. Of the failed surgeries 73% involved the use of the 0.5 mm implants. This result may indicate that larger implants are associated with a higher likelihood of success, although it is difficult to make conclusions due to the numerous variables involved in this surgical procedure. Other than a temporary cataract formation no deleterious side effects of the surgical procedure were observed. In each animal the incision site was covered in connective tissue by the time of sacrifice although no suture was used to close the incision site.

3.2. Implant Activity *in vivo*

The ability of the implant to remain in a stable position after implantation was evaluated by viewing the implant through fundus examination at regular intervals. Figure 1B, taken from an umimplanted eye, shows the fundus abnormalities associated with the RCS disorder. Figure 1C,D shows fundus photographs taken from the implanted fellow eye of the same animal. In all successful surgeries the implant retained in a stable position in the subretinal space for at least 5 months. To evaluate the functional status of the implant we recorded the electrical response produced by the MPA in response to infrared light in a subset of implanted rats. In previous implantation studies, a decreased implant response was indicative of implant deterioration in the subretinal space (Chow et al. 2001a). A typical implant spike recorded from the corneal surface is shown in figure 2A. Figure 2B shows implant spike amplitudes for each animal followed bimonthly and illustrates how the response remained stable up to 6 months post implantation. Additional implanted rats examined at 7 months had similar implant spike amplitudes. This suggests that the intrinsic functional properties of the MPA remain intact following implantation, a result that is consistent with parallel studies carried out in normal adult cats (Chow et al. 2001b). No response could be recorded in the contralateral unimplanted eye to infrared or white light in RCS and S344ter rats at 12 months of age (data not shown).

3.3. Evaluation of Retinal Tissue

Figure 3A shows that the presence of the implant did not appear to cause further degeneration in these animals as retinal tissue overlying the implant and adjacent to the implant are comparable. In some portions of the RCS rat retina we observed a proliferation of RPE cells similar to results described in a previous studies (Villegas-Perez et al., 1998). Since this proliferation was found in both the implanted and the fellow eye, it is presumably not a reaction to the presence of the implant. As a further measure of retinal tissue reaction to the presence of the implant, figure 3B,C shows retinal sections labeled with GFAP. GFAP is a sensitive marker to retinal insult, being upregulated in Muller cells with retinal detachment, ischemia, and photoreceptor

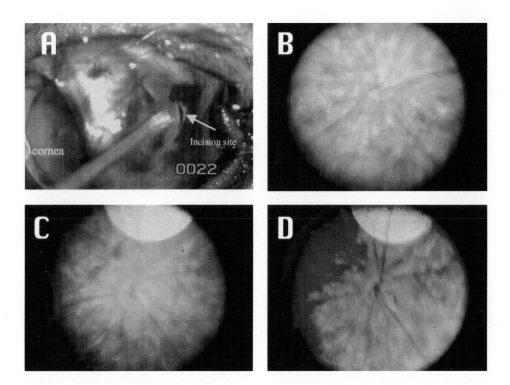

Figure 1. A. The incision site just before a 1.0 mm microphotodiode was placed through the incision and into the subretinal space. The eye globe is rotated ventrally and held in place with anchored suture threads. B. A fundus photograph of the contralateral unimplanted eye of RCS rat 2016 at 15 months of age. Note pigmentary changes associated with this model of photoreceptor degeneration. C and D. Fundus photographs from the same RCS rat at 15 months of age and at 8 months post implantation taken at two planes of focus: the posterior most part of the retina (C) and an anterior plane, focusing on the surface of the implant (D).

A   B

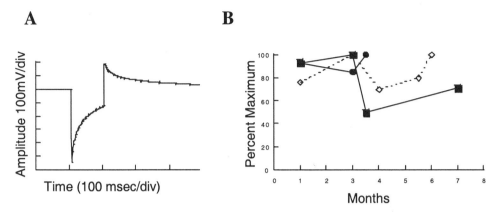

Figure 2. A) A typical implant spike recorded from the corneal surface of the eye in response to a 100 msec flash of IR light plotted as spike amplitude (100 μV/div) versus time (100 msec/div). B) Normalized amplitude (percent of maximal response) of implant spike plotted as a function of time (months) during the post-operative period. Each symbol represents a different rat and responses are normalized to the maximal response for that animal. The implant response remained fairly stable during the post-operative period.

degeneration (Bignami and Dahl, 1979; Eisenfeld et al., 1984; Erickson et al., 1987; Lewis et al, 1994). The extent of labeling in areas immediately overlying the retina are comparable to adjacent areas and areas outside of the implant area (3C). The lack of GFAP increase is somewhat difficult to interpret due to the very prominent, localized increase observed in normal implanted feline (Pardue et al., 1999) and rat (Zrenner et al., 1999) retina. Perhaps the overall high levels of GFAP in the degenerating retina are already at a maximal level before implantation, thus masking any specific implant effect. However, a more extensive examination of retinal tissue in regions adjacent to and overlying the implant is needed to fully evaluate the effects of the implant on retinal health.

4. SUMMARY

These results establish the feasibility of using rodent models of photoreceptor degeneration to evaluate a subretinal MPA device. Surgical techniques have been developed to implant a small MPA device into the rat subretinal space with relative reproducibility. In both rat models of photoreceptor degeneration, the retina detached readily and there was no evidence of abnormal adherence of the degenerate retina to the

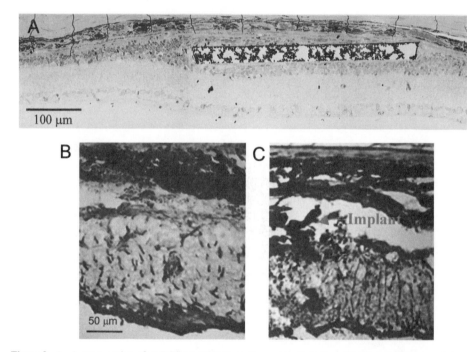

Figure 3. A. A cross section of an RCS rat retina 5 months post implantation showing no differences between retinal tissue immediately overlying the implant and areas adjacent to the implant. B). Retinal cross from an RCS rat retina 7months post-implantation sections labeled with GFAP in retinal areas overlying the implant and in areas adjacent to the implant (C).

RPE. Electrophysiological and immunohistochemical results show that the implant maintains the ability to generate activity *in vivo* without inducing further degeneration or inflammation in the rat eye. Future studies will focus on determining the nature of information transmitted to cortical visual centers using cortical recording and behavioral tests as well as improving and testing new implant designs.

5. ACKNOWLEDGEMENTS

This work was supported by a Merit Review from VA Rehabilitation Research & Development and by a Fight for Sight post-doctoral award. We are grateful to Deborah Ross, CRA for assistance with fundus photography.

REFERENCES

Aramant, R., Seiler, M.J., and Ball, S.L., 1999, Successful cotransplantation of intact sheets of fetal retina with retinal pigment epithelium, *Invest. Ophthalmol. Vis. Sci.* **40**:1557-1564.

Bignami, A. and Dahl, D., 1979, The radial glia of Müller in the rat retina and their response to injury. An immunofluorescence study with antibodies to the glial fibrillary acidic (GFA) protein, *Exp. Eye Res.*.**28**:63-69.

Chow, A.Y., Pardue, M.T., Chow, V.Y., Peyman, G.A., Liang, C., Perlman, J.I. and Peachey, N.S., 2001a, Implantation of semiconductor-based photodiodes into the cat subretinal space, *IEEE Trans. Rehab. Eng.* in press.

Chow, A.Y., Pardue, M.T., Perlman, J.I., Ball, S.L., Chow, V.Y., Hetling, J.R., Peyman, G.A., Liang, C., Stubbs, E.B. and Peachey, N.S., 2001 b, Subretinal implantation of semiconductor-based photodiodes. I. Biocomptibility and function of novel implant designs and laser ablation of overlying retina, *Exp. Eye Res.* submitted.

Dowling, J.E., and Sidman, R.L., 1962, Inherited retinal dystrophy in the rat, *J. Cell Biol.* **14**:73-109.

Eckmiller, R., 1975, Electronic stimulation of the vertebrate retina, *IEEE Trans. Biomed. Eng.* **22**:305-311.

Eisenfeld A.J., Bunt-Milam A.H., and Sarthy P.V., 1984, Müller cell expression of glial fibrillary acidic protein after genetic and experimental photoreceptor degeneration in the rat retina, *Invest. Ophthalmol. Vis. Sci.* **25**:1321-1328.

Erickson P.A., Fisher S.K., Guérin C.J., Anderson D.H., and Kaska D.D., 1987, Glial fibrillary acidic protein increases in Müller cells after retinal detachment, *Exp. Eye Res.* **44**:37-48.

Flannery, J.G., Farber, D.B., Bird, A.C, and Bok, D., 1989, Degenerative changes in a retina affected with autosomal dominant retinitis pigmentosa, *Invest. Ophthalmol. Vis. Sci.* **30**:191-211

Ghosh, F., Arnér, K., Ehinger, B., 1998, Transplantation of full-thickness embryonic rabbit retina using pars plana vitrectomy, *Retina* **18**:136-142.

Gouras, P., Du, J., Kjeldbye, H., Yamamoto, S., Zack, D.J., 1994, Long-term photoreceptor transplants in dystrophic and normal mouse retina, *Inves.t Ophthalmol. Vis Sci.* **35**:3145-53.

Grumet, A.E., Wyatt, J.L., and Rizzo, J.F., 2000, Multi-electrode stimulation and recording in the isolated retina, *J. Neurosci. Meth.* **101**:31-42.

Huang, J.C., Ishida, M., Hersh, P., Sugino, I.K., and Zarbin, M.A., 1998, Preparation and transplantation of photoreceptor sheets, *Curr. Eye Res.* **17**:573-85.

Humayun, M.S., del Cerro, M., deJuan, E. Jr., Dagnelie, G., Radner, W., Sadd, S.R., and del Cerro, C., 2000,Human retinal transplantation, *Invest. Ophthalmol. Vis. Sci.* **41**:3100-3106.

Humayun M.S., de Juan E. Jr., Weiland, J.D., Dagnelie, G., Katona, S., Greenberg, R., Suzuki, S., 1999, Pattern electrical stimulation of the human retina, *Vis. Res.* **39**:2569-2576.

Humayun M.S., de Juan E. Jr., Dagnelie G., Greenburg R.J., Propst R.H., and Philips H., 1996, Visual perception elicited by electrical stimulation of retina in blind humans, *Arch. Ophthalmol.* **114**:40-46.

Humayun M., Sato Y., Propst R., and de Juan E. Jr., 1995, Can potentials from the visual cortex be elicited electrically despite severe retinal degeneration and a markedly reduced electroretinogram? *Ger. J. Ophthalmol.* **4**:57-64.

Juliusson, B., Bergström, A., van Veen, T., and Ehinger, B., 1993, Cellular organization in retinal transplants using cell suspensions or fragments of embryonic retinal tissue, *Cell Transplant.* **2**:411-418.

Lewis G.P., Guérin C.J., Anderson D.H., Matsumoto B., and Fisher S.K., 1994, Rapid changes in the expression of glial cell proteins caused by experimental retinal detachment, *Amer. J. Ophthalmol..* **118**:368-376.

Narfström K., and Nilsson S.E., 1986, Progressive retinal atrophy in the Abyssinian cat: electron microscopy, *Invest. Ophthalmol. Vis. Sci.* **27**:1569-1576.

Pardue, M.T., Ball, S.L., Hetling, J.R., Chow, V.Y., Chow, A.Y., and Peachey, N.S., 2001, Visual evoked potentials to infrared stimulation in normal cats and rats, *Doc. Ophthalmol.* in press.

Pardue M.T., Stubbs E.B. Jr., Perlman J.I., Chow A.Y., Narfström K., and Peachey N.S., 1999, Immunocytochemical analysis of the cat retina following subretinal implantation of microphotodiode-based retinal prostheses. ARVO Abstracts, *Invest. Ophthalmol.. Vis. Sci.* **40**:S731.

Peachey N.S., and Chow A.Y., 1999, Subretinal implantation of semiconductor-based photodiodes: Progress and challenges, *J. Rehab. Res. & Devel.* **36**:371-378.

Santos A., Humayun M.A., de Juan E., Jr., Greenburg R.J., March M.J., Klock I.B., and Milam A.H., 1997, Preservation of the inner retina in retinitis pigmentosa, *Arch. Ophthalmol.* **115**:511-515.

Seiler, M.J., Aramant, R.B., and Ball, S.L., 1999, Photoreceptor function of retinal transplants implicated by light-dark shift of S-antigen and rod transducin, *Vis. Res.* **39**:2589-2596.

Seiler, M.J. and Aramant, R.B., 1998, Intact sheets of fetal retina transplanted to restore damaged rat retinas, *Invest. Ophthalmol. Vis. Sci.* **39**:2121-2131.

Seiler, M.J., Aramant, R.B, and Bergstrom, A., 1995, Co-transplantation of embryonic retina and retinal pigment epithelial cells to rabbit retina, *Curr. Eye Res.* **14**:199-207.

Silverman, M.S. and Hughes, S.E., 1989, Transplantation of photoreceptors to light-damaged retina,*Invest. Ophthalmol. Vis. Sci.* **30**:1684-1690.

Stone J.L., Barlow W.E., Humayun M.S., de Juan E., Jr., and Milam A.H., 1992, Morphometric analysis of macular photoreceptors and ganglion cells in retinas with retinitis pigmentosa, *Arch. Ophthalmol.* **110**:1634-1639.

Villegas-Pérez, M.P., Lawrence, J.M., Vidal-Sanz, M., Lavail, M.M., and Lund, R.D., 1998, Ganglion cell loss in RCS rat retina: a result of compression of axons by contracting intraretinal vessels linked to the pigment epithelium, *J. Comp. Neurol.* **392**:58-77.

Zrenner, E., Stett A., Weiss S., Aramant R.B., Guenther E., Kohler K., Miliczek K.D., Seiler M.J., and Haemmerle H., 1999, Can subretinal microphotodiodes successfully replace degenerated photoreceptors? *Vision Res.* **39**:2555-2567.

THE ROLE OF FATTY ACIDS IN THE PATHOGENESIS OF RETINAL DEGENERATION

Daniel C. Garibaldi, Zhenglin Yang, Yang Li, Zhengya Yu and Kang Zhang[*]

1. INTRODUCTION

Retinal dystrophies are a heterogeneous group of ocular disorders that result in varying degrees of progressive peripheral or central visual loss.[1] Though these dystrophies have traditionally been distinguished by their histopathology, ophthalmoscopic findings, and associated physiologic deficits, advances in genetics and molecular biology have permitted researchers to elucidate the genetic determinants of these clinical and histopathological changes.   Techniques such as genetic linkage analysis, positional cloning, and the candidate gene approach have been utilized to associate genomic loci and specific mutations with known disease phenotypes.

Stargardt-like macular dystrophy (STGD3, MIM 600110) and autosomal dominant macular dystrophy (adMD), are inherited forms of macular degeneration characterized by decreased visual acuity, macular atrophy, and extensive fundus flecks. [2, 3, 4, 5] Previous studies have suggested that these clinically similar phenotypes may in fact result from mutations in a common disease gene.[3, 4, 5]   Recent work demonstrates that affected members in independent families with STGD3 and adMD have a single five base-pair deletion within the protein-coding region of a novel retinal photoreceptor-specific gene on chromosome 6q14.[6] Analysis of this gene revealed homology to the *ELO* family of yeast proteins that function in the elongation of very long chain fatty acids, leading to its given name *ELOVL4*, and implicating the biosynthesis and modification of fatty acids in the pathogenesis of inherited macular degeneration.[6] In this chapter, we review previous studies of the association between fatty acid metabolism and retinal degeneration. The potential role of ELOVL4 in the synthesis of polyunsaturated fatty acids (PUFAs) in the retina is discussed, future research directions is proposed based on models of retinal degeneration caused by PUFA deficiency.

---

*Cole Eye Institute, The Cleveland Clinic Foundation, Cleveland, OH

New Insights Into Retinal Degenerative Diseases
Edited by Anderson *et al.*, Kluwer Academic/Plenum Publishers, 2001

2. FATTY ACIDS

Fatty acids are comprised of a hydrocarbon chain with a terminal carboxyl group. They may be saturated (containing no double bonds) or unsaturated (containing one or more double bonds, inserted by specific fatty acid desaturase enzymes).[7] Polyunsaturated fatty acids (PUFAs) are defined as those fatty acids of at least 18 carbons in length containing two or more double bonds, and are named according to the number of double bonds and location of the first double bond in relation to the methyl terminal carbon. Thus, arachnidonic acid (AA C20:4:n-6) is 20 carbons long with 4 double bonds, the first double bond located 6 carbons from the methyl terminal carbon. PUFAs are divided into the "n-3" and "n-6" families based on their formation, by desaturation and elongation reactions, from the two essential fatty acids linoleic acid (LA 18:2:n-6) and α-linolenic acid (ALA 18:3:n-3, Figure 1).[8, 9]   The n-6 family of biologically active molecules include prostaglandins, thromboxanes, prostacyclins and leukotrienes while the n-3 derivatives, termed the eicosonoids, include prostaglandins and leukotrienes.[10]

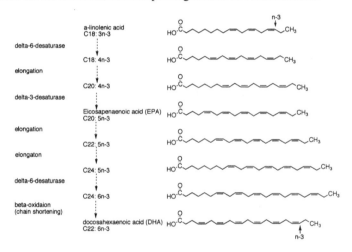

Figure 1. The biosynthetic pathway of long chain fatty acids in the retina.   Biosynthesis of polyunsaturated long chain fatty acids, such as DHA, requires dietary consumption of the essential α-linolenic acid and a subsequent series of three elongation steps and three desaturation steps.

Polyunsaturated fatty acids play integral roles in the structure and function of biologic organisms, serving as the building blocks of cellular structures and acting as signaling molecules.[8, 9] Early studies demonstrated that certain PUFAs were "essential" to growth and development, as Burr and Burr in 1929 found that a fat-free environment yielded stunted growth in rats and that this effect was reversed through the supplementation of LA, ALA or AA to the rats' diet.[11]

More recent work has investigated the function of PUFAs in fetal and neonatal neural and visual development.[12, 13] Much of this research has focused on the n-3 PUFA docosahexaenoic acid (DHA 22:6:n-3), as it has been shown that phospholipids of the retina and brain are highly enriched in DHA. Moreover, DHA comprises up to 60% of the fatty acids in phospholipids in the outer segments of retinal photoreceptors,[14, 15, 16, 17, 18, 19] supporting a potential role for this molecule in the process of visual transduction. In contrast, mammalian retinal pigment epithelium (RPE) contains larger concentrations of shorter chain mono- and dienoic fatty acids with only small proportions of DHA.[20, 21, 22, 23] Recent work suggests that this concentration gradient is essential to retinoid targeting within the eye through the influence of DHA on the interaction between 11-cis-retinal and interphotoreceptor retinoid-binding protein (IRBP).[24] In this model, proposed by Chen et al., the high concentration of DHA in photoreceptor cells reduces the affinity of the IRBP for 11-cis-retinal, inducing the exchange of 11-cis-retinal for all-trans-retinol. Thus, 11-cis-retinal is targeted to the photoreceptor cell, where it is required for rhodopsin regeneration.

Numerous studies have investigated the relationship between dietary n-3 PUFA consumption and the specific fatty acid composition of the brain and retina. It has been demonstrated that increasing the dietary intake of n-3 PUFAs will increase the levels of n-3 fatty acids in the rod outer segments of retinal photoreceptors.[25, 26] Small alterations in the specific PUFAs ingested in the diet have been shown to modify the fatty acid composition of the cell membranes in the plasma, brain and developing retina.[27] This has been demonstrated in human premature infants where the relative content of n-3 and n-6 dietary fatty acids affects the composition of n-3 and n-6 PUFAs in plasma and erythrocytes.[28, 29, 30, 31, 32] Others have found that breastfed human term infants have higher cerebral DHA concentrations than formula-fed infants,[33, 34] and have attributed this observation to the absence of DHA in formulas (DHA comprises approximately 0.1-1.0% of fatty acids in breast milk, depending on diet).[35]

3. MODELS OF N-3 FATTY ACID DEFICIENCY

To elucidate the role of PUFAs, specifically DHA, in the development and function of the retina, many researchers have examined animal models of n-3 deficiency. Animals grown in an environment free of dietary n-3 PUFAs are known to develop abnormal electroretinograms (ERGs) in the setting of decreased retinal levels of DHA-phospholipids.[36, 37, 38] This adverse effect of n-3 deficiency on visual function has been reproduced in multiple species including rats,[39, 40, 41] guinea pigs,[42] monkeys,[37, 38] and humans.[30, 43] While these experiments have not been directly replicated in humans, researchers have studied premature human infants as models of n-3 fatty acid deficiency, as they are deprived in the final weeks to months of intrauterine nutrition during which DHA accumulation is high.[44, 45] In these infants, low n-3 fatty acids diets have been associated with a reduction in ERG sensitivity and maximal amplitude,[30, 43] and supplementation with n-3 PUFAs has been shown to improve visual evoked potentials (VEP) and forced choice preferential looking activity (FPL).[46] In full term infants, studies have suggested that retinal function, as measured by ERG and VEP, may mature at an accelerated pace when infants' diets are supplemented with AA and DHA.[30, 43, 47, 48] Though this improvement in physiologic testing has been supported by several studies,[49, 50] others have found that supplementation with DHA yields no improvement in visual acuity development relative to placebo.[51, 52, 53]

Recently, attention has been given to the observed reduction of serum and retinal PUFAs, especially DHA, in retinal degeneration phenotypes. This association is most striking in patients with Retinitis Pigmentosa (RP), and decreased plasma PUFAs have been observed in nearly all types of RP.[54, 55, 56, 57, 58] Lower blood levels of PUFAs have also been observed with Usher's Syndrome type I, which combines an RP-like phenotype with hearing loss.[59] Hoffman et al. described a positive correlation between the plasma level of PUFAs and the severity of the RP phenotype in patients who inherited RP in an X-linked fashion.[60]

Decreased levels of DHA have also been observed in various animal models of RP. One such model is the progressive rod cone degeneration (*prcd*) phenotype. Both miniature poodles[61] and Abyssinian cats[62] with this phenotype have been shown to have decreased plasma levels of DHA. Researchers have also noted decreased DHA levels in rod outer segments of animals with RP-like phenotypes including the *prcd* dog,[63] transgenic pigs with a Pro247Leu rhodopsin mutation, and *rds* heterozygous mice with a peripherin mutation.[64]

These results beg the question of whether the observed decreases in plasma and ROS levels of DHA may represent a causal factor in the development of retinal degeneration. Researchers have examined this issue through the repletion of PUFAs, specifically DHA, in models of n-3 deficiency in attempts to demonstrate recovery of visual function in the setting of restored DHA levels. Numerous studies have revealed that one can replete brain and retinal DHA content in models of n-3 deficiency using dietary supplementation of various PUFAs, including ALA, EPA, and DHA itself.[42, 65, 66, 67, 68, 69] However, does the ability to restore DHA content confer an ability to recover visual function and if so, does this establish an etiologic role for decreased retinal n-3 in models of retinal degeneration?

The results in this area of research have been mixed. Weisinger et al.[42] demonstrated that guinea pigs fed prolonged n-3 deficient diets after weaning had statistically significant loss of ERG amplitudes in the setting of markedly decreased retinal DHA levels relative to controls. After ten weeks of repletion of dietary DHA, they observed complete functional recovery, suggesting that a critical level of retinal DHA during development is required for normal retinal function. These results, however, were not supported by studies of the *prcd* dog. In these animals, though researchers were able to increase plasma and ROS levels of DHA through dietary supplementation, photoreceptor loss continued.[63] Similarly, though dietary supplementation with DHA restored the fatty acid composition of the adult primate retina, it did not reverse the observed ERG abnormalities.[70] In humans, DHA supplementation failed to restore visual function in patients with both X-linked[60] and autosomal dominant retinitis pigmentosa.[71] Weisinger's group attempted to alleviate the conflicting results of these studies by suggesting that restoration of function was evident only with early and prolonged DHA repletion and that there may be a critical developmental age after which functional losses may be irreversible.[42]

4. VISUAL DYSFUNCTION IN SETTINGS OF DECREASED PUFA LEVELS

What is the physiologic basis for the observed association between decreased levels of n-3 PUFAs, especially DHA, and visual dysfunction? Though the precise role of PUFAs in the retina is unknown, it has long been postulated that the unique fatty acid composition of the retina is essential to the development and function of the visual system. The link between lipid environment and retinal function is supported by studies focused on the function of the ABCA4 protein, a photoreceptor-specific ATP-binding cassette transporter implicated in the pathophysiology of numerous retinal dystrophies including RP, cone-rod dystrophy and Stargardt disease. Initial molecular, genetic, and histopathologic studies suggested that ABCA4 functions to transport retinoids across photoreceptor outer segment membranes.[72, 73, 74] This was supported by *in vitro* studies demonstrating that the ATPase activity of ABCA4 was stimulated 5-fold when retinal is used as substrate,[75] and studies of an *ABCA4* knockout mouse demonstrating impaired photoreceptor function in the setting of light-dependent increases in all-*trans*-retinal and N-retinylidene-phosphatidylethanolamine and an accumulation of A2E (a pyridium bis-retinoid compound) in photoreceptor and RPE cells.[76]

Recently, researchers have shown that the function of ABCA4 *in vitro* is strongly influenced by its surrounding lipid environment.[77] They demonstrated that the V_{max} of ABCA4 reconstituted into ROS phospholipid is four times higher than ABCA4 in soybean phospholipid and over six times higher than ABCA4 in brain polar lipid and that retinal increased the V_{max} 2.5-5 fold for ABCA4 in ROS and brain lipids and only 1.5-2 fold for ABCA4 in soybean phospholipids. Furthermore, they showed that environments enriched for DHA- and PE-containing phospholipids resulted in a 2-3-fold increase in basal and retinal-stimulated ABCA4 ATPase activity.[77] These results suggest that the lipid composition of the retina, consisting of high levels of DHA and PE, is essential to the normal function of ABCA4.

Other studies have examined the relationship between the specific phospholipid milieu of retinal membranes and the function of rhodopsin.[78] In developmental studies, researchers have observed that increases in rhodopsin content were associated with increases in retinal DHA and that these levels could be modified by dietary fatty acid content.[27] Functional studies have suggested that delipidated rhodopsin cannot be regenerated and that this property is restored by supplementation with phospholipids.[79] Similarly, it has been shown that n-3 deficient rats exhibit decreased rhodopsin regeneration,[65] and that reductions in membrane DHA *in vitro* can adversely affect the conversion of rhodopsin to its active state.[80, 81] Littman and Mitchell[81] provided a model to explain such results by proposing that a highly unsaturated fatty acid environment within photoreceptor outer segments' membranes may create the optimal environment for rhodopsin transformation upon exposure to light.

Further evidence of the influence of DHA on the function of rhodopsin has been provided by studies demonstrating the relationship between DHA levels, light exposure, and light-induced damage. Rats exposed to prolonged light[82] or reared in a high light environment[83] demonstrate reduced levels of DHA in rod outer segments, suggesting an influence of light on the fatty acid composition of the retina. Conversely, reduced levels of DHA, as induced by n-3 PUFA deficiency, have been shown to exert a protective effect against light induced degeneration.[84, 85, 86, 87] Some have postulated that this protection is due to a decreased amount of oxidative material available for peroxidation induced damage, supported by studies demonstrating decreased retinal damage when rats are pretreated with antioxidants prior to light exposure.[88, 89] However, in settings of n-3 PUFA deficiency, DHA is replaced by n-6 docosapentaenoic acid, as a result of a shared Δ-6 desaturase,[30, 69, 90, 91, 92, 93,] maintaining a relatively constant combined level of DHA and 22:5n-6.[42, 94] Thus the total oxidative potential of retinal lipids remains unchanged in n-3 deficient states, suggesting that the observed decreased light mediated damage is not solely due to loss of potentially oxidizable DHA.

An alternative mechanism for the decreased susceptibility to light mediated damage observed in models of n-3 fatty acid deficiency involves the alteration of rhodopsin function in the setting of decreased DHA retinal content. If decreases in retinal DHA interfere with normal rhodopsin function, this would potentially explain both the abnormal ERGs observed in n-3 deficiency and the decrease in light induced photoreceptor damage. To this end, Bush et al.[65] demonstrated that photon capture in the retinas of n-3 deficient rats is reduced secondary to a decrease in rhodopsin regeneration, suggesting that n-3 PUFAs are essential to the normal function of rhodopsin.

5. RELATIONSHIP BETWEEN FATTY ACID SYNTHESIS AND HEREDITARY RETINAL DEGENERATION

Stargardt macular dystrophy (STGD) is a juvenile onset macular dystrophy characterized by decreased visual acuity, macular atrophy, and extensive flecks. It may be inherited in an autosomal recessive or autosomal dominant fashion. Stargardt-like macular dystrophy (STGD3, MIM 600110) is an autosomal dominant form of the disease (Figure 2).[6] Genetic mapping data suggest that a mutation in the STGD3 gene may be responsible for another macular dystrophy known as autosomal dominant macular dystrophy (adMD), which has striking clinical resemblance to STGD3.[6]

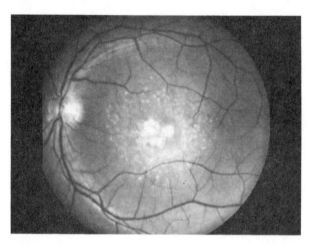

Figure 2. Fundus photograph from a STGD3 patient with a visual acuity of 20/240 showing a large area of macular atrophy with characteristic flecks surrounding the region of atrophy.

Recently, attention has turned to the potential role of fatty acids in the pathogenesis of hereditary retinal degeneration, specifically STGD3 and adMD. This stems from the observation, by Zhang et al.,[6] that affected members of families with STGD3 and adMD share a five base-pair deletion in exon 6 of the ELOVL4 gene on chromosome 6q14. This gene was found to share identity with members of the ELO gene family in yeast that have been shown to function in long chain fatty acid elongation (Figure 3),[95, 96] and with a human gene, *HELO1*, thought to function in PUFA elongation (Figure 3),[97] thus implicating defects in fatty acid synthesis and modification in the pathogenesis of STGD3 and adMD.[6]

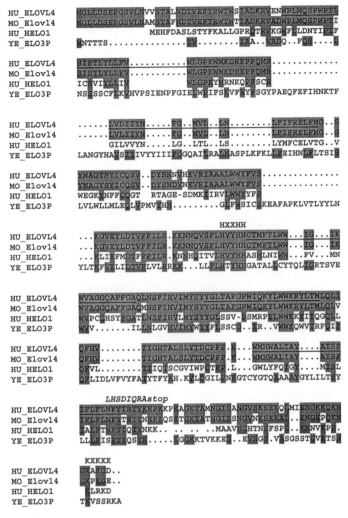

Figure 3. Comparison of human ELOVL4 and mouse Elovl4 protein sequences with the product of the yeast ELO3 gene, Elo3p and human HELO1. Five putative transmembrane domains are underlined. Conservative and identical residues are indicated in dark gray and light gray, respectively. Carboxy-terminal dilysine signal responsible for ER retention and the dioxy iron-binding HXXHH motif are indicated above the protein sequence. The truncated COOH-terminus caused by the 5-bp deletion in ELOVL4 is shown in bold italics above the human ELOVL4 sequence.

5.1. The ELO family in yeast is involved in fatty acid chain elongation

Three members of the ELO family, namely ELO1, ELO2 and ELO3, have been identified in yeast, and all appear to play essential roles in fatty acid chain elongation.

ELO1, the first member to be described, has been shown to participate in microsomal fatty acid elongation between C14 and C16.[95] ELO2 and ELO3, identified secondary to their homology to ELO1, function in elongation of chains up to 24 and 26 carbons, respectively.[96] Members of the ELO family share common biochemical features, including at least five membrane-spanning domains and a single histidine-box motif.[95, 96] This histidine box motif (HXXHH) is characteristic of diiron-oxo proteins such as fatty acid desaturases, in which the histidine residues help to coordinate the reception of electrons in reduction reactions.[98, 99] Thus the ELO family members possess biochemical features that would allow them to participate in the catalysis of reduction reactions occurring during fatty acid elongation.[100] Each gene is thought to encode a single enzyme component of complex enzymatic systems that function in the synthesis of very long chain saturated and monounsaturated fatty acids.[96]

Recently, Leonard et al.[97] identified a human gene, named *HELO1*, with significant sequence identity to the yeast ELO family. The predicted amino acid sequence of this gene, showed 29.1% identity with ELO2 and 27.7% identity with ELO3, implicating a role for the encoded protein HELO1p in fatty acid elongation. However, unlike the ELO proteins, which have been shown to function in the elongation of saturated and monounsaturated long chain fatty acids, HELO1p, when expressed in yeast, was shown to participate in both polyunsaturated and monounsaturated fatty acid elongation. Analysis of HELO1 expression in human tissues revealed highest levels in the adrenal gland and testes, with approximately 50% lower expression in the brain.

5.2. A Mutation in ELOVL4, a human homologue of the ELO family, results in hereditary retinal degeneration phenotypes

ELOVL4 was identified using a positional cloning approach. First, genetic linkage analysis limited the minimal genetic region for the disease gene for STGD3 and adMD to a 0.6 cM interval on chromosome 6q14.[6] Analysis of this region revealed a candidate gene coined ELOVL4, as the most likely disease gene for STGD3 and adMD. Sequencing of ELOVL4 cDNA revealed that it encoded a putative 314 amino acid protein that shared 35% identity with members of the yeast ELO family and 38% identity with HELO1. This putative protein was shown to possess the three characteristics shared by all members of the ELO family: five membrane-spanning domains, a single histidine-box motif, and a signal sequence required for localization of transmembrane proteins to the endoplasmic reticulum.[101, 102, 103] Thus, amino acid sequence of the protein encoded by ELOVL4 implicate a role in PUFA elongation.

Mutation analysis of ELOVL4 revealed a five base-pair deletion starting at position 797 of exon 6 (797-801delAACTT). This frameshift mutation results in a premature stop codon, causing the loss of a 51 amino acid fragment at the protein's C-terminus that includes the dilysine targeting signal responsible for localization to the endoplasmic reticulum. Zhang et al. demonstrated that this mutation was present in all affected members of four unrelated STGD3 kindreds and one five-generation family with adMD and was not present in chromosomes of normal control individuals.

ELOVL4 was found to be expressed in both rod and cone photoreceptors of the adult retina. Northern blot studies of an ELOVL4-cDNA probe revealed transcription in human retina, with ten-fold lower transcription in human brain and no detectable transcription in a human RPE cell line. This retinal specific expression of ELOVL4 was confirmed using RT-PCR analysis. *In situ* hybridization studies of rhesus monkey and mouse retinal sections demonstrated strong signal in the photoreceptor inner segments, supporting the observation that ELOVL4 is expressed solely in the photoreceptor cells of adult retina.[6]

Several lines of evidence strongly suggest that the observed deletion in ELOVL4 is the disease-causing mutation in several families with STGD3 and adMD. First, the localization of ELOVL4 to the retinal photoreceptors establishes it as a candidate gene for retinal degeneration phenotypes. Second, the mutation is deleterious, resulting in the elimination of the terminal-targeting signal required for the localization of the encoded protein to the endoplasmic reticulum. Finally, linkage disequilibrium implicates ELOVL4 as the disease-causing gene in the families studied, as all affected members in five distinct families segregated the 797-801delAACTT mutation while it was not present in more than 1100 control alleles.[6]

5.3. A model for the role of defects in fatty acid synthesis in the pathogenesis of hereditary retinal degeneration

Though the specific function of the protein encoded by ELOVL4 has yet to be elucidated by *in vitro* expression studies, one can infer, through its shared sequence identity to the yeast ELO family and the human HELO1 gene, a role for ELOVL4 in PUFA elongation in humans. Thus, by demonstrating that a mutation in ELOVL4 results in both STGD3 and adMD phenotypes, Zhang et al. implicate, for the first time, the biosynthesis of fatty acids in the pathogenesis of hereditary retinal degeneration. How then, does a defect in an enzyme involved in PUFA elongation result in phenotypes characterized by decreased visual acuity in the setting of macular atrophy? Furthermore, why do the observed phenotypes resemble those resulting from mutations in ABCA4, the disease-causing gene for autosomal recessive Stargardt macular dystrophy?

The loss of visual acuity and electrophysiologic abnormalities resulting from ELOVL4 mutations may potentially be attributed to loss of the native fatty acid composition of retinal membranes that are normally enriched in PUFAs such as DHA. Numerous studies have demonstrated that systemic n-3 deficiency leads to a decrease in the n-3 fatty acid content of the retina, with a resultant adverse effect on visual function. It is reasonable to suggest that the 797-801delAACTT mutation in ELOVL4 will be deleterious to its putative function in fatty acid elongation, leading to a local deficiency of very long chain fatty acids in the retina. This would likely lead to ERG abnormalities and visual loss, as demonstrated by multiple animal models of n-3 deficiency.

Mutations in ELOVL4 may contribute directly to visual dysfunction through an alteration of the normal lipid environment of the retina, shown to be essential for the function of rhodopsin in visual transduction. This may also contribute to an alteration of the normal DHA concentration gradient from the photoreceptor cells to the RPE, resulting in a disturbance in interphotoreceptor retinoid-binding protein mediated 11-cis-retinal targeting to the photoreceptor cell.

The recent studies of the ABCA4 protein, however, allow for an alternative model, that may help to explain the similarities between recessive and dominant forms of Stargardt macular dystrophy caused by discrete genetic mutations. Studies of *ABCA4* knockout mice demonstrated that ABCA4 acts as an outwardly directed flippase for N-retinylidene-PE and that loss of ABCA4 function results in defects in dark adaptation in the setting of lipofuscin accumulation in the RPE. Ahn et al.,[77] demonstrated that the function of the ABCA4 protein is strongly dependent on its surrounding lipid environment, with the kinetics of its basal and retinal stimulated ATPase activity enhanced by exposure to a lipid milieu similar to that of the ROS.

One might thus postulate that mutations in ELOVL4 would lead to defects in ABCA4 function through the loss of the normal fatty acid composition of the retina. This could lead to retinal degeneration and visual abnormalities through the model established by Weng et al.,[76] whereby photoreceptor death stems from the loss of RPE support as a result of the accumulation of A2-E in RPE lysosomes. This model may explain the phenotypic similarities between recessive and dominant Stargardt macular dystrophy, as it suggests that defects in ABCA4 function play a role in both the pathophysiology of recessive Stargardt macular dystrophy and Stargardt-like phenotypes not associated with mutations in the ABCA4 gene.

The dominant phenotype of the 797-801delAACTT mutation in *ELOVL4* may result from the dominant-negative effect of the mutant ELOVL4 on its interacting/associated proteins, perhaps by poisoning its enzymatic complex. Alternatively, the dominant phenotype can be explained by haploid-insufficiency due to inactivation of one copy of *ELOVL4*. Thus, one may speculate that inactivation of both alleles of *ELOVL4* may lead to earlier onset, more severe, and more diffuse retinal degeneration, such as RP and Leber's Congenital Amaurosis (LCA).

6. SUMMARY

Benefits of the identification of the *ELOVL4* gene will not be limited to patients with STGD3 or adMD. Study of the *ELOVL4* gene product will aid in elucidating the pathogenesis of both recessive and dominant Stargardt macular dystrophy and related phenotypes. It will also help to clarify the potential role of defects in fatty acid synthesis in the development of macular degeneration. It will be interesting and important to investigate the role of ELOVL4 and its pathway in age-related macular degeneration, as well as other forms of inherited retinal degeneration such as RP and LCA. This line of research may yield potential treatments for certain retinal phenotypes through the use of early and sustained dietary fatty acid supplementation for those individuals inheriting a specific defect in the synthesis of very long chain and polyunsaturated fatty acids.

REFERENCES

1. D. C. Garibaldi and K. Zhang, Molecular genetics of macular degeneration, Int. Ophthalmol. Clin. 39(4), 117- 142 (1999).
2. G. W. Cibis, M. Morey, and D. J. Harris, Dominantly inherited macular dystrophy with flecks (Stargardt), Arch Ophthalmol. 98, 1785-1789 (1980).
3. A. O. Edwards, A. Miedziak, T. Vrabec, et al., Autosomal dominant Stargardt-like macular dystrophy: Clinical characterization, longitudinal follow-up, and evidence for a common ancestry in families linked to chromosome 6q14, Am. J. Ophthalmol. 127, 426-435 (1999).
4. I. B. Griesinger, P. A. Sieving, and R. Ayyagari, Autosomal dominant macular atrophy at 6q14 excludes CORD7 and MCDR1/PBCRA loci, Invest. Ophthalmol. Vis. Sci. 41,248-255(2000).
5. E. M. Stone, B. E. Nichols, A. E. Kimura, et al., Clinical features of a Stargardt-like dominant progressive macular dystrophy with genetic linkage to chromosome 6q, Arch. Ophthalmol. 112, 765-772 (1994).
6. K. Zhang, M. Kniazeva, M. Han, et al., A five base-pair deletion in the ELOVL4 gene is associated with two related forms of autosomal dominant macular dystrophy, Nat. Genet., 27, 89-93 (2001).
7. S.S. Pandian, O. E. Eremin, S. McClinton, et al, Fatty acids and prostate cancer: current status and future challenges, J. R. Coll. Surg. Edinb. 44, 352-61 (1999).

8. I. Gill and R. Valivety, Trends Biotechnol. 15, 401-409 (1997).
9. P. Broun, S. Gettner and C. Somerville, Annu. Rev. Nutr. 19, 197-216 (1999).
10. M. A. Crawford, Background to essential fatty acids and their prostanoid derivatives, Br. Med. Bul. 39, 210-213 (1983).
11. G. O. Burr and M. M. Burr, On the nature and role of fatty acids essential in nutrition, J. Biol. Chem. 86, 587-622 (1930).
12. S. E. Carlson, S. H. Werkman, J. M. Peeples, et al., Arachnidonic acid status correlates with first year growth in preterm infants, Proc. Natl. Acad. Sci. U. S. A. 90, 1073-1077 (1993).
13. R. Uauy, E. Birch, D. Birch and P. Peirano, Visual and brain function measurements in studies of n-3 fatty acid requirements of infants, J. Pediatr. 120, S168-180 (1992).
14. S. J. Fliesler and R. E. Anderson, Chemistry and metabolism of lipids in the vertebrate retina, Prog Lipid Res, 22, 79-131 (1983).
15. N. Salem Jr., Omega-3 fatty acids: molecular and biochemical aspects, in New Protective Roles for Selected Nutrients, edited by G. A. Spiller and J. Scala (Alan R. Liss, New York, 1989), pp. 109-228.
16. M. I. Aveldano de Cladironi and N. G. Bazan, Composition and biosynthesis of molecular species of retina phosphoglycerides, Neurochem. 1:381-392 (1980).
17. M. I. Aveldano and N. G. Bazan, Molecular species of phosphatidylcholine, -ethanolamine, -serine, and –inositol in microsomal and photoreceptor membranes of bovine retina, J. Lipid. Res. 24,620-627 (1983).
18. H. G. Choe and R. E. Anderson, Unique molecular species composition of glycerolipids of frog rod outer segments, Exp. Eye Res. 51, 159-165 (1990).
19. A. M. Stinson, R. D. Wiegand and R. E. Anderson, Fatty acid and molecular species compositions of phospholipids and diacylglycerols from rat retinal membranes, Exp. Eye Res. 52, 213-218 (1991).
20. R. E. Anderson, P. M. Lissandrello, M. B. Maude, and M. T. Matthes, Lipids of bovine retinal pigment epithelium, Exp. Eye Res. 23, 149-157 (1976).
21. D. W. Batey, J. F. Mead, and C. D. Eckhert, Lipids of the retinal pigment epithelium in RCS dystrophic and normal rats, Exp. Eye Res. 43, 751-757 (1986).
22. S. C. Braunagel, D. T. Organisciak, and H-M Wang, Isolation of plasma membranes from the bovine retinal pigment epithelium, Biochim. Biophys. Acta 426, 183-194 (1985).
23. S. C. Braunagel, D. T. Organisciak, and H-M Wang, Characterization of pigment epithelial cell plasma membranes from normal and dystrophic rats, Invest. Ophthalmol. Vis. Sci. 29, 1066-1075 (1988).
24. Y. Chen, L. A. Houghton, J. T. Brenna, and N. Noy, Docosahexaenoic acid modulates the interactions of the interphotoreceptor retinoid-binding protein with 11-cis-retinal, J. Biol. Chem. 271, 20507-20514 (1996).
25. D. S. Lin, G. J. Anderson, W. E. Connor, and M. Neuringer, Effect of dietary n-3 fatty acids upon the phospholipid molecular species of the monkey retina, Invest. Ophthalmol. Vis. Sci. 35, 794-803 (1994).
26. M. Suh, A. A. Wierzbicki, and M. T. Clandinin, Dietary fat alters membrane composition in rod outer segments in normal and diabetic rats: impact on content of very-long-chain (C≥24) polyenoic fatty acids, Biochim. Biophys. Acta 1214, 54-62 (1994).
27. M. Suh, A. A. Wierzbicki, E. L. Lien and M. T. Clandinin, Dietary 20:4n-6 and 22:6n-3 modulates the profile of long- and very-long-chain fatty acids, rhodopsin content, and kinetics in developing photoreceptor cells, Pediatr. Res. 48, 524-530 (2000).
28. S. M. Innis, K. D. Foote, M. J. MacKinnon, and O. J. King, Plasma and red blood cell fatty acids of low-birth-weight infants fed their mother's expressed breast milk or preterm infant formula, Am. J. Clin. Nutr. 51, 994-1000 (1990).
29. B. Koletzko, E. Schmidt, H. J. Bremer, et al., Effects of dietary long-chain polyunsaturated fatty acids on the essential fatty acid status of premature infants, Eur. J. Pediatr. 148, 669-675 (1989).
30. R. Uauy, D. G. Birch, E. E. Birch, et al., Effect of dietary omega-3 fatty acids on retinal function of very-low-birth-weight neonates, Pediatr. Res. 28, 485-492 (1990).
31. M. L. Pita, M. R. Fernandez, C. DeLucchi, et al., Changes in the fatty acid pattern of red blood cell phospholipids induced by type of fat, dietary nucleotides, and postnatal age in preterm infants, J. Pediatr. Gastroenterol. Nutr. 7, 740-747 (1988).
32. S. E. Carlson, P. G. Rhodes, and M. G. Ferguson, Docosahexaenoic acid status of preterm infants at birth and following feeding human milk or formula, Am. J. Clin. Nutr. 44, 798-804 (1986).
33. J. Farquharson, F. Cockburn, W. A. Patrick, et al., Infant cerebral cortex phospholipid fatty-acid composition and diet, Lancet 340, 810-813 (1992).
34. M. Makrides, M. A. Neumann, R. W. Byard, et al., Fatty acid composition of brain, retina, and erythrocytes in breast- and formula-fed infants, Am. J. Clin. Nutr. 60, 189-194 (1994).
35. S. M. Innis, Human milk and formula fatty acids, J. Pediatr. 120, S56-S61 (1992).

36. T. G. Wheeler, R. M. Benolken, and R. E. Anderson, Visual membranes: specificity of fatty acid precursors for the electrical response to illumination, Science 188, 1312-1314 (1975).
37. M. Neuringer, W. E. Connor, C. Van Petten, and L. Barstad, Dietary omega-3 fatty acid deficiency and visual loss in infant rhesus monkeys, J. Clin. Invest. 73, 272-276 (1984).
38. M. Neuringer, W. E. Connor, D. S. Lin, L. Barstad, and S. Luck, Biochemical and functional effects of prenatal and postnatal omega-3 fatty acid deficiency on retina and brain in rhesus monkeys, Proc. Natl. Acad. Sci. USA 83, 4021-4025 (1986).
39. R. M. Benolken, R. E. Anderson, and T. G. Wheeler, Membrane fatty acids associated with the visual response in visual excitation, Science 182, 1253-1254 (1973).
40. J-M Bourre, M. Francois, A. Youyou, et al., The effects of dietary α-linolenic acid on the composition of nerve membranes, enzymatic activity, amplitude of electrophysiological parameters, resistance poisons and performance of learning tasks in rats, J. Nutr. 119, 1880-1992 (1989).
41. T. G. Wheeler, R. M. Benolken, and R. E. Anderson, Visual membranes: specificity of fatty acid precursors for the electrical response to illumination, Science 188, 1312-1314 (1975).
42. H. S. Weisinger, A. J. Vingrys, B. V. Bui, and A. J. Sinclair, Effects of dietary n-3 fatty acid deficiency and repletion in the guinea pig retina, Invest. Ophthalmol. Vis. Sci. 40, 327-328 (1999).
43. D. G. Birch, E. E. Birch, D. R. Hoffman, and R. D. Uauy, Retinal development in very-low-birth-weight infants fed diets differing in omega-3 fatty acids, Invest. Ophthalmol. Vis. Sci. 33, 2365-2376 (1992).
44. M. T. Clandinin, J. E. Chappell, T. Heim, et al., Fatty acid utilization in perinatal de novo synthesis of tissues, Early Hum. Dev. 5, 355-366 (1981).
45. M. Neuringer and W. E. Connor, n-3 Fatty acids in the brain and retina: evidence for their essentially, Nutr. Rev. 44, 285-294 (1986).
46. E. E. Birch, D. G. Birch, D. R. Hoffman, and R. Uauy, Dietary essential fatty acid supply and visual acuity development, Invest. Ophthalmol. Vis. Sci. 33, 3242-3253 (1992).
47. E. Birch, D. Birch, D. Hoffman, L. Hale, M. Everett, R. Uauy, Breast feeding and optimal visual development, J. Pediatr. Ophthalmol. Strabismus 30, 33-38 (1993).
48. S. E. Carlson, S. H. Werkman, P. G. Rhodes, E. A. Tolley, Visual-acuity development in healthy preterm infants: effects of marine-oil supplementation, Am. J. Clin. Nutr. 58, 35-42 (1993).
49. S. E. Carlson, A. J. Ford, S. H. Werkman, J. M. Peeples, W. W. K. Koo, Visual acuity and fatty acid status of term infants fed human milk and formulas with and without docosahexaenoate and arachnidonate from egg yolk lecithin, Pediatr. Res. 39, 882-888 (1996).
50. E. E. Birch, D. R. Hoffman, R. Uauy, D. G. Birch, C. Prestidge, Visual acuity and the essentiality of docosahexaenoic acid and arachnidonic acid in the diet of term infants, Pediatr. Res. 44, 201-209 (1998).
51. N. Auestad, M. B. Montalto, R. T. Hall, et al., Visual acuity, erythrocyte fatty acid composition, and growth in term infants fed formulas with long chain polyunsaturated fatty acids for one year, Pediatr. Res. 41, 1-10 (1997).
52. M. H. Jorgensen, G. Holmer, P. Lund, O. Hernell, K. F. Michaelsen, Effect of formula supplemented with docosahexaenoic acid and gamma-linolenic acid on fatty acid status and visual acuity in term infants, J. Pediatr. Gastroenterol. Nutr. 26, 412-421 (1998).
53. M. Makrides, M. A. Neumann, K. Simmer, R. A. Gibson, A critical appraisal of the role of dietary long-chain polyunsaturated fatty acids on neural indices of term infants: a randomized, controlled trial, Amer. Acad. Pediatr. 105, 32-38 (2000).
54. R. Anderson, M. Maude, R. Lewis, D. Newsome, and G. Fishman, Abnormal plasma levels of polyunsaturated fatty acid in autosomal dominant retinitis pigmentosa, Exp. Eye. Res. 44, 155-159 (1987).
55. C. Converse, H. Hammer, C. Packard, and J. Shepherd, Plasma lipid abnormalities in retinitis pigmentosa and related conditions, Trans. Ophthalmol. Sci. UK. 103, 508-512 (1983).
56. J. Gong, B. Rosner, D. Rees, E. Berson, C. Weigel-DiFranco, and E. Schaefer, Plasma docosahexaenoic acid levels in various genetic forms of retinitis pigmentosa, Invest. Ophthalmol. Vis. Sci. 33, 2596-2602 (1992).
57. D. Hoffman, R. Uauy, and D. Birch, Red blood cell fatty acid levels in patients with autosomal dominant retinitis pigmentosa, Exp. Eye Res. 57, 359-368 (1993).
58. R. Holman, D. Bibus, G. Jeffrey, P. Smethurst, and J. Crofts, Abnormal plasma lipids of patients with retinitis pigmentosa, Lipids 29, 61-65 (1994).
59. N. Bazan, B. Scott, T. Reddy, and M. Pelias, Decreased content of docosahexaenoate and arachidonate in plasma phospholipids in Usher's syndrome, Biochem. Biophys. Res. Comm. 141, 600-604 (1986).
60. D. Hoffman and D. Birch, Docosahexaenoic acid in red blood cells of patients with X-linked retinitis pigmentosa, Invest. Ophthalmol. Vis. Sci. 36, 1009-1018 (1995).

61. R. E. Anderson, M. B. Maude, R. A. Alvarez, G. M. Acland, and G. D. Aguirre, Plasma lipid abnormalities in the miniature poodle with progressive rod-cone degeneration, Exp. Eye Res. 52, 349-355 (1991).

62. R. E. Anderson, M. B. Maude, S. E. G. Nilsson, and K. Narfstrom, Plasma lipid abnormalities in Abyssinian cat with a hereditary rod-cone degeneration, Exp. Eye Res. 53, 415-417 (1991).

63. G. Aguirre, G. Acland, M. Maude, and R. Anderson, Diets enriched in docosahexaenoic acid fail to correct progressive rod-cone degeneration (prcd) phenotype, Invest. Ophthalmol. Vis. Sci. 38, 2387-2407 (1997).

64. R. E. Anderson, M. B. Maude, R. A. Alvarez, G. Acland, and G. D. Aguirre, A hypothesis to explain the reduced blood levels of docosahexaenoic acid in inherited retinal degenerations caused by mutations in genes encoding retina-specific proteins, Lipids 34, S235-S237 (1999).

65. R. A. Bush, A. Malnoe, C. E. Reme, T. P. Williams, Dietary deficiency of n-3 fatty acids alters rhodopsin content and function in the rat retina, Invest. Ophthalmol. Vis. Sci. 35, 91-100 (1994).

66. G. J. Anderson, W. E. Connor, and J. D. Corliss, Docosahexaenoic acid is the preferred dietary n-3 fatty acid for the development of the brain and retina, Pediatr. Res. 27, 89-97 (1990).

67. A. Youyou, G. Durand, G. Pascal, M. Piciotti, O. Dumont, and J. M. Bourre, Recovery of altered fatty acid composition induced by a diet devoid of n-3 fatty acids in myelin, synaptosomes, mitochondria, and microsomes of developing rat brain, J. Neurochem. 46, 224-228 (1986).

68. P. Homayoun, G. Durand, G. Pascal, and J. M. Bourre, Alteration in fatty acid composition of adult rat brain capillaries and choroids plexus induced by a diet deficient in n-3 fatty acids: slow recovery after substitution with a nondeficient diet, J. Neurochem. 51, 45-48 (1988).

69. J-M. Bourre, G. Durand, G. Pascal, and A. Youyou, Brain cell and tissue recovery in rats made deficient in n-3 fatty acids by alteration of dietary fat, J. Nutr. 119, 15-22 (1989).

70. N. G. Bazan and T. S. Reddy, Retina, in: Handbook of Neurochemistry, edited by A. Lajtha (Plenum Press, New York, 1985), pp. 507.

71. D. R. Hoffman, R. Uauy, and D. G. Birch, Metabolism of omega-3 fatty acids in patients with autosomal dominant retinitis pigmentosa, Exp. Eye Res. 60, 279-289 (1995).

72. R. Allikmets, N. Singh, H. Sun, et al., A photoreceptor cell-specific ATP-binding transporter gene (ABCA4) is mutated in recessive Stargardt macular dystrophy, Nat. Genet. 15, 236-246 (1997).

73. M. Illing, L. L. Molday, and R. S. Molday, The 220-kDa rim protein or retinal rod outer segments is a member of the ABC transporter superfamily, J. Biol. Chem. 272, 10303-10310 (1997).

74. H. Sun and J. Nathans, Stargardt's ABCA4 is localized to the disc membrane of retinal rod outer segments, Nat. Genet. 17, 15-16 (1997).

75. H. Sun, R. S. Molday, and J. Nathans, Retinal stimulates ATP hydrolysis by purified and reconstituted ABCA4, the photoreceptor-specific ATP-binding cassette transporter responsible for Stargardt disease, J. Biol. Chem. 274, 8269-8281 (1999).

76. J. Weng, N. L. Mata, S. M. Azarian, R. T. Tzekov, D. G. Birch, and G. H. Travis, Insights into the function of Rim protein in photoreceptors and etiology of Stargardt's disease from the phenotype in ABCA4 knockout mice, Cell 98, 13-23 (1999).

77. J. Ahn, J. T. Wong, and R. S. Molday, The effect of lipid environment and retinoids on the ATPase activity of ABCA4, the photoreceptor ABC transported responsible for Stargardt macular dystrophy, J. Biol. Chem. 27, 20399-20405, (2000).

78. T. S. Wiedmann, R. D. Pates, J. M. Beach, A. Salmon, M. F. Brown, Lipid-protein interactions mediate the photochemical function of rhodopsin, Biochemistry 27, 6469-6474 (1988).

79. H. Shichi, Biochemistry of visual pigments. II. Phospholipid requirements and opsin conformation for regeneration of bovine rhodopsin, J. Biol. Chem. 246, 6178-6182 (1971).

80. E. E. Dratz and L. L. Holte, The molecular spring model for the function of docosahexaenoic acid (22:6n-3) in biological membranes, in: The Third International Congress on Essential Fatty Acids and Eicosanoids, edited by A. J. Sinclair and R. A. Gibson (AOCS Press, Champaign, IL, 1993), pp. 122-127.

81. B. J. Littman and D. C. Mitchell, A role for phospholipid polyunsaturation in modulating membrane protein function, Lipids 31, S193-S197 (1996).

82. R. D. Wiegand, N. M. Gusto, L. M. Rapp, and R. E. Anderson, Evidence for rod outer segment peroxidation following constant illumination of the rat retina, Invest. Ophthalmol. Vis. Sci. 24, 1433-1435 (1983).

83. J. S. Penn and R. E. Anderson, Effect of light history on rod outer segment membrane composition in the rat, Exp. Eye Res. 44, 767-778 (1987).

84. D. T. Organisciak, R. M. Darrow, Y-L. Jiang, and J. C. Blanks, Retinal light damage in rats with altered levels of rod outer segment docosahexaenoate, Invest. Ophthalmol. Vis. Sci. 37, 2243-2257 (1996).

85. R. A. Bush, C. E. Reme, A. Malnoe, Light damage in the rat retina: the effect of dietary deprivation of n-3 fatty acids on acute structural alterations, Exp. Eye Res. 53, 741-752 (1991).

86. C. A. Koutz, R. D. Wiegand, L. M. Rapp, and R. E. Anderson, Effects of dietary fat on the response of the rat retina to chronic and acute light stress, Exp. Eye Res. 60, 307-316 (1995).
87. C. E. Reme, A. Malnoe, H. H. Jung, Q. Wei, and K. Munz, Effect of dietary fish oil on acute light-induced photoreceptor damage in the rat retina, Invest. Ophthalmol. Vis. Sci. 35, 78-90 (1994).
88. D. T. Organisciak, H-M. Wang, and W. K. Noell, Aspects of the ascorbate protective mechanism in retinal light damage in rats with normal and reduced ROS docosahexaenoic acid, in: Degenerative Retinal Disorders: Clinical and Laboratory Investigations, edited by J. G. Hollyfield, R. E. Anderson, and M. M. Lavail (Alan R. Liss, New York, 1987) pp. 455-468.
89. W. K. Noell, D. T. Organisciak, H. Ando, M. A. Braniecki, and C. Durlin, Ascorbate and dietary protective mechanisms in retinal light damage of rats: electrophysiological, histological and DNA measurements, in: Degenerative Retinal Disorders: Clinical and Laboratory Investigations, edited by J. G. Hollyfield, R. E. Anderson, and M. M. Lavail (Alan R. Liss, New York, 1987) pp. 469-483.
90. R. R. Brenner and R. O. Peluffo, Effect of saturated and unsaturated fatty acids on the desaturation in vivo of palmitic, stearic, oleic, linoleic, and linolenic acids, J. Biol. Chem. 241, 5213-5219 (1966).
91. R. R. Brenner and R. O. Peluffo, Regulation of unsaturated fatty acid biosynthesis, Biochim. Biophys. Acta 176, 471-479 (1969).
92. S. E. Carlson, J. D. Carver, and S. G. House, High fat diets varying in ratios of polyunsaturated to saturated fatty acid and linoleic to linolenic acid: a comparison of rat neural and red cell membrane phospholipids, J. Nutr. 116, 718-725 (1986).
93. S. E. Carlson, P. G. Rhodes, O. S. Rao, and D. E. Goldgar, Effect of fish oil supplementation on the n-3 fatty acid content of red blood cell membrane in preterm infants, Pediatr. Res. 21, 507-510 (1987).
94. H. S. Weisinger, A. J. Vingrys, and A. J. Sinclair, The effect of docosahexaenoic acid on the electroretinogram of the guinea pig, Lipids 31, 65-70 (1996).
95. D. A. Toke and C. E. Martin, Isolation and characterization of a gene affecting fatty acid elongation in Saccharomyces cerevisiae, J. Biol. Chem. 271, 18413-18422 (1996).
96. C. S. Oh, D. A. Toke, S. Mandala, and C. E. Martin, ELO2 and ELO3, homologues of the Saccharomyces cerevisiae ELO1 gene, function in fatty acid elongation and are required for sphingolipid formation, J. Biol. Chem. 272, 17376-17384 (1997).
97. A. E. Leonard, E. G. Bobik, J. Dorado, et al., Cloning of a human cDNA encoding a novel enzyme involved in the elongation of long-chain polyunsaturated fatty acids, Biochem. J. 350, 765-770 (2000).
98. B. G. Fox, J. Shanklin, J. Ali, T. M. Loehr, and L. J. Sanders, Resonance raman evidence for an Fe-O-Fe center in stearoyl-ACP desaturase. Primary sequence identity with other diiron-oxo proteins, Biochemistry 33, 12776-12786 (1994).
99. J. Shanklin, E. Whittle, and B. G. Fox, Eight histidine residues are catalytically essential in membrane-associated iron enzyme, stearoyl-CoA desaturase, and are conserved in alkane hydroxylase and xylene monooxygenase, Biochemistry 33, 12787-12794 (1994).
100. P. Tvrdik, R. Westerberg, S. Silve, et al., Role of a new mammalian gene family in the biosynthesis of very long chain fatty acids and sphingolipids, J. Cell. Biol. 149, 707-717 (2000).
101. M. R. Jackson, T. Nilsson, and P. A. Peterson, Identification of a consensus motif for retention of transmembrane proteins in the endoplasmic reticulum, EMBO J. 9, 3153-3162 (1990).
102. M. R. Jackson, T. Nilsson, and P. A. Peterson, Retrieval of transmembrane proteins to the endoplasmic reticulum, J. Cell. Biol. 121, 317-333 (1993).
103. D. L. Cinti, L. Cook, M. N. Nagi, and S. K. Suneja, The fatty acid chain elongation system of mammalian endoplasmic reticulum, Prog. Lipid Res. 31, 1-51 (1992).

CIRCADIAN SIGNALING IN THE RETINA
The role of melatonin in photoreceptor degeneration

Allan F. Wiechmann[*]

1. INTRODUCTION

The neurohormone, melatonin, is a primary output signal of the circadian clock in the retina and pineal gland. Melatonin enhances the susceptibility of rat photoreceptors to light-induced cell death. It also appears to enhance the sensitivity of the retina and central visual system to light. Thus melatonin, synthesized at night, may act as a paracrine hormone, and bind to receptors in the retina and brain to increase the sensitivity of the visual system to facilitate dark adaptation and other diurnal events that occur in the retina. Because one function of melatonin may be to increase the sensitivity of the retina to light as part of a dark-adaptation mechanism, an undesirable consequence of this may be an increased sensitivity by the photoreceptors to the damaging effects of light.

2. PINEAL-RETINA RELATIONSHIPS

Melatonin is synthesized by pinealocytes and retinal photoreceptors. Both the pineal gland and retina likely evolved from primitive photoreceptive organs that produced melatonin (Wiechmann, 1986). It is theorized that the two lateral eyes became more specialized for phototransduction, whereas the third eye became specialized for secretion of melatonin into the circulation. However, the lateral eyes retained the capacity for melatonin production, to serve as a paracrine signal to regulate diurnal events in the eye and sensitivity to light.

3. MELATONIN BIOSYNTHESIS IN THE RETINA

Both the pineal gland and retina synthesize the indolamine hormone, melatonin. Melatonin (N-acetyl-5-methoxytryptophan) is a lipophilic molecule and is synthesized

[*]Allan F. Wiechmann, University of Oklahoma Health Sciences Center, Oklahoma City, Oklahoma, 73190

from tryptophan in a series of four enzymatic steps. A cyclic rhythm of melatonin has been measured in the retina of several species (Cahill and Besharse, 1992; Pang et al., 1980; Wiechmann et al., 1986), with peak levels occurring during the dark period. There is compelling evidence for the identification of the photoreceptors as the sites of retinal melatonin synthesis (Guerlotte et al., 1996; Wiechmann, 1996; Cahill and Besharse, 1992; Niki et al., 1998; Zwaliska and Iuvone, 1992; Green at al., 1995).

4. ROLE OF RETINAL MELATONIN

The cyclic changes that occur in the rate of retinal melatonin synthesis suggest that melatonin may be involved in other diurnal events in the retina. Studies have suggested that melatonin is involved in photoreceptor outer segment disc shedding and phagocytosis (Besharse and Dunis, 1983; Ogino et al., 1983; White and Fisher, 1989), and photomechanical movements (Pierce and Besharse, 1985). In amphibians, diurnal movements of the RPE pigment granules and photoreceptors may protect the photoreceptors from light damage during the day, and optimize their sensitivity at night.

4.1. Melatonin-Dopamine Antagonism

Some functions of melatonin in the retina may be mediated indirectly through antagonism of dopamine. Dopamine blocks the melatonin-induced cone elongation in amphibians (Pierce and Besharse, 1985), and melatonin inhibits the calcium-dependent release of dopamine from amacrine cells (Dubocovich, 1983). Furthermore, the high levels of retinal dopamine that occur in the light inhibit the activity of NAT (Iuvone and Besharse, 1986). These studies suggest that dopamine and melatonin serve as chemical messengers of day and night, respectively, and exert some of their influences by a mutual antagonism. Information may be exchanged between the melatonin-synthesizing photoreceptors and the dopaminergic amacrine cells via paracrine signaling. Since melatonin appears to be rapidly metabolized within the *Xenopus* retina, the actions of retinal melatonin are likely restricted to the retina (Cahill and Besharse, 1989).

4.2. Melatonin and Light-Dark Adaptation

Melatonin may play an important role in dark adaptation. Melatonin appears to enhance the sensitivity of the central visual system to light (Reuss and Kiefer, 1989; Semm and Vollrath, 1982), and we have previously shown that melatonin increases the sensitivity of horizontal cells to light in the salamander retina (Wiechmann et al., 1988). This is supported by the report that horizontal cells in the rat retina appear to express melatonin receptor RNA and protein (Fujieda et al., 1999). Melatonin, synthesized at night, may bind to receptors in the retina and brain to increase the sensitivity of the visual system when photic input is low, and thus facilitate dark adaptation.

The presence of D_1 receptors (positively-coupled to cyclic AMP synthesis) on horizontal cells (Krizaj and Witkovsky, 1993) and the dopamine-induced uncoupling of horizontal cell gap junctions (Lasater et al., 1984) offers one possible explanation on the role of melatonin in light sensitivity. Melatonin released by photoreceptors at night may bind to melatonin receptors on amacrine cells to inhibit dopamine release. Alternatively, melatonin may bind to receptors on GABA-ergic neurons in the INL, stimulating these cells to inhibit dopamine release from nearby amacrine cells (Boatright et al., 1994). Resultant lower dopamine levels may then cause an increase in receptive field size due to horizontal cell coupling. This would result in lower visual acuity, but would potentially increase the sensitivity of the retina to light during the dark period, since more second-order neurons would respond to a light stimulus. The hyperpolarization of horizontal cells (and enhanced dark adaptation) that occurs in response to a reduction in endogenous retinal dopamine levels supports this hypothesis (Dowling, 1991; Witkovsky et al., 1988).

In combination with this pathway, melatonin could bind to receptors on horizontal cells to directly increase horizontal cell coupling. This hypothesis is supported by a recent report that melatonin receptor RNA and protein appears to be expressed in horizontal cells of the rat retina (Fujieda et al., 1999). Furthermore, melatonin may postsynaptically regulate horizontal cell activity by inhibiting the cAMP response to D_1 receptor activation in chick retina (Iuvone and Gan, 1995). Melatonin may therefore modulate dopaminergic transmission by a combination of reducing dopamine release from amacrine cells, and inhibiting postsynaptic responses to D_1 receptor activation on horizontal cells. These observations, combined with our report that melatonin increases horizontal cell sensitivity to light in salamander retina (Wiechmann et al., 1988), support a role for a direct action of melatonin on horizontal cells.

4.3. Melatonin and Light-Induced Photoreceptor Death

We have previously shown that an injection of melatonin immediately prior to continuous light exposure increases the degree of light-induced photoreceptor cell death in albino rats (Wiechmann and O'Steen, 1992). This work has been recently confirmed and extended (Sugawara et al., 1998), in which it was shown that a melatonin receptor antagonist, luzindole, protects photoreceptors from light-induced damage, thus demonstrating that the effect of melatonin is mediated through a retinal melatonin receptor. Because one function of melatonin may be to increase the sensitivity of the retina to light as part of a dark-adaptation mechanism, an undesirable consequence of this may be an increased sensitivity to the damaging effects of light.

There is accumulating evidence that supports a direct action of melatonin on retinal photoreceptors. Melatonin induces membrane conductance changes in isolated frog rod photoreceptors (Cosci et al., 1997), and we have reported that melatonin appears to bind with low affinity to structures in the outer plexiform layer (OPL), perhaps on photoreceptor terminals or horizontal cell processes in frog retina (Wiechmann et al., 1986). Since photoreceptors are the sites of melatonin synthesis, it is reasonable to expect that melatonin receptors located on these cells would have a lower affinity for melatonin, since the concentration of melatonin directly adjacent to these cells is likely to be quite high. Furthermore, we have observed Mel_{1b} and Mel_{1c} melatonin receptor RNA expression in *Xenopus* photoreceptor inner segments (Wiechmann and Smith, 2001). Conversely, the human mt_1 (Mel_{1a}) receptor has been detected in the inner nuclear layer (INL) and presumptive horizontal cells of the rat retina, but not in photoreceptors (Fujieda et al., 1999). However, it has been clearly demonstrated that the human Mel_{1a} receptor is very poorly expressed in human retina, whereas the Mel_{1b} receptor is highly expressed in human retina (Reppert et al., 1995). Perhaps the Mel$_{1b}$ receptor is expressed in photoreceptors of the human retina, while the Mel$_{1a}$ receptor is expressed exclusively in neurons of the inner retina. The possibility of melatonin receptor expression in photoreceptor cells may be worthy of further study.

The Equivalent Light Hypothesis suggests that many forms of blindness, such as retinitis pigmentosa and macular degeneration, are due to chronic rod activation from constitutively active mutant phototransduction proteins (Fain and Lisman, 1993; Lisman and Fain, 1995). The mutated protein(s) would produce a constant 'equivalent light' that triggers photoreceptor degeneration. This hypothesis further predicts that continuous real or equivalent light may produce photoreceptor degeneration by interfering with circadian processes, such as protein synthesis and outer segment disc shedding and lead to the loss of photoreceptors including those not expressing the mutant gene. Melatonin, a primary output of the circadian clock may therefore play a formerly unsuspected major role in regulating the susceptibility of photoreceptors to inherited and environmentally-induced degeneration.

5. LOCALIZATION OF MELATONIN RECEPTOR BINDING

The distribution of ^{125}I-melatonin binding sites in the mammalian brain is primarily restricted to the hypothalamus and pituitary, compared to the wider distribution in non-mammals. ^{125}I-Melatonin binding in non-mammals is associated primarily with visual and other sensory processing structures (Mazzucchelli et al., 1996; Rivkees at al., 1989; Weaver at 1., 1989: Wiechmann and Wirsig-Wiechmann, 1993). This dichotomy in distribution of melatonin receptors supports the concept of a phylogenetically early role for melatonin in modulation of visual function.

In the retina, ^{125}I-melatonin binding occurs in the inner plexiform layer (IPL) of many species (Blazynski and Dubocovich, 1991; Laitinen and Saavedra, 1990; Wiechmann and Wirsig-Wiechmann, 1991). Since melatonin inhibits dopamine release from the retina, and high-affinity melatonin binding occurs in the IPL of the retina, a candidate for the site of action of melatonin is the dopaminergic amacrine cell. Another strong candidate cell for melatonin receptor expression is the GABA-ergic neuron in the INL. GABA$_A$ receptor antagonists block melatonin-induced suppression of dopamine release (Boatright et al., 1994), suggesting that the effect of melatonin on dopamine release may not be by direct action on the dopaminergic cells, but mediated through GABA. Ganglion cells also appear to express some melatonin receptor subtype RNA expression, although receptor protein does not appear to be present, and melatonin does not appear to bind to these cells in the retina (Blazynski and Dubocovich, 1991; Fujieda et al., 1999). Melatonin receptors produced by these ganglion cells may perhaps be transported through their axons in the optic nerve to their presynaptic terminals located in the brain. The observation that the human Mel$_{1a}$ receptor is expressed in presumptive horizontal cells in the rat retina makes these cells somewhat unexpected candidates for direct target sites of melatonin action (Fujieda et al., 1999). However, our report that melatonin increases the sensitivity of horizontal cells to light in salamander retina (Wiechmann et al., 1988), supports this hypothesis.

In an early study by us, we used ^{3}H-melatonin to identify low-affinity melatonin binding elements in the retinal pigment epithelium (RPE) and outer plexiform layer (OPL) of the *Rana pipiens* retina (Wiechmann et al., 1986). This supports the observation that melatonin directly affects photoreceptor voltage responses to light (Cosci et al., 1997), and supports our recent evidence of melatonin receptor RNA expression in *Xenopus* photoreceptors (Wiechmann and Smith, 2001). It is becoming increasingly apparent that melatonin is likely to influence retinal physiology by acting on multiple cell types, including RPE, ganglion, amacrine, horizontal, and photoreceptor cells.

6. STRUCTURE AND FUNCTION OF MELATONIN RECEPTORS

The relatively recent cloning of the melatonin receptor represents an enormous advancement towards the elucidation of the function of melatonin. The melatonin receptor cDNA was first cloned from cultured *Xenopus* melanophores (Ebisawa et al., 1994), and then from mammalian brain (Reppert et al., 1994, 1995a,b). As predicted from earlier studies, the cDNAs encode a seven-pass transmembrane G-protein coupled receptor. Also, several subtypes of the melatonin receptor have been identified in several species (Reppert et al., 1995a,b).

6.1. Melatonin Receptor Signaling Mechanisms

Since melatonin inhibits cyclic AMP accumulation in most tissues (Godson et al., 1997; Iuvone and Gan, 1994), the G protein coupled to the melatonin receptor is thought to be an inhibitory (G_i) G protein. In cultures of chicken retinal neurons, melatonin inhibits forskolin-induced cAMP accumulation (Iuvone and Gan, 1994). Cultures of human and rat RPE cells may express melatonin receptors, since melatonin inhibits forskolin-stimulated cyclic AMP synthesis in these cells (Nash and Osborne, 1995). RNA encoding all three melatonin receptor subtypes has been identified in frog RPE (Wiechmann et al., 1999, Wiechmann and Smith, 2001).

Melatonin exerts opposite effects on $GABA_A$ receptor-mediated currents in different brain regions through activation of different receptor subtypes (Wan et al., 1999). In the rat suprachiasmatic nucleus, melatonin potentiates $GABA_A$ receptor-mediated current via the Mel_{1a} receptor subtypes, but inhibits the $GABA_A$ current in the hippocampus via the Mel_{1b} receptor. Also, since differences in $GABA_A$ receptor responses occur within the same transfected cells, via two different melatonin receptor subtypes (Mel_{1a} and Mel_{1b}), different second-messenger systems must be involved (Wan et al., 1999). Therefore, although most studies have linked melatonin receptor activation to inhibition of adenylate cyclase, different cell types that express melatonin receptors may potentially utilize different intracellular signaling mechanisms.

All three melatonin receptor subtypes (Mel_{1a}, Mel_{1b}, and Mel_{1c}) have been shown to reduce cAMP accumulation via inhibition of adenylate cyclase activity (Godson et al., 1997; Iuvone and Gan, 1994). However, cells that express Mel_{1b} or Mel_{1c} receptors appear to also inhibit cGMP synthesis (Jockers et al., 1997, Petit et al., 1999). Therefore, cells that express only Mel_{1a} receptors may inhibit cAMP synthesis only, whereas cells that also express Mel_{1b} or Mel_{1c} receptors may inhibit both cAMP and cGMP synthesis (Jockers et al., 1997, Petit et al., 1999). This suggests that inhibition of cAMP accumulation may be a general feature of melatonin receptor signaling, but that this may be complemented by modulation of cGMP synthesis, depending on the expression of additional receptor subtypes (Petit et al., 1999). Furthermore, the Mel_{1a} melatonin receptor appears to be coupled to several different G proteins (Brydon et al., 1999). In addition to inhibiting adenylate cyclase activity, Mel_{1a} receptor stimulation also appears to potentiate phospholipase activation (Godson and Reppert, 1997). It has also been shown that in some tissues, stimulation of the Mel_{1a} receptor does not result in inhibition of cAMP synthesis (Conway et al, 1997). These studies suggest that melatonin activation of its receptor may stimulate multiple signaling pathways.

6.2. Localization of Melatonin Receptor Expression

In situ hybridization studies of the melatonin receptor on the brain and retina have revealed differential expression patterns of the receptor subtypes, which may provide clues to their distinct functions (Reppert et al, 1994, 1995b). In the chicken retina, a band of receptor RNA hybridization has been observed in the ganglion cell layer (GCL) and inner nuclear layer (INL; Reppert et al, 1995b), and we have observed Mel_{1b} and Mel_{1c} receptor RNA hybridization over the INL, GCL, and photoreceptors of the frog retina (Wiechmann and Smith, 2001), The pattern of hybridization is more suggestive of a GABA-ergic amacrine and/or horizontal cell localization in the INL, rather than just dopaminergic cells. Recently, mt_1 (Mel_{1a}) receptor immunoreactivity has been localized to dopaminergic and GABA-ergic amacrine cells of the guinea pig retina (Fujieda et al., 2000). These results provide strong support for the hypothesis that melatonin has direct actions on both dopaminergic and GABA-ergic neurons of the inner retina.

7. MELATONIN AS AN ANTIOXIDANT

An interesting paradox to the enhanced susceptibility to photoreceptor cell death exerted by melatonin is the potential protective effect that melatonin may have in the retina because of its antioxidant properties. Melatonin may act as an antioxidant in the retina to protect photoreceptor outer segment membranes from free radicals generated by light (Marchiafava and Longoni, 1999; Siu et al., 1998). At physiological concentrations in living rod photoreceptors, melatonin is 100 times more potent than vitamin E in inhibiting light-induced oxidative processes (Marchiafava and Longoni, 1999). The enhancement of light-induced photoreceptor cell death in response to injected melatonin appears to be receptor-mediated (Wiechmann and O'Steen, 1992; Sugawara et al., 1998), whereas the antioxidative protection is probably not a receptor-mediated function.

Some theories regarding the pathology of age-related macular degeneration (ARMD) have suggested that cumulative effects of solar radiation, drugs, and other chemicals that generate free radicals may contribute to the progression of this disease (Christen, 1994; Gerster, 1991). Since melatonin is produced by the retinal photoreceptors at nighttime, it could potentially have an antioxidative protective function at night, but a receptor-mediated detrimental effect in the light.

8. CONCLUSIONS

There is an emergent realization that melatonin has a fundamentally important role in the circadian entrainment of the neuroendocrine system. Melatonin entrains the biological clock, and regulates seasonal reproduction (Reiter, 1980). Melatonin makes retinal photoreceptors more susceptible to light-induced cell death, and there is a huge and controversial body of literature suggesting that melatonin is an immunoenhancing compound that can aid in the treatment of some cancers, and help defend against immunodeficiency diseases (Fraschini et al., 1998). Further study of melatonin certainly is needed to understand its myriad functions.

Studies suggest that inappropriate (ie., daytime) exposure of retinal cells to melatonin may be detrimental to photoreceptor survival. The photosensitizing properties of melatonin may potentially contribute to increased sensitivity to the damaging effects of light on photoreceptors. For example, a case report described a woman with toxic optic neuropathy that had taken Zoloft (a serotonin re-uptake inhibitor; SSRI) and melatonin supplementation (Lehman and Johnson, 1999). After discontinuing the melatonin supplementation, the patient's visual acuity and color vision returned to normal. Since SSRI's increase serotonin levels at neuronal synapses, and serotonin is a precursor of melatonin, the combined effects of Zoloft and melatonin supplementation may have produced a melatonin/dopamine imbalance, resulting in the toxic optic neuropathy.

Understanding the molecular mechanisms that convey the actions of melatonin may help to identify defects of the melatonin pathway in some diseases of vision, and to assess whether alterations in normal melatonin levels may contribute to a disruption of normal retinal function. For example, ARMD is likely the result of cumulative effects of genetic, environmental and chemical factors. Chronic exposure to melatonin at inappropriate times of day and lighting conditions may increase the risk of susceptibility to this debilitating disease.

REFERENCES

Besharse, J.C., and Dunis, D.A., 1983, Methoxyindoles and photoreceptor metabolism: activation of rod shedding, *Science* **219**:1341-1342.

Blazynski, C., and Dubocovich, M.L., 1991, Localization of 2-[^{125}I]iodomelatonin binding sites in mammalian retina, *J. Neurochem.* **56**:1873-1880.

Boatright, J.H., Rubim, N.M., and Iuvone, P.M., 1994, Regulation of endogenous dopamine release in amphibian retina by melatonin: The role of GABA, *Vis. Neurosci.* **11**:1013-1018.

Brydon, L., Roka, F., Petit, L, de Coppet. P., Tissot, M., Barrett, P., Morgan., P.J., Nanoff, C., Strosberg, D.A., and Jockers, R., 1999, Dual signaling of human Mel1a melatonin receptors via G_{i2}, G_{i3}, and $G_{q/11}$ proteins, *Mol. Endocrinol.* **13**:2025-2038.

Cahill, G.M., and Besharse, J.C., 1989, Retinal melatonin is metabolized within the eye of *Xenopus laevis, Proc. Natl. Acad. Sci. U.S.A.* **86**:1098-1102.

Cahill, G.M., and Besharse, J.C., 1992, Light-sensitive melatonin synthesis by *Xenopus* photoreceptors after destruction of the inner retina, *Vis. Neurosci.* **8**:487-490.

Christen, W.G., 1994, Antioxidants and eye disease, *Am. J. Med.* **97**:14S-17S.

Conway, S., Drew, J.E., Canning, S.J., Barrett, P., Jockers, R., Strosberg, A.D., Guardiola-Lemaitre, B., Delagrange, P., and Morgan, P.J., 1997, Identification of mel$_{1a}$ melatonin receptors in the human embryonic kidney cell line HEK293: evidence of G protein-coupled melatonin receptors which do not mediate the inhibition of stimulated cyclic AMP levels, *FEBS Lett.* **407**:121-126.

Cosci, B., Longoni, B., and Marchiafava, P.L., 1997, Melatonin induces membrane conductance changes in isolated retinal rod receptor cells, *Life Sci.* **60**:1885-1889.

Dowling, J.E., 1991, Retinal neuromodulation: the role of dopamine, *Vis. Neurosci.* **7**:87-97.

Dubocovich, M.L., 1983, Melatonin is a potent modulator of dopamine release in the retina, *Nature* **306**:782-784.

Ebisawa, T., Karne, S., Lerner, M.R., and Reppert, S.M., 1994, Expression cloning of a high-affinity melatonin receptor from *Xenopus* dermal melanophores, *Proc. Natl. Acad. Sci. U.S.A.* **91**:6133-6137.

Fain, G.L., and Lisman, J.E., Photoreceptor degeneration in vitamin A deprivation and retinitis pigmentosa: the equivalent light hypothesis, *Exp. Eye Res.* **57**:335-340.

Fraschini, F., Demartini, G., Esposti, D, and Scaglione, F., 1998, Melatonin involvement in immunity and cancer, *Biol. Signals Recept.* **7**:61-72.

Fujieda, H., Hamadanizadeh, S.A., Wankiewicz, E., Pang, S.F., and Brown, G.M., 1999, Expression of mt_1 melatonin receptor in rat retina: Evidence for multiple cell targets for melatonin, *Neuroscience* **93**:793-799.

Fujieda, H., Scher, J., Hamadanizadeh, S.A., Wankiewicz, E., Pang., S.F., and Brown, G.M., 2000, Dopaminergic and GABAergic amacrine cells are direct targets of melatonin: Immunocytochemical study of mt_1 melatonin receptor in guinea pig retina, *Vis. Neurosci.* **17**:63-70.

Gerster, H., 1991, Review: antioxidant protection of the ageing macula, *Age and Ageing,* **20**:60-69.

Godson, C., and Reppert, S.M., 1997, The Mel_{1a} melatonin receptor is coupled to parallel signal transduction pathways, *Endocrinology* **138**:397-404.

Green, C.B., Cahill, G.M., and Besharse, J.C., 1995, Tryptophan hydroxylase is expressed by photoreceptors in *Xenopus laevis* retina, *Vis. Neurosci.* **12**:663-670.

Guerlotte, J., Greve, P., Bernard, M., Grechez-Cassiau, A., Morin, F., Collin, J.-P., and Voisin, P., 1996, Hydroxyindole-O-methyltransferase in the chicken retina: Immunocytochemical localization and daily rhythm of mRNA, *Eur. J. Neurosci.* **8**:710-715.

Iuvone, P.M., and Besharse, J.C., 1986, Dopamine receptor-mediated inhibition of serotonin N-acetyltransferase activity in retina, *Brain Res.* **369**:168-176.

Iuvone, P.M., and Gan, J., 1994, Melatonin receptor-mediated inhibition of cyclic AMP accumulation in chick retinal cultures, *J. Neurochem.* **63**:118-124.

Iuvone, P.M., and Gan, J., 1995, Functional interaction of melatonin receptors and D1 dopamine receptors in cultured chick retinal neurons, *J. Neurosci.* **15**:2179-2185.

Jockers, R., Petit, L., Lacroix, I., de Coppet, P., Barret, P., Morgan, P.J., Guardiola, B., Delagrange, P., Marullo, S., and Strosberg, A.D., 1997, Novel isoforms of Mel1c melatonin receptor modulating intracellular cyclic guanosine 3'-5'-monophosphate levels, *Mol. Endocrinol.* **11**:1070-1081.

Krizaj, D., and Witkovsky, P., 1993, Effects of submicromolar concentrations of dopamine on photoreceptor to horizontal cell communication, *Brain Res.* **627**:122-128.

Laitinen, J.T., and Saavedra, J.M., 1990, The chick retinal melatonin receptor revisited: Localization and modulation of agonist binding with guanine nucleotides, *Brain Res.* **528**:349-352.

Lasater, E.M., Dowling, J.E., and Ripps, H., 1984, Pharmacological properties of isolated horizontal and bipolar cells from the skate retina, *J. Neurosci.* **4**:1966-1975.

Lehman, N.L., Johnson, L.N., 1999, Toxic optic neuropathy after concomitant use of melatonin, Zoloft, and a high-protein diet, *J. Neuro-Opthalmol.* **19**:232-234.

Lisman, J., and Fain, G., 1995, Support for the equivalent light hypothesis for RP, *Nature Med.* **1**:1254-1255.

Marchiafava, P.L., and Longoni, B., Melatonin as an antioxidant in retinal photoreceptors, *J. Pineal Res.* **26**:184-189.

Mazzucchelli, C., Capsoni, S., Angeloni, D., Fraschini, F., and Stankov, B., 1996, Expression of the melatonin receptor in *Xenopus laevis:* A comparative study between protein and mRNA distribution, *J. Pineal Res.* **20**:57-64.

Nash, M.S., and Osborne, N.N., 1995, Pertussis toxin-sensitive melatonin receptors negatively coupled to adenylate cyclase associated with cultured human and rat retinal pigment epithelial cells, *Invest. Ophthalmol. Vis. Sci.* **36**:95-102.

Niki, T., Hamanda, T., Ohtomi, M., Sakamoto, K., Suzuki, S., Kako, K., Hosoya, Y., Horikawa, K., and Ishida, N., 1998, The localization of the site of arylalklamine N-acetyltransferase circadian expression in the photoreceptor cells of mammalian retina, *Biochem. Biophys. Res. Commun.* **248**:115-120.

Ogino, N., Matsumura, M., Shirakawa, H., and Tsukahara, I., 1983, Phagocytic activity of cultured retinal pigment epithelial cell from chick embryo: inhibition by melatonin and cyclic AMP, and its reversal by taurine and cyclic GMP, *Ophthalmic Res.* **15**:72-89.

Pang, S.F., Yu, H.S., Suen, H.C., and Brown, G.M., 1980, Melatonin in the retina of rats: a diurnal rhythm, *J. Endocrinol.* **87**:89-93.

Petit, L., Lacroix, I., de Coppet, P., Strosberg, A.D., and Jockers, R., 1999, Differential signaling of human Mel1a and Mel1b melatonin receptors through the cyclic guanosine 3'-5'-monophosphate pathway, *Biochem. Pharmacol.* **58**:633-639.

Pierce, M.E., and Besharse, J.C., 1985, Circadian regulation of retinomotor movements. I. Interaction of melatonin and dopamine in the control of cone length, *J. Gen. Physiol.* **86**:671-689.

Reiter, R.J., 1980, The pineal gland and its hormone in the control of reproduction in mammals, *Endocr. Rev.* **1**:109-131.

Reppert, S.M., Weaver, D.R., and Ebisawa, T., 1994, Cloning and characterization of a mammalian melatonin receptor that mediates reproductive and circadian responses, *Neuron* **13**:1177-1185.

Reppert, S.M., Godson, C., Mahle, C.D., Weaver, D.R., Slaugenhaupt, S.A., and Gusella, J.F., 1995, Molecular characterization of a second melatonin receptor expressed in human retina and brain: The Mel$_{1b}$ melatonin receptor, *Proc. Natl. Acad. Sci. U.S.A.* **92**:8734-8738.

Reppert, S.M., Weaver, D.R., Cassone, V.M., Godson, C., and Kolakowski, L.F., 1995, Melatonin receptors are for the birds: Molecular analysis of two receptor subtypes differentially expressed in chick brain, *Neuron* **15**:1003-1015.

Reuss, S., and Kiefer, W., 1989, Melatonin administered systematically alters the properties of visual cortex cells in cat: further evidence for a role in visual information processing, *Vision Res.* **29**:1089-1093.

Rivkees, S.A., Cassone, V.M., Weaver, D.R., and Reppert, S.M., 1989, Melatonin receptors in chick brain: Characterization and localization, *Endocrinology* **125**:363-368.

Semm, P., and Vollrath, L., 1982, Alterations in the spontaneous activity of cells in the guinea pig pineal gland and visual system produced by pineal indoles, *J. Neural Trans.* **53**:265-275.

Sui, A.W., Reiter, R.J., and To, C.H., 1998, Pineal indolamines and vitamin E reduce nitric oxide-induced lipid peroxidation in rat retinal homogenates, *J. Pineal Res.* **24**:239-244.

Sugawara, T., Sieving, P.A., Iuvone, P.M., and Bush, R.A., 1998, The melatonin antagonist luzindole protects retinal photoreceptors from light damage in the rat, *Invest. Ophthalmol. Vis. Sci.* **39**:2458-2465.

Wan, Q., Man, H.-Y., Liu, F., Braunton, J., Niznik, H.B., Pang, S.F., Brown, G.M., and Wang, Y.T., 1999, Differential modulation of GABA$_A$ receptor function by Mel$_{1a}$ and Mel$_{1b}$ receptors, *Nature Neurosci.* **2**:401-403.

Weaver, D.R., Rivkees, S.A., and Reppert, S.M., 1989, Localization and characterization of melatonin receptors in rodent brain by in vitro autoradiography, *J. Neurosci.* **9**:2581-2590.

White, M.P., and Fisher, L.J., 1989, Effects of exogenous melatonin on circadian disc shedding in the albino rat retina, *Vision Res.* **29**:167-179.

Wiechmann, A.F., 1986, Melatonin: Parallels in pineal gland and retina, *Exp. Eye Res.* **42**:507-527.

Wiechmann, A.F., 1996, Hydroxyindole-*O*-methyltransferase is expressed in a subpopulation of photoreceptors in the chicken retina, *J. Pineal Res.* **20**:217-225.

Wiechmann, A.F., and O'Steen, W.K., 1992, Melatonin increases photoreceptor susceptibility to light-induced damage, *Invest. Opthalmol. Vis. Sci.* **33**:1894-1902.

Wiechmann, A.F., and Smith, A.R., 2001, Melatonin receptor RNA is expressed in photoreceptors and displays a cyclic rhythm in *Xenopus* retina, (submitted).

Wiechmann, A.F., and Wirsig-Wiechmann, C.R., 1991, Localization and quantification of high-affinity melatonin binding sites in *Rana pipiens* retina, *J. Pineal Res.* **10**:174-179.

Wiechmann, A.F., and Wirsig-Wiechmann, C.R., 1993, Distribution of melatonin receptors in the brain of the frog *Rana Pipiens* as revealed by *in vitro* autoradiography, *Neuroscience* **52**:469-480.

Wiechmann, A.F., Bok, D., and Horwitz, J., 1986, Melatonin binding in the frog retina: autoradiographic and biochemical analysis, *Invest. Ophthalmol. Vis. Sci.* **27**:153-163.

Wiechmann, A.F., Campbell, L.D., and Defoe, D.M., 1999, Melatonin receptor RNA expression in *Xenopus* retina, *Mol. Brain Res.* **63**:297-303.

Wiechmann, A.F., Yang, X.-L., Wu, S.M., and Hollyfield, J.G., 1988, Melatonin enhances horizontal cell sensitivity in salamander retina, *Brain Res.* **453**:377-380.

Witkovsky, P., Stone, S., and Besharse, J.C., 1988, Dopamine modifies the balance of rod and cone inputs to horizontal cells of the *Xenopus* retina, *Brain Res.* **449**:332-336.

Zwaliska, J.B., and Iuvone, P.M., 1992, Melatonin synthesis in chicken retina: Effect of kainic acid-induced lesions on the diurnal rhythm and D2 dopamine receptor-mediated regulation of serotonin N-acetyltransferase activity, *Neurosci. Lett.* **135**:71-74.

LOCALIZATION AND FUNCTIONAL ANALYSIS OF BESTROPHIN

George Hoppe, Lihua Y. Marmorstein, and Alan D. Marmorstein*

1. SUMMARY

Best Macular Dystrophy (BMD) is an inherited autosomal dominant disorder that exhibits symptoms and histopathology reminiscent of age-related macular degeneration (AMD) the leading cause of blindness in the western world. The vitelliform macular dystrophy 2 gene (*VMD2*), mutated in BMD, encodes a previously unknown protein named bestrophin. Bestrophin is predicted to be a transmembrane protein with 4 membrane spanning alpha helical domains. On the basis of northern blot and in situ hybridization, bestrophin was predicted to be uniquely expressed in the retinal pigment epithelium (RPE). Using bestrophin specific antibodies, we have confirmed the RPE specific localization of bestrophin in the eye, and determined that bestrophin is a basolateral plasma membrane protein. Here we attempt to combine our existing knowledge on structure, cellular localization, and the clinical consequences of mutations of bestrophin, in order to gain insights into its possible function in retina.

2. INTRODUCTION

Best Macular Dystrophy (BMD) represents one of a number of single gene disorders that exhibit symptoms and histopathologies reminiscent of age-related macular degeneration (AMD) (10, 15) the leading cause of blindness in the western world. AMD affects nearly 30% of those over the age of 75 (13), impacting on the quality of life of those affected by causing a debilitating loss of central vision. BMD is an autosomal dominant disease with a juvenile age of onset (11). Clinically, the disease is characterized by an "egg yolk" or vitelliform lesion in the macula easily visible during a routine fundus examination (6, 10). It is thought that the vitelliform lesion may be due to the abnormal deposition of lipofuscin in the retinal pigment epithelium (25). Over time the vitteliform lesion will break up, and may appear as a "bulls eye."

---

* Cole Eye Institute, Cleveland Clinic – i31, 9500 Euclid Avenue, Cleveland, OH 44195

New Insights Into Retinal Degenerative Diseases
Edited by Anderson *et al.*, Kluwer Academic/Plenum Publishers, 2001

A defining characteristic of BMD, however, is a light peak to dark trough ratio of the electrooculogram (EOG) of less than 1.5, without aberrations in the clinical electroretinogram (6). The depressed light response in the EOG of an individual with a vitelliform lesion in the absence of any abnormality in the a and b waves of the ERG is diagnostic for BMD. Even otherwise asymptomatic carriers of BMD-associated mutations, as assessed by pedigree, will exhibit an altered EOG (3, 14).

Histopathologically BMD is poorly characterized with data available from only a small number of donor eyes. In all cases the disease has been shown to manifest as a generalized RPE abnormality associated with abnormal lipofuscin accumulation, though not to the extent seen in Stargardt's disease (8, 19, 25). An electron-dense, granular, PAS positive, mucopolysacharide negative material has been observed in the inner segments of degenerating photoreceptors. In addition, regions of geographic atrophy of the RPE, drusen, and the deposition of an abnormal fibrillar material beneath the RPE have been observed. Occasional breaks in Bruch's membrane with accompanying neovascularization have also been reported, although BMD is not noted for extensive choroidal neovascularization as is Sorsby's Fundus Dystrophy. Many of these features are also found in AMD (4, 7, 12, 23).

The presence of both drusen-like basolateral depositions beneath RPE and autofluorescent accumulations in RPE is typical for AMD. Thus, BMD bares a close pathophysiological resemblance to AMD. Understanding the cellular and biochemical mechanisms by which mutations in *VMD2* gene cause retinal degeneration in Best disease will shed light on pathophysiological pathways of AMD. Specifically, elucidation of bestrophin function and its alterations due to mutatagenesis will help to find the cause for the formation of drusen and accumulation of lipofuscin.

3.   MATERIALS AND METHODS

3.1. Western Blot

Porcine tissues (retina, cerebrum, cerebellum, and medulla) were first homogenized and then lysed in nonreducing Laemmli sample buffer. Proteins were resolved on 10% SDS-PAGE, and transfered to PVDF. Western blots were performed using monoclonal antibody E6–6, alkaline phosphatase-conjugated secondary antibodies, and tetranitroblue tetrazolium/5-bromo-4-chloro-3-indolyl phosphate as a substrate.

3.2. Immunocytochemistry

Human donor eyes obtained form the Cleveland Eye Bank were opened posterior to the limbus and fixed in 4% paraformaldehyde in PBS containing 1mM $MgCl_2$ and 0.1mM $CaCl_2$. Tissues were rinsed in several changes of PBS and processed for paraffin microscopy using conventional procedures. Sections were deparaffinized with xylene and hydrated through graded ethanols, then subject to antigen retrieval in 0.01M sodium citrate, pH 6.0 under 15 lbs of pressure at 121°C for 1 minute. Tissue sections were incubated with 3% BSA in PBS for 30 min, then incubated with monoclonal antibody E6-6 ascites diluted 1:4000 in PBS or a 1:4000 dilution of E6-6 ascites preincubated with the immunizing peptide at 100ug/ml for 1 hour at room temperature. After washing with PBS, sections were treated with goat-anti-mouse IgG conjugated to peroxidase ABC (avidin-biotin-peroxidase complex, Vector Labs, Burlingame, CA, 1:200 dilution) for 1h. Sections were washed with PBS then incubated in 0.05% 3,3'-diaminobenzidine and 0.03% hydrogen peroxide in the phosphate buffer. Sections were viewed on a Nikon Microphot2 microscope. Images were acquired using a SPOT2 CCD camera and Metamorph software and processed in Adobe Photoshop 5.0.

3.3. Protein Sequence Analysis

Sequence similarity searches were performed using BLAST v 2.1 and Position Specific Iterated (PSI-) BLAST software available at National Center for Bioinformatics (NCBI) Server (http://www.ncbi.nlm.nih.gov/BLAST/). Protein sequence alignment was performed using Clustal W v 1.81 software available at European Bioinformatics Institute (EBI) Server (http://www2.ebi.ac.uk/clustalw). Similarity searches of short amino acid motifs (less then 10 aa) were performed using GenCore v 5 software (http://www2.ebi.ac.uk/bic_sw/) at EBI Server, which employs the ultra-fast rigorous Smith/Waterman algorithm.

4. *VMD2* GENE, mRNA, AND PREDICTED PROTEIN

In 1998, Petrukhin *et al.* (20) identified the gene mutated in Best disease, *VMD2*, on chromosome 11q13 encoding bestrophin. Although mutations in *VMD2* in individuals with AMD are rare, mutations have been reported in up to 1.5% of individuals with AMD. Based on northern blot analysis, a 2.2 kb transcript of *VMD2* is abundantly yet uniquely expressed in the retina based on northern blotting and in the RPE based on in situ hybridization. Using RT-PCR we were able to detect bestrophin mRNA in several RPE cell lines, i.e, D407, ARPE19, and RPE-J cells, however the protein was not detected in D407 or ARPE-19 cells on western blots.

Bestrophin is a 585 amino acid protein with a predicted mass of 68 kDa and a theoretical isoelectric point of 6.9. Computer assisted structural analysis predicts that bestrophin is a transmembrane protein with 4 membrane spanning alpha helical domains (Fig. 1). There are no obvious targeting signals present in the amino acid sequence that would provide clues to its subcellular localization, and bestrophin has no obvious sites for post-translational modifications such as N-glycosylation.

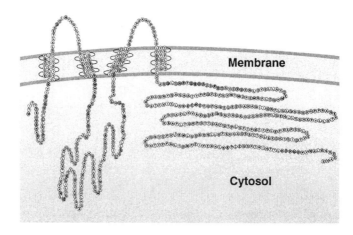

Figure 1. A model of bestrophin structure and membrane orientation. Shaded circles indicate BMD associated mutations.

5. CHARACTERIZATION OF BESTROPHIN

We have produced one polyclonal and several monoclonal antibodies, which were raised by immunizations with the synthetic peptide KDHMDPYWALENRDEAHS, corresponding to the amino acid sequence at the C-terminus of human bestrophin (Fig. 1). Western blot analysis and inmmunohistochemistry demonstrated immunopositive material in the retinas of human, monkey, and pig, but not cow, rat, or mouse. Mouse and rat have very low homology to the human sequence in their C-terminus. BLAST analysis of a pig EST against *VMD2* showed a very high homology between the pig and human C-terminal amino acid sequences (Fig. 2).

```
Human bestrophin (NP_004174.1)          TTLKDHMDPYWALENRDEAHS
Porcine bestrophin (AW480265.1)         AILKDHRDPYWALENRDEAHS
```

Figure 2. Homology between C-terminus of human bestrophin and the predicted amino acid sequence of a porcine homologue of bestrophin identified through BLAST analysis of publicly available EST databases. Identical amino acids are shaded.

Western blot analysis of bestrophin distribution among ocular tissues revealed its abundant and exclusive RPE localization confirming *in situ* hybridization data. Due to limited availability of samples of human tissues an extensive tissue distribution analysis of bestrophin is ongoing. Petrukhin et al. (20), have shown expression of bestrophin mRNA in brain by RT-PCR. Our preliminary experiments have failed to detect immunoreactive protein in the lysates of pig brain (Fig. 3), however, the possibility remains that expression of this protein might be limited to a specific region(s) of brain (e.g., choroid plexus, hypothalamus, hippocampus, etc.).

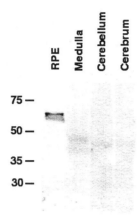

Figure 3. Expression of bestrophin in porcine brain and RPE. Lysates of pig RPE, cerebrum, cerebellum, and medulla were subjected to SDS-PAGE separation and Western blot analysis using monoclonal antibody E6-6. Note that 100 ug protein/lane of brain lysates were loaded compared to 10 ug protein/lane for RPE lysate.

Anti-bestrophin antibodies were used to probe the cellular and subcellular localization of bestrophin in the eye. Our data indicate that bestrophin is a basolateral plasma membrane protein *in situ* (Fig. 4).

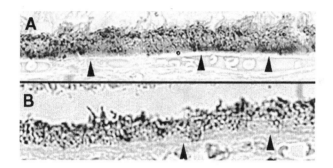

Figure 4. Immunohistochemical localization of bestrophin in human retina. Sections of human eye from a 55 year old male donor were stained for bestrophin with monoclonal antibody E6-6 (A), or to control for specificity with monoclonal antibody E6-6 preincubated with the peptide antigen (B). Bestrophin is concentrated along the basal surface of the cells. Bruch's membrane is indicated by the arrows as a point of reference.

6. POSSIBLE FUNCTION OF BESTROPHIN

The amino acid sequence of bestrophin offers few clues to its possible physiological role in the RPE, i.e., it contains no known motifs, nor has it any homology to proteins with known functions. However, several theoretical considerations based on the analysis of genomic localization of the gene, the amino acid sequence of bestrophin, as well as the characteristic pathophysiological features of Best disease, might prove useful in identifying the function of bestrophin.

6.1. Functional Implications Based On Genomic Structure

The 3-prime untranslated region (UTR) of *VMD2* contains a region of antisense complementarity to the 3' UTR of the ferritin heavy-chain (FHC) gene, *FTH1*. Petrukhin and colleagues (20) have proposed that the overlap of bestrophin's 3' UTR with the 3' UTR of the FHC gene could exert an anti-sense effect on FHC resulting in the downregulation of FHC gene expression in BMD. Deficiency in the iron storage protein, ferritin, could in turn lead to free iron overload, lipid peroxidation and subsequently to protein damage and enhanced lipofuscin formation (1). Preliminary data however does not support this hypothesis.

```
                    *    **   *     *    *   ** *   *                    *
human        MTITYTSQVANAR-LGSFSRLLLCWRGSIYKLLYGEFLIFLLCYYIIRFI  49
Drosophila   MTITYTGEVATCRGFGCFLKLLLRWRGSIYKLVWLDLLAFLTIYYAINMV  50
C. elegans   MTVTYSLDVASSS-FFCLYKLLFRWKGSIWKSVWAELVVWLCLYAVLSVI  49

                     *                    *  *   *      *** *   *  *
human        YRLALTEEQQLMFEKLTLYCDSYIQLIPISFVLGFYVTLVVTRWWNQYEN  99
Drosophila   YRFGLNPAQKETFEAIVQYCDSYRELIPLSFVLGFYVSIVMTRWWNQYTS 100
C. elegans   YRCLLTMKQRATFEDLCIFFDTYSNFIPITFMLGFYVSAVFTRWWQIFDN  99

             ** *   *                              * *    ** *     *
human        LPWPDRLMSLVSGFVEGKDEQGRLLRRTLIRYANLGNVLILRSVSTAVYK 149
Drosophila   IPWPDPIAVFVSSNVHGQDERGRMMRRTIMRYVCLCLTMVLANVSPRVKK 150
C. elegans   IGWIDTPCLWITQYIKGETERAKCVRRNCIRYSILTQAMVYRDVAASVRK 149

human        RFPSAQHLVQAGFMTPAEHKQLEKLSLPHN---MFWVPWVWFANLSMKAW 196
Drosophila   RFPGLNNLVEAGLLNDNEKTIIETMNKAFPRPSKHWLPIVWAASIITRAR 200
C. elegans   RFPTFNHLVTAGLMTEKEMAEFESIPSPHA---KYWQPMHWLFSMITLAR 196

                 *        *  *          *  ** *    *     *  *        *
human        LGGRIRDPILLQSLLNEMNTLRTQCGHLYAYDWISIPLVYTQVVTAVYS 246
Drosophila   KEGRIRDDFAVKTIIDELNKFRGQCGLLISYDTISVPLVYTQVVTLAVYS 250
C. elegans   DEGMISSDIIYVDLMEKMRQFRVNILSLTLFDWVPVPLVYTQVVHLAVRS 246

                                             *       *             *
human        FFLTCLVGRQFLN---PAKAYPGHELDLVVPVFTFLQFFFYVGWLKVAEQ 293
Drosophila   YFLTCCMGQQWTDGKVVGNTTYLNKVDLYFPVFTTLQFFFYMGWLKVAES 300
C. elegans   YFLIALFGRQYLHPESNRLNDFKQTIDLYVPIMSLLQFIFFIGWMKVAEV 296

                ***  **** *   ***   ***
human        LINPFGEDDDDFETNWIVDRNLQVSLLAVDEMHQDLPRMEPDMYWNKPEP 343
Drosophila   LINPFGEDDDDFEVNWMVDRNLQVSYLIVDEMHHDHPELLKDQYWDEVFP 350
C. elegans   LLNPLGEDDDDFECNWILDRNLQVGLMVVDTAYNRYPTLEKDQFWEDAIA 346

human        -QPPYTAASAQFRRASFMGSTFNISLNK-------EEMEFQPNQEDEE- 383
Drosophila   NELPYTIAAERFRENHPEPSTAKIEVPKNAAMPSTMSSVRIDEMADDASG 400
C. elegans   -EPLYTAESAMRPLNPQVGSCADMPTEE-------EPFMVRPRRRTLSR 387

human        --------------------------DAHAG----IIGRFLG------ 395
Drosophila   IHFSAGNGKMRLDSSPSLVSVSGTLSRVNTVAS----ALKRFLSRDDSRP 446
```

```
C. elegans        MSHWDG-------------------DMEDTDVVPVVG---------- 405


human             -----LQSHDHHPPRANSRTKLLWPKRESLLHEGLPKNHKAAKQNVRGQE 440
Drosophila        GSATPSQDQPYKFPASASSASLSGAVVGSATSAGKPAGSLRITQQVIEEV 496
C. elegans        --------LKHTRDNSNYASGESLAFSNSFANGGRKLSEMFRRMRAGSRI 447


human             -----DNKA----------------WKLKAVDAFKSAPLYQRPGYYSAP 468
Drosophila        -----DEQATITSMRANDPRPNVMDIFAQTSSGAGTSGPLQPPPAHSEPV 541
C. elegans        GDRYR-------------------KRNSSAQDFENGMAKKNSIDENA 475


human             QTPLSP----------TPMFFP------------------LEPSAPSKL 489
Drosophila        DIPSRPPSYNRAQSQYEPNLFPPGGVDALLSTSAPAGGSPLLLSNAATAP 591
C. elegans        DIHSNRLDQASGTPKSGRLWSS------------MPQTQLEEMLKNKNF 512


human             HSVTGIDTK--DKSLKTVSSGAKKSFELLS------------------- 517
Drosophila        SSPVGESSKSLYDPQKGASRETVESMDLRSSTDLLGDAAVQPEDEGDDFD 641
C. elegans        NSPVKYNTDGMKDRELQNPTPITDHIDLPLHVAS--------------- 546


human             ------ESDGALMEHPEVSQVRRKTVEFNLTDMPEIPENHLKEPLEQSPT 561
Drosophila        KLKAEREKEKLMRQQKNLARTISTAPGMEATAVPMVPMVPVNVAVQQAQL 691
C. elegans        ------SQSWFNESLPVIKEEEEAKRKSNTDTESPKSSKHSSMSIRRSEL 590


human             NIHTTLKDHMDPYWALENRDEAHS------ 585
Drosophila        QPVASSADLLAGGDQFSNSTMKSEDAINGS 721
C. elegans        RRSSSSGSDLGKSGKRERKKSE-------- 612
```

Figure 5. Similarities between human and Drosophila bestrophin, and a model *C. elegans* RPF protein. Accession numbers: NP_004174.1 - human bestrophin, AAF32327.1 - Dbest [*Drosophila melanogaster*], AAB70976.1 - hypothetical protein C01B12.3 [*Caenorhabditis elegans*]. Conserved amino acids are shaded. Amino acids mutated in Best disease patients are marked with an asterisk. Predicted transmembrane domains are shown in bold font.

6.2. Functional Implications Based On Protein Sequence

Very few functional properties of bestrophin can be deduced from its primary structure. Generally, a tetra-spanning membrane protein with long cytosolic tail generally could be regarded as (i) a transmembrane channel/transporter with ability to be regulated by the cytosolic environment, (ii) a membrane receptor with ability to transduce a signal from extracellular milieu to cytoplasmic compartment (iii) a structural protein.

Bestrophin shares significant homology with a large gene family in *C. elegans* (Fig. 5). The RFP family of proteins is named for the presence of a conserved arginine (R), phenylalanine (F), proline (P), amino acid sequence motif. However, the function of the RFP genes is unknown. In addition, an exhaustive search of genome and EST databases revealed bestrophin homologues in a variety of species. The homologies are shown in Figure 5 between human and drosophila bestrophin, and a representative member of the *C. elegans* RFP family. 48 different mutations associated with BMD, predominantly missense mutations, have been described (http://www.uni-wuerzburg.de/humangenetics/vmd2.html). For the most part, these mutations affect amino acids in the first 50% of the protein, and occur in 4 distinct clusters (2). Using these conserved regions to search for motifs in other proteins that bear similarities has revealed one intriguing feature of the bestrophin sequence. The very acidic motif GEDDDDXE, is conserved between many bestrophin homologues in many species, and also is present in many Na^+/Ca^{2+} exchangers. It is also a "hot spot" for BMD associated mutations.

6.3. Functional Implications Based On Pathophysiology

The hallmark of Best disease is an abnormal EOG in which the Arden index is depressed below 150 in the absence of changes in the clinical ERG (6). The voltage changes associated with the EOG have been previously attributed to changes in the transepithelial potential of the RPE due to a hyperpolarization of the basolateral membrane mediated by basolateral Cl⁻ channels (21). Two major mechanisms of Cl⁻ flux have been described for the basolateral plasma membrane of RPE cells. The first is a cyclic AMP sensitive conductance apparently mediated by the Cystic Fibrosis membrane transporter (CFTR) (22). CFTR is not however thought to be involved in generation of the light peak, which is the altered response in BMD(18). The light peak is thought to be generated by an as yet unidentified Ca^{2+} dependent Cl⁻ channel (9). Our recent findings demonstrating that bestrophin is a basolateral plasma membrane protein (16) suggest the possibility that bestrophin either is the putative Ca^{2+} dependent Cl⁻ channel, or that it plays a role in modulating the activity of RPE ion channels.

Histopathologic examinations of eyes with BMD have demonstrated large accumulations of lipofuscin in the RPE and the deposition of material between the RPE and Bruch's membrane (8, 19, 25). Little is known about the origin of the basal deposits, and the chemistry of lipofuscin formation is not entirely clear. Indeed it is not known whether the composition of lipofuscin associated with BMD is similar to that observed in AMD or Stargardt's disease (7, 12, 17). Lipofuscin formation has been linked by many studies to the visual cycle, transport/oxidation of retina specific polyunsaturated fatty acids, and deficient lysosomal processing of photoreceptor outer segments (5, 17, 24, 26). Given the basolateral plasma membrane localization of bestrophin, it is unlikely that the protein exerts a direct effect on phagocytic uptake of photoreceptor outer segments, as this occurs at the apical plasma membrane. It is also unlikely that bestrophin functions directly in the recycling of retinoids (though it may promote uptake of retinoids from the bloodstream). Since bestrophin is not present in lysosomes, it is unlikely to be directly affecting lysosomal degradation either. However, the altered EOG of individuals with BMD suggests that bestrophin could exert effects on all three pathways by altering the ionic homeostasis of the RPE cell. If bestrophin is an ion channel or modulates the activity of specific ion channels the resultant changes in channel activity could result in alterations in the kinetics of these pathways which over time could accelerate the process of lipofuscin formation.

7. CONCLUDING REMARKS

Analysis of bestrophin's DNA and protein sequences has yielded few insights into the role of bestrophin in the RPE cell. Our recent finding that bestrophin is a basolateral plasma membrane protein has allowed us to rule out direct roles in certain pathways such as phagocytic uptake and lysosomal degradation of photoreceptor outer segments, as well as the recycling of retinoids. By combining our knowledge of localization with that of the pathophysiology of BMD we can postulate that bestrophin functions as an ion channel or as a modulator of RPE ion transporters. Future studies will be guided by this hypothesis with the goal not only of understanding the normal function of bestrophin, but of understanding how mutation of bestrophin results in BMD, and whether the pathway(s) in wich bestrophin participates are involved in the pathogenesis of AMD.

REFERENCES

1. Aust, S. D. Ferritin as a source of iron and protection from iron-induced toxicities. *Toxicol Lett* 82-83: 941-4, 1995.
2. Bakall, B., T. Marknell, S. Ingvast, M. J. Koisti, O. Sandgren, W. Li, A. A. Bergen, S. Andreasson, T. Rosenberg, K. Petrukhin, and C. Wadelius. The mutation spectrum of the bestrophin protein--functional implications. *Hum Genet* 104: 383-9, 1999.
3. Bard, L. A., and H. E. Cross. Genetic counseling of families with Best macular dystrophy. *Trans Am Acad Ophthalmol Otolaryngol* 79: OP865-73, 1975.
4. Bird, A. C., N. M. Bressler, S. B. Bressler, I. H. Chisholm, G. Coscas, M. D. Davis, P. T. de Jong, C. C. Klaver, B. E. Klein, R. Klein, and et al. An international classification and grading system for age-related maculopathy and age-related macular degeneration. The International ARM Epidemiological Study Group. *Surv Ophthalmol* 39: 367-74, 1995.
5. Brunk, U. T., U. Wihlmark, A. Wrigstad, K. Roberg, and S. E. Nilsson. Accumulation of lipofuscin within retinal pigment epithelial cells results in enhanced sensitivity to photo-oxidation. *Gerontology* 41: 201-12, 1995.
6. Cross, H. E., and L. Bard. Electro-oculography in Best's muscular dystrophy. *Am J Ophthalmol* 77: 46-50, 1974.
7. Curcio, C. A., N. E. Medeiros, and C. L. Millican. The Alabama Age-Related Macular Degeneration Grading System for donor eyes. *Invest Ophthalmol Vis Sci* 39: 1085-96, 1998.
8. Frangieh, G. T., W. R. Green, and S. L. Fine. A histopathologic study of Best's macular dystrophy. *Arch Ophthalmol* 100: 1115-21, 1982.
9. Gallemore, R., B. Hughes, and S. Miller. Light-induced responses of the retinal pigment epithelium. In: *The retinal pigment epithelium*, edited by M. Marmor and T. Wolfensberger. New York Oxford: Oxford University Press, 1998, p. 175-198.
10. Gass, D. J. M. Heredodystrophic disorders affecting the pigment epithelium and retina. In: *Stereoscopic atlas of macular diseases, diagnosis and treatment*, edited by D. J. M. Gass. St. Louis: Mosby, 1997, p. 303-313.
11. Godel, V., G. Chaine, L. Regenbogen, and G. Coscas. Best's vitelliform macular dystrophy. *Acta Ophthalmol Suppl* 175: 1-31, 1986.
12. Green, W. R., and C. Enger. Age-related macular degeneration histopathologic studies. The 1992 Lorenz E. Zimmerman Lecture. *Ophthalmology* 100: 1519-35, 1993.
13. Leibowitz, H. M., D. E. Krueger, L. R. Maunder, R. C. Milton, M. M. Kini, H. A. Kahn, R. J. Nickerson, J. Pool, T. L. Colton, J. P. Ganley, J. I. Loewenstein, and T. R. Dawber. The Framingham Eye Study monograph: An ophthalmological and epidemiological study of cataract, glaucoma, diabetic retinopathy, macular degeneration, and visual acuity in a general population of 2631 adults, 1973-1975. *Surv Ophthalmol* 24: 335-610, 1980.
14. Maloney, W. F., D. M. Robertson, and S. M. Duboff. Hereditary vitelliform macular degeneration: variable fundus findings within a single pedigree. *Arch Ophthalmol* 95: 979-83, 1977.
15. Marmor, M. F., and K. Small. Dystrophies of the retinal pigment epithelium. In: *The retinal pigment epithelium*, edited by M. F. M. a. T. J. Wolfensberger. New York: Oxford Press, 1998, p. 330-333.

16. Marmorstein, A. D., L. Y. Marmorstein, M. Rayborn, X. Wang, J. G. Hollyfield, and K. Petrukhin. Bestrophin, the product of the best vitelliform macular dystrophy gene (VMD2), localizes to the basolateral plasma membrane of the retinal pigment epithelium [In Process Citation]. *Proc Natl Acad Sci U S A* 97: 12758-63, 2000.

17. Mata, N. L., J. Weng, and G. H. Travis. Biosynthesis of a major lipofuscin fluorophore in mice and humans with ABCR-mediated retinal and macular degeneration. *Proc Natl Acad Sci U S A* 97: 7154-9, 2000.

18. Miller, S., J. Rabin, T. Strong, M. Iannuzzi, A. Adams, F. Collins, W. Reenstra, and P. McCray, Jr. Cystic fibrosis (CF) gene product is expressed in retina and retinal pigment epithelium. In: *Invest Ophthalmol Vis Sci*, 1992, p. 1009.

19. O'Gorman, S., W. A. Flaherty, G. A. Fishman, and E. L. Berson. Histopathologic findings in Best's vitelliform macular dystrophy. *Arch Ophthalmol* 106: 1261-8, 1988.

20. Petrukhin, K., M. J. Koisti, B. Bakall, W. Li, G. Xie, T. Marknell, O. Sandgren, K. Forsman, G. Holmgren, S. Andreasson, M. Vujic, A. A. Bergen, V. McGarty-Dugan, D. Figueroa, C. P. Austin, M. L. Metzker, C. T. Caskey, and C. Wadelius. Identification of the gene responsible for Best macular dystrophy. *Nat Genet* 19: 241-7, 1998.

21. Quinn, R. H., J. N. Quong, and S. S. Miller. Adrenergic receptor activated ion transport in human fetal retinal pigment epithelium [In Process Citation]. *Invest Ophthalmol Vis Sci* 42: 255-64, 2001.

22. Sheppard, D. N., and M. J. Welsh. Structure and function of the CFTR chloride channel. *Physiol Rev* 79: S23-45, 1999.

23. van der Schaft, T. L., C. M. Mooy, W. C. de Bruijn, F. G. Oron, P. G. Mulder, and P. T. de Jong. Histologic features of the early stages of age-related macular degeneration. A statistical analysis. *Ophthalmology* 99: 278-86, 1992.

24. Wassell, J., S. Davies, W. Bardsley, and M. Boulton. The photoreactivity of the retinal age pigment lipofuscin. *J Biol Chem* 274: 23828-32, 1999.

25. Weingeist, T. A., J. L. Kobrin, and R. C. Watzke. Histopathology of Best's macular dystrophy. *Arch Ophthalmol* 100: 1108-14, 1982.

26. Winkler, B. S., M. E. Boulton, J. D. Gottsch, and P. Sternberg. Oxidative damage and age-related macular degeneration. *Mol Vis* 5: 32, 1999.

PATHOGENICITY OF MYELIN BASIC PROTEIN IN ANTERIOR UVEITIS

Pathogenicity of MBP in uveitis

Grazyna Adamus, Brad Sugden, Maria Manczak, Anatol Arendt, and Paul A. Hargrave [*]

1. INTRODUCTION

Uveitis is a group of inflammatory eye diseases that may result from infections, malignancies, trauma, or incompletely understood processes that are presumably mediated by the body's immune system. Uveitis can affect any portion of the uveal track, thus the neighboring tissues such as the retina and vitreous are often affected. Approximately 50% of patients with uveitis have an underlying systemic disease including multiple sclerosis (MS) where central nervous system (CNS) inflammation is associated with the inflammation of the eye. The most common form of uveitis is anterior uveitis (AU), which may precede, occur with, or follow the systemic disease. [1-5].

Animal models have been developed to study the pathogenesis of uveitis, because of the difficulties obtaining patient's tissues from the inflamed eye for experiments. There are animal models for experimental uveitis that provoke inflammation of different tissues of the eye and represent different forms of uveitis. Most of those models represent acute forms of uveitis induced with proteins specific for photoreceptor cells or endotoxin. In our studies we use myelin basic protein (MBP) to induce AU in Lewis rats. The hallmark of MBP-induced uveitis is anterior uveitis, which is associated with the inflammation of the spinal cord [6-8] and uveitis can relapse upon re-immunization. Possible target antigens are myelinated neurons in the CNS and iris, which exist there in quite abundant quantity [9]. One of the main components of myelin, myelin basic protein, is expressed in significant quantities (20-30% of total protein) in both the CNS and peripheral nervous system (PNS). The major proteins of the CNS myelin

---

[*] G. Adamus, B. Sugden and M. Manczak, Neurological Sciences Institute, Oregon Health Sciences University, 3181 SW Sam Jackson Park Road, Portland, OR 97201; P.A. Hargrave and A. Arendt, Department of Ophthalmology, University of Florida School of Medicine, Gainesville, FL 32610

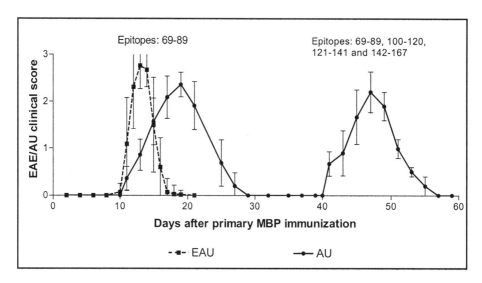

Figure 1. Kinetics of acute and recurrent uveitis in Lewis rats induced by immunization with 25 μg myelin basic protein in complete Freund's adjuvant supplemented with 150 μg Mycobacterium tuberculosis. The initial episode is associated with the recognition of epitope 69-89 whereas the recurrent episode is associated with more diverse T cell specificities.

are MBP and proteolipid protein (PLP), which are produced by oligodendrocytes [10]. Other myelin proteins include myelin-associated protein (MAG) and myelin oligodendrioglia glycoprotein (MOG). The myelin of the iris nerves is produced by Schwann cells that express MBP, glycoprotein P_0 and small amounts of PLP. The role of PLP, P_0, MOG and MAG in the induction of uveitis is unknown. In MS, the immune attack is targeted mainly to the myelin-forming cell of the CNS, the oligodendrocytes [11]. Therefore, it is possible that the Schwann cells are the targets in uveitis associated with MS.

The onset of MBP-induced AU usually coincides with the onset of EAE but has a longer duration and persists after clinical signs of EAE subside (Figure 1). Of particular interest, these recovered rats develop a total resistance to further attempts to induce active EAE but are susceptible to reinduction of uveitis. The mechanism of susceptibility of the eye to recurrent uveitis is not well understood although minor epitopes of MPB are likely to play a role in progression to relapse by expansion of the T cell repertoire with time. MBP-induced recurrent ocular disease was associated with the recognition of new MBP epitopes in Lewis rats, especially peptides 100-120, 121-141 and 142-167 [12].

2. MBP PEPTIDES INDUCE AU

In this study, we examined the pathogenic potential of all MBP peptides to induce acute and recurrent anterior uveitis in Lewis rats. The question was whether all those peptides found to stimulate T cells after immunization with the intact MBP were capable of eliciting EAE and AU, including recurrent AU (R-AU). The initial priming of Lewis rats with MBP always led to the predominance of T cells specific for the major epitope 69-89 early after immunization. Thus, we tested whether those determinants of MBP, cryptic after primary immunization, can be immunogenic in the course of EAE and in AU. For this, Lewis rats were immunized with MBP synthetic peptides covering the entire protein sequence of guinea pig MBP in complete Freund's adjuvant supplemented with 150 µg *Mycobacterium tuberculosis*. Then, rats were scored for the development of clinical EAE and AU [12]. Table 1 summarizes the results from acute (first episode) and recurrent (second episode) anterior uveitis induced in Lewis rats with MBP peptides.

Table 1. Pathogenicity of MBP peptide in anterior uveitis

Antigen (MBP Peptide)	Acute AU		Recurrent AU	
	Incidence	Mean maximal score	Incidence	Mean maximal score
Intact MBP	5/5	2.80±0.6	5/5	2.70±0.4
1-20	4/5	1.30±0.9	5/5	1.45±0.9
21-40	0/5		NT	
41-60	0/5		3/5	1.15±1.0
55-69	0/5		5/5	1.40±0.4
69-89	5/5	2.70±0.4	4/5	1.50±1.2
87-99	5/5	1.50±0.9	5/5	1.65±0.8
100-120	5/5	3.10±0.5	5/5	1.75±1.5
121-141	3/5	1.80±1.2	5/5	1.70±1.2
142-167	5/5	3.20±0.4	5/5	2.40±0.7
CFA alone	0/3		0/5	

NT – not determined

A total of 5 peptides from the carboxyl terminus of guinea pig MBP induced neurological disease that coincided with the onset of ocular disease 10-13 days after immunization with similar incidence of AU as in rats immunized with intact MBP. The major encephalitogenic peptide 69-89 was also uveitogenic. Other peptides 100-120 and 142-167 were also strong inducers of both EAE and AU. In contrast, the peptide corresponding to MBP residues 1-20 was uniquely capable of only inducing AU but not EAE. A control group of rats injected with saline/CFA did not produce EAE or AU. The relapsing disease was induced in Lewis rats by immunization with the MBP peptides, including pathogenic and non-pathogenic epitopes after full recovery from the first episode of AU induced with MBP. Rats were evaluated for the development of EAE and AU.

All rats remained resistant to EAE and did not show any signs of paralysis. The severity and the incidence were comparable to the acute disease induced with peptides. However, we found that injection of MBP peptides induced AU, some of them with an accelerated onset of AU in 7-8 days after they received the second dose when compared to day 10-11 in the initial episode. This early onset suggests the presence of memory cells generated after the initial immunization. In addition to well-characterized encephalitogenic and uveitogenic dominant peptide epitopes, there were other MBP peptides capable of inducing R-AU. MBP peptides 69-89, 100-120 and 142-167 induced AU with accelerated onset. In control experiments, rats were re-injected with saline/CFA only and they did not develop any signs of neurological or ocular disease indicating that this disease is triggered by specific antigen.

3. PEPTIDE 1-20 IS UVEITOGENIC BUT NOT ENCEPHALITOGENIC

The peptide 1-20 was uniquely capable of inducing AU but not EAE. Rats immunized with this peptide developed mild clinical AU on day 11 and disease peaked on day 18 after immunization. Similar to induction of the initial episode of AU with the epitope 1-20, the recurrent disease could also be produced with this peptide. However, the accelerated onset was not observed like in the case of major uveitogenic peptide, suggesting the absence or low density of previously activated T cells specific for the injected epitope. To further investigate the pathogenicity of 1-20 T cells we developed a T cell line specific for this peptide. We compared its pathogenic potential to two other selected cell lines specific for uveitogenic peptides 69-89 and 142-167. A cell line specific to non-ocular antigen, ovalbumin (OVA) was developed for control experiments.

As expected the injection of OVA-specific T cells did not elicit any neurological or ocular inflammation. However, T cell lines specific for MBP peptides 69-89 and 142-167 were capable of eliciting both diseases in contrast to the T cell line specific for the MBP peptide 1-20 that had only uveitogenic potential. The latter confirmed the results from the active immunization with this peptide. As seen in Figure 2, the activated T cells specific for MBP 1-20 were capable of transferring AU.

T cells specific for the immunodominant MBP epitope at residue 69-89 utilize $V\alpha2$ and $V\beta8.2$ elements in their T cell receptor (TCR) gene [13]. T cells obtained from the iris/ciliary body and spinal cord of the first episode demonstrated the same biased V gene expression [14]. However, during recurrent AU we observed a shifting of TCR usage by infiltrating T cells from the $V\beta8.2$ to the more diverse repertoire [12]. In this study, we examined the TCR V gene usage of T cells specific for peptide 1-20 by RT-PCR analysis. RNA was isolated from T cell line obtained from rats immunized with peptide 1-20 and stimulated with the same peptide *in vitro*. The up-regulation of $V\beta$ mRNA was examined using primers specific for rat V genes. $V\beta$ analysis demonstrated the expression of $V\beta4$, $V\beta8.6$ and V19 genes (data not shown).

It is not clear why peptide 1-20 was uniquely uveitogenic in acute and recurrent AU. In earlier studies of EAE, it has been shown that peptide 1-20 can be encephalitogenic under a special protocol of immunization, when the animals were treated with pertussis toxin to open up the blood brain barrier [15]. In our experiments the active immunization with the peptide without additional manipulations and the passive transfer of 1-20 specific T cells produced the same results; the rats developed AU but not EAE. It is possible that the T cells specific for peptide 1-20 preferentially target the eye rather than spinal cord in recipient rats. Perhaps the effector cells phenotype enables the migration across the blood ocular barrier more easily than across the blood brain barrier and the entry to the eye in sufficient numbers to cause disease.

The analysis of pathogenic determinants important for the induction of uveitis provides further evidence that MBP-specific T cells also contribute to the pathogenesis of anterior uveitis. Elucidation of the mechanism by which AU occurs and relapses provides important information for treatment of ocular autoimmune diseases, preventing relapses, and tissue destruction.

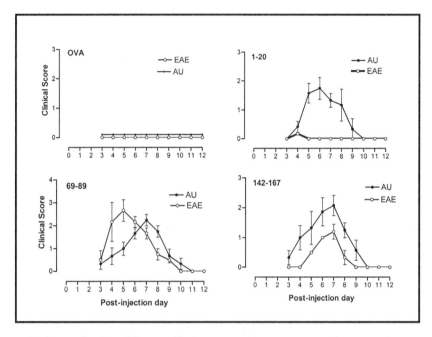

Figure 2. Passive transfer of T cell lines specific for MBP peptides 1-20, 69-89 (major encephalitogenic and uveitogenic epitope), 142-167 and ovalbumin (control), respectively, to naïve Lewis rats.

4. ACKNOWLEDGMENTS

This work was supported by the grant EY 12477 from the National Institutes of Health (G.A.).

REFERENCES

1.   Rothova A, Buitenhuis HJ, Meenken C, Brinkman CJJ, linssen A, Alberts C, et al. Uveitis and systemic disease. Br. J. Ophthalmol. 76:137-41 (1992).
2.   Rodriguez A, Calonge M, Pedroza-Seres M, Akova YA, Messmer EM, D'Amico DJ, et al. Referral Patterns of uveitis in a tertiary eye care center. Arch. Ophthalmol 114:593-99 (1996).
3.   Suttorp-Schulten MSA, Rothova A. The possible impact of uveitis in blindness: a literature survey. Br J. Ophthalmol 80:844-48 (1996).
4.   McCluskey PJ, Towler HM, Lightman S. Management of chronic uveitis. BMJ 320:555-8 (2000).
5.   Towler HM, Lightman S. Symptomatic intraocular inflammation in multiple sclerosis. Clin Experiment Ophthalmol. 28:97-102 (2000).
6.   Shikishima K, Lee WR, Behan WMH, Foulds WS. Uveitis and retinal vasculitis in acute experimental allergic encephalomyelitis in the Lewis rat: an ultrastructural study. Exp. Eye Res. 56:167-75 (1993).

7.   Verhagen C, Mor F, Cohen IR. T cell immunity to myelin basic protein induces anterior uveitis in Lewis rats. J. Neuroimmunol. 53:65-71 (1994).

8.   Adamus G, Amundson D, Vainiene M, Ariail K, Machnicki M, Weinberg A, et al. Myelin basic protein specific T-helper cells induce experimental anterior uveitis. J. Neurosci. Res. 44:513-18 (1996).

9.   Huhtala A. Origin of myelinated nerves in the rat iris. Exp. Eye Res. 22:259-65 (1976).

10.   Lemke G. Unwrapping the genes of myelin. Neuron 1:535-43 (1988).

11.   Steinman L. Multiple sclerosis: A coordinated immunological attack against myelin in the central nervous system. Cell 85:299-302 (1996).

12.   Adamus G, Manczak M, Sugden B, Arendt A, Hargrave PA, Offner H. Epitope recognition and T cell receptors in recurrent autoimmune anterior uveitis in Lewis rats immunized with myelin basic protein. J. Neuroimmunol. 108:122-30 (2000).

13.   Burns FR, Li X, Shen N, Offner H, Chou YK, Vandenbark A, et al. Both rat and mouse T cell receptors specific for the encephalitogenic determinant of myelin basic protein use similar $V\alpha$ and $V\beta$ chain genes even though the MHC and encephalitogenic determinants being recognized are different. J. Exp. Med. 169:27-39 (1989).

14.   Buenafe AC, Offner H, Machnicki M, Elerding H, Adlard K, Jacobs R, et al. EAE TCR motifs and antigen recognition in myelin basic protein-induced anterior uveitis in Lewis rats. J. Immunol. 161:2052-59 (1998).

15.   Mor F, Cohen IR. Pathogenicity of T cells responsive to diverse cryptic epitopes of myelin basic protein in the Lewis rat. J. Immunol. 155:3693-99 (1995).

A2E INHIBITS MITOCHONDRIAL FUNCTION, CAUSES THE RELEASE OF PRO-APOPTOTIC PROTEINS AND INDUCES APOPTOSIS IN MAMMALIAN CELLS

Christian Grimm, Marianne Suter, Andreas Wenzel, Marja Jäättela, Peter Esser, Norbert Kociok, Marcel Leist, Christoph Richter, and Charlotte E. Remé[*]

1. SUMMARY

Age-related macular degeneration (AMD) is the leading cause of severe visual impairment in humans living in industrialized countries. A precondition for AMD appears to be the accumulation of the age pigment lipofuscin in lysosomes of retinal pigment epithelial (RPE) cells. Here we show that A2E (*N*-retinyl-*N*-retinylidene ethanolamine) the major fluorophore of lipofuscin, induces apoptosis in RPE and other cells at concentrations found in human retina. Apoptosis is accompanied by appearance of the pro-apoptotic proteins cytochrome c and apoptosis inducing factor (AIF) in the cytoplasm and the nucleus but does not involve activation of caspase-3. Biochemical examinations show that A2E inhibits mitochondrial function by specifically targeting cytochrome oxidase (COX). With both, isolated mitochondria and purified COX, A2E inhibits oxygen consumption synergistically with light. Inhibition is reversed by addition of cytochrome c or cardiolipin. Loss of RPE cell viability through inhibition of mitochondrial function might constitute a pivotal step towards the progressive degeneration of the central retina. We present a working hypothesis that suggests that

[*] Christian Grimm, Andreas Wenzel, Charlotte E Remé, Department of Ophthalmology, University Hospital Zurich, Frauenklinikstr. 24, CH-8091 Zurich, Switzerland. Marianne Suter, Christoph Richter, Institute of Biochemistry, Swiss Federal Institute of Technology (ETH), Universitätstr. 16, CH-8092 Zurich, Switzerland. Maria Jäätela, Danish Cancer Society, Apoptosis Laboratory, Strandboulevarden 49, DK-2100 Copenhagen, Denmark. Peter Esser, Norbert Kociok, Eye Clinic, University of Cologne, Joseph Stelzmannstr. 9, D-50931 Cologne, Germany. Marcel Leist, Department of Neurobiology, H. Lundbeck A/S, Ottiliavej 9, 2500 Valby, Denmark.

A2E, once released from lysosomes (by lysosomal rupture or by 'overflow' of the lysosomal capacity) can target mitochondria and inhibit mitochondrial function. This causes the release of the pro-apoptotic proteins and the induction of cell death.

2. INTRODUCTION

Age-related macular degeneration (AMD) constitutes one of the leading causes of severe visual impairment and affects 10 - 20% of people at an age of 65 and older in industrialized nations[1]. Due to the slow progression of the disease in humans and the lack of suitable animal models, the molecular events occurring during the disease process are poorly understood. It is suggested that both genetic components[2-5] and exogenous enhancing factors[6,7] contribute to the pathogenesis of AMD. One of these exogenous factors might be light[6]. Exposure to visible light greatly enhances the formation of A2E (N-retinyl-N-retinylidene ethanolamine) in a transgenic mouse model lacking a functional Rim protein (RmP), an ABC transporter for retinoids in the disc membranes of rod photoreceptors[8,9]. A2E is the major fluorophore of lipofuscin, a lysosomal deposit accumulating with age in the cells of the RPE[10,11]. A2E was also detected in lysosomes of cells exposed in culture to the free compound or to A2E coupled to low-density lipoprotein particles[12,13]. In cell cultures, A2E inhibits hydrolytic activities in lysosomes[14,15] and mediates blue light-induced damage to RPE cells[16]. A2E can be synthesized from 2 retinals and 1 ethanolamine[11], both components of photoreceptor outer segment membranes, where 11-*cis*-retinal serves as the chromophore of the visual pigment rhodopsin and phosphatidylethanolamine is an abundant membrane phospholipid. In the course of photoreceptor renewal, rod outer segment tips are shed and phagocytozed by the underlying cells of the retinal pigment epithelium (RPE)[17-19]. During this process, A2E or its precursor molecules may be incorporated as components of lipofuscin into phagolysosomes of the RPE cells. Since lipofuscin and A2E are stable, these molecules accumulate with age and may increasingly stress the capacity of the lysosomes. It is believed that this accumulation of lipofuscin in RPE cells is a predicament for the development of AMD. Similarly, the formation of Drusen and other deposits in the Bruch's membrane might be initial steps in the pathogenesis of AMD. It is not known how Drusen are formed, but they might be the consequence of progressive death and/or exocytosis of RPE cells in the retina[14,20]. A role of A2E in the pathogenesis of AMD is supported by human studies showing that AMD is frequently associated with mutations in the gene for RmP (ABCR)[2] suggesting, on the basis of the animal data, an enhanced accumulation of A2E in RPE cells of these patients.

Several studies showed in *in vitro* systems using cultured RPE cells, that A2E has cytotoxic activity and the capacity to disrupt lysosomal function[12,15,16]. However, cellular mechanisms leading to the death of the cells remain to be elucidated. Evidence from human autopsy eyes suggests that cells are lost by an apoptotic process during the course of AMD[21]. Apoptosis plays major roles during development and homeostasis by removing surplus, unwanted or damaged cells[22]. Furthermore, apoptosis is strongly involved in many disease processes such as cancer, autoimmune diseases, acquired immune deficiency syndrome and tissue degeneration (for review, see[23]). Intracellular death signaling and execution of cell death strongly depends on the apoptotic stimulus applied[24,25]. A major control center in the regulation of apoptosis appears to be mitochondria[26,27]. Destabilization of mitochondrial membranes causes the release of apoptosis–inducing proteins such as cytochrome c and apoptosis inducing factor (AIF). Cytoplasmic appearance of cytochrome c often but not always[28] induces activation of caspases, cysteine proteases active in apoptosis. Accumulation of AIF in nuclei of cells induces cell death independently of caspase activation[29,30]. Present knowledge suggests that mitochondria function as cellular sensors of stress into which different apoptosis induction pathways converge, and that mitochondria act as central executioners of apoptosis[31].

In this study, we show that the application of A2E induces apoptotic cell death in a variety of mammalian cells in culture, including RPE. Apoptosis is not accompanied by detectable caspase activation but is preceded by a decline in mitochondrial activity and the cytoplasmic and nuclear accumulation of cytochrome c and AIF. Experiments in highly characterized systems demonstrate that A2E targets mitochondrial function by the specific inhibition of cytochrome oxidase (COX). We propose a model for the cytotoxicity of A2E that involves mitochondrial dysfunction and the mobilization of pro-apoptotic proteins by A2E. Our findings may also suggest new strategies to retard or overcome AMD.

3. MATERIAL AND METHODS

3.1. Animals

Eight day-old BALB/c mice were obtained from the Animal Unit of the University of Constance, female Wistar rats were from the Animal Unit of the Institute of Biochemistry, Swiss Federal Institute of Technology (ETH), Zurich. All experiments were performed in accordance with international guidelines to minimize pain and discomfort (NIH-guidelines and European Community Council Directive 86/609/EEC) and conformed to the ARVO statement for care and use of animals in research.

3.2. Cell culture, viability assays and immunocytochemistry

RPE cells were prepared from porcine eyes obtained from a local slaughter house essentially as described[32]. The purity of the culture was verified by immunohistochemical staining with anti-cytokeratin antibodies (Sigma c-2931). Experiments were performed with passage 1-2 RPE cells.

Murine cerebellar granule cells (CGC) were isolated from 8 days old BALB/c mice and cultured as described[33]. Contamination with non-neuronal cells (β-III-tubulin negative) was < 5 %. Dissociated neurons were plated on poly-L-lysine–coated dishes at a density of about 0.25×10^6 cells/cm^2 and cultured in Eagle's Basal Medium (BME, Gibco) supplemented with 10 % heat-inactivated fetal calf serum, 20mM KCl, 2mM L-glutamine and penicillin-streptomycin, and cytosine arabinoside (10μM; added 48h after plating). CGC were used without further medium changes after 5 days in culture. The cells were exposed to A2E in their original medium in the presence of 2μM (+)-5-methyl-10,11-dihydro-5H-dibenzo[a,d]cyclohepten-5,10-imine (MK801; Biotrend Chemikalien GmbH, GER) and 2mM Mg^{2+} to prevent NMDA receptor activation and excitotoxicity[34,35].

Apoptosis and secondary lysis were routinely quantified by double staining of neuronal cultures with 1µg/ml H-33342 (membrane permeant, blue-fluorescent chromatin stain; Molecular Probes, Eugene, OR, USA) and 0.5µM SYTOX (non-membrane permeant, green-fluorescent chromatin stain; Molecular Probes, Eugene, OR, USA) as described[36,37]. Apoptotic cells were characterised by scoring condensed nuclei. In addition, mitochondrial activity was quantified by their capacity to reduce MTT (3-(4,5-dimethylthiazole-2-yl)-2,5-diphenyltetrasodium bromide).

For Immunocytochemistry, CGC were grown on glass-bottomed culture dishes, exposed to A2E or solvent, fixed with 4% paraformadehyde and permeabilized with 0.1% Triton X-100 (for cytochrome c) or 0.1% $NaDodSO_4$ (for AIF). Staining of cytochrome c (anti-cytochrome c antibody; clone 6H2.B4, Pharmingen, Hamburg, Germany) was performed as described [38,39]. Rabbit anti-AIF serum [29] was used at a dilution of 1:500. CGC were embedded in phosphate-buffered saline containing 50% glycerol and 0.5µg/ml H-33342, and imaged by confocal microscopy as described[40].

3.3. Synthesis, purification, and HPLC analysis of A2E

A2E was synthesized from all-*trans* retinal and ethanolamine as described[11] and purified chromatographically on silica gel 60 thin layer chromatography plates using the primary developing system as described[41]. A2E was detected on the plates by their fluorescence upon illumination with 366nm light. The material containing A2E was scraped off, eluted with chloroform/methanol/water (30/25/4), dried, and re-chromatographed. For HPLC, total A2E was loaded onto a reversed phase column (Nucleosil 100-5 C18, Macherey-Nagel, Oensingen, CH) and eluted with a gradient of methanol (+ 0.1% trifluoroacetic acid) in water (86% - 100% in 20min). A2E was detected at 436nm with a peak maximum at 16 min.

3.4. Cytotoxicity of A2E and apoptosis assays

A2E (25µM) or staurosporin (0.5µM) was applied to RPE cells in cell culture medium and 0.5% DMSO. Incubation was at 37°C with 5% CO_2 for 24hr in darkness. TUNEL staining was done as described[40].

For the analysis of DNA fragmentation, cells were scraped from the plates after incubation with A2E (50µM; 0.5% DMSO; 16h; 5% CO_2; 37°C; darkness) and DNA prepared as described[40]. 15µg of total DNA were analyzed by electrophoresis on a 1.5% agarose gel and stained with SybrGreen (Molecular Probes, Eugene, OR, USA).

3.5. Oxygen consumption measurements

Liver mitochondria were obtained from overnight-starved rats weighing 200g by differential centrifugation [42]. COX was isolated from rat liver.

Oxygen consumption was measured at 28°C with a Clark-type oxygen electrode (Yellow Spring Instruments, Yellow Spring, OH) under continuous stirring as described[40]. A2E was added to the mitochondrial suspension before rotenone (5µM) and succinate (0.4 to 1.2mM). Cytochrome c (1µM) or cardiolipin (1µg/ml) were added as indicated. COX (263µg/ml) assays were performed in 40mM phosphate buffer, pH 7.0, and 50µM EDTA with 1mM ascorbate and 0.4mM tetramethyl-*p*-phenylenediamine.

4. RESULTS AND DISCUSSION

4.1. Mitochondrial targeting and apoptosis induction by A2E

A2E has originally been isolated from aged human eyes. We sacrificed Wistar rats of different age and measured A2E concentrations in cells of the retinal pigment epithelium. A2E accumulated linearly with age showing that also in rats, A2E can be considered an age pigment (Fig. 1).

Synthetic A2E was toxic to all cell types tested including pig RPE cells (Fig. 2), human fibroblasts, the neuroblastoma lines CHP-100 and SH-SY5Y, and murine cerebellar granule cells (CGC). Apoptosis induction was dose–dependent and an A2E concentration of 25µM killed approximately 50% of cultured RPE cells. In old human eyes, 830 pmol A2E were isolated on average[11]. Assuming that A2E is present exclusively in RPE cells and evenly distributed, a concentration of about 15µM A2E in such cells of old humans can be calculated. Our experimental concentrations are therefore close to the A2E concentrations found in humans.

The concentrations of A2E which induced apoptosis either in only few or in almost all cells differed usually less than by a factor of 2 to 4 (Fig.3 and data not shown). This suggests that a specific cell type can tolerate a particular A2E concentration but that already a small amount of A2E exceeding this threshold dose is sufficient to efficiently induce cell death. Such a phenomenon could be explained by the presence of an intracellur 'buffering system' which could neutralize the toxicity of A2E. However, once this putative system is saturated, A2E could react with intracellular components critical for the control of cell survival. Mitochondria are regarded as sensitive control centers of apoptosis and participate in the regulation and execution of apoptosis induced by many agents, including lipophilic cations[34]. Since A2E is a lipophilic cation, we tested the effect of A2E on mitochondrial function. When incubated with A2E, CGC accumulated A2E in mitochondria[40] which are very abundant in these cells. Significant apoptotic death was induced at A2E concentrations below 15µM. Interestingly, already lower A2E concentrations decreased mitochondrial function as measured by the MTT assay (Fig. 3). Therefore, the decline in mitochondrial activity preceded cell death (Fig. 3).

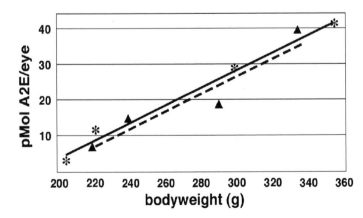

Figure 1. Age-dependent accumulation of A2E in RPE cells of Wistar rats. Rats were sacrificed at the indicated bodyweight (indicative of age), and A2E prepared from eyecups (without retina). Two independent series are shown: stars and solid line; triangles and dotted line.

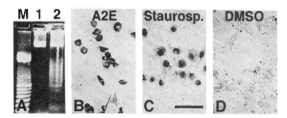

Figure 2. A2E induces apoptosis in RPE cells. (A) Pig RPE cells were exposed to 50μM A2E in 0.5% DMSO (lane 2) or to solvent alone (lane 1). DNA was extracted after 24h and analyzed by agarose gel electrophoresis. M: marker. (B-D) TUNEL staining of cells exposed for 24 hours to 25μM A2E (B), 0.5μM staurosporin (C), or 0.5% DMSO (D). Scale bar: 100μm. (Reprinted with permission from J Biol Chem, 2000, 275, 39625).

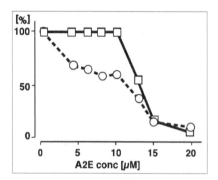

Figure 3. A2E decreases mitochondrial activity and causes loss of cell viability. CGC were incubated in medium containing A2E concentrations as indicated. After 45h the capacity of the cell population to reduce MTT (mitochondrial activity) was determined (circles). Data are standardized to untreated control cells as 100 % reference value. In parallel cultures, cell viability was determined by staining with the chromatin dyes H-33342 and SYTOX. The percentage of cells with intact plasma membrane and non-condensed chromatin (squares) was determined. Data represent means of three independent experiments. (In part reprinted with permission from J Biol Chem, 2000, 275, 39625).

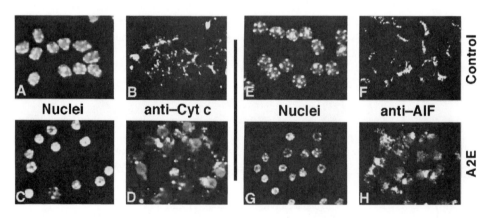

Figure 4. A2E triggers release of cytochrome c and of apoptosis inducing factor. (A-D) CGC cultures were exposed to 20μM A2E (C, D) or solvent (A, B) for 20 h, fixed, stained for chromatin structure (H-33342; A, C) and cytochrome c (B, D), and imaged by confocal microscopy. Optical sections were obtained at the level of the neuronal somata, where mitochondria (B) lie in a close circle around the nucleus. (E-H) CGC cultures were exposed to 20μM A2E (G, H) or solvent control (E, F), fixed after 24h, stained for chromatin structure (H-33342; E, G) and apoptosis inducing factor (F, H). A2E leads to the appearance of cytochrome c and AIF in the cytosol and the nucleus. Scale bar: 25μm. (In part reprinted with permission from J Biol Chem, 2000, 275, 39625).

In response to A2E treatment, the mitochondria released the pro-apoptotic proteins cytochrome c (Fig. 4, panels A to D) and AIF (Fig. 4, panels E to H) into the cytoplasm and nucleus. The detachment of cytochrome c from the inner mitochondrial membrane has important consequences for mitochondria: First, mitochondria experience oxidative stress because the reduction of the respiratory chain members upstream of the cytochrome c binding sites results in enhanced superoxide formation[43] and second, interruption of electron flow impairs mitochondrial ATP synthesis. Both events are highly relevant for apoptosis because oxidative stress and decreased energy levels weaken mitochondria, cause leakiness of their inner membrane and favor further release of pro-apoptotic proteins[26,44]. Interestingly, A2E induced cell death did apparently not involve caspase activity since death was not preventable by the pan caspase inhibitor Z-VAD.fmk or other established caspase inhibitors (data not shown). Similarly, A2E treatment did not activate caspase-3 in CGC (Fig. 5) or RPE (data not shown) cells as measured by DEVD cleavage assays[45]. It has been shown in other systems that mitochondrial release and nuclear translocation of AIF can be correlated with large-scale DNA fragmentation and apoptosis induction in the absence of caspase activation[29,30].This suggests that AIF translocation from mitochondria to the nucleus might be a crucial step in the apoptotic response to A2E exposure. Release of cytochrome c often but not always results in the activation of the caspase cascade. Activation of caspases can be entirely prevented in cells defective in ATP generation, despite high cytosolic cytochrome c concentrations[28,35,36]. This may explain the lack of inhibition of A2E-induced cell death by the pan-caspase inhibitor Z-VAD.fmk. Thus, caspases may play at most only a minor role in A2E-induced apoptosis. There are also *in vitro* precedents for apoptosis independent of caspase activation in the retina[46] or tumor cells[47].

The presence of A2E in lysosomes but not in mitochondria reported by others [12,15], and the mitochondrial localization found in our study may suggest that after damaging lysosomes A2E can be released and subsequently taken up by mitochondria. Alternatively, A2E may initially accumulate in and be retained by mitochondria, which when damaged may be engulfed by lysosomes.

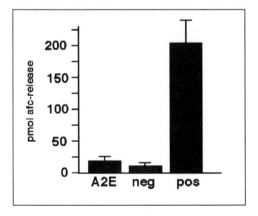

Figure 5. A2E–induced apoptosis does not involve Caspase–3 activation. CGC cells were exposed to A2E (25µM; 20 hours), solvent as negative control or colchicine (16 hours). Caspase-3 activity was measured by the afc-release (from DEVD-afc) pro min und pro mg protein.

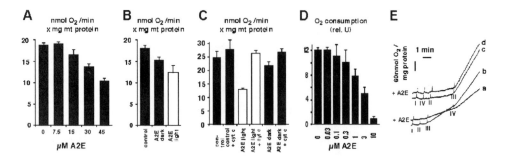

Figure 6. A2E inhibits respiration of mitochondria and of purified COX. Oxygen consumption of mitochondria isolated from rat liver was measured in the absence and presence of A2E in a Clarke-type electrode in the dark (aluminum foil-wrapped) or under light (70W Tungsten lamp, 40cm distance). (A) Concentration dependence of A2E–mediated inhibition of oxygen consumption by mitochondria in the dark. (B) Inhibition by 10μM A2E in the dark and in light. (C) Reversal of A2E (15μM) inhibition by cytochrome c (1μM). (D) Concentration dependence of A2E–mediated inhibition of oxygen consumption by COX in the dark. (E) Cardiolipin (1μg/ml) mediated reversal of A2E (3μM) induced inhibition of oxygen consumption by COX; the following additions were made: I, ascorbate/TMPD; II, COX; III, cytochrome c; IV, cardiolipin; representative recordings are shown. The columns in A-D represent mean values ± SD of four independent experiments

4.2. Inhibition of mitochondrial function by A2E

Since A2E targeted mitochondria, we measured mitochondrial activity in dependence of A2E concentration. A2E suppressed oxygen consumption by mitochondria and isolated cytochrome oxidase (COX) in a dose-dependent manner (Fig. 6, panels A, D), A2E being more potent in the light than in the dark (Fig. 6, panels B, C). The A2E-induced inhibition is apparently due to detachment of cytochrome c from mitochondria because inhibition could be overcome by added cytochrome c or cardiolipin (Fig. 6, panels C, E and data not shown), a compound that facilitates the binding of cytochrome c to COX-containing membranes[48]. This restoration of respiratory activity may be due to a direct interaction of A2E and cardiolipin at the level of COX since oxygen consumption of purified COX was unaffected by A2E when cardiolipin was added to the reaction prior to COX and cytochrome c (Fig. 6 panel E, trace d versus trace a). Alternatively, the anionic cardiolipin may bind to and inactivate the cationic A2E in solution. However, we did not detect complex formation between the two compounds by UV/VIS spectroscopy (data not shown). These results strongly suggest that A2E inhibits mitochondrial respiration because it prevents cytochrome c binding to the inner mitochondrial membrane and thereby interrupts electron flow between cytochrome bc_1 and COX.

5. HYPOTHESIS

At least initial steps of A2E synthesis may occur in rod photoreceptors. Precursors or final forms of A2E may reach the RPE through disc shedding of photoreceptors and/or may be specifically transported through the extracellular matrix. In the RPE, A2E is incorporated into lysosomes which may act as a cellular buffer system for the toxic 'waste' A2E. A2E accumulates with increasing age and may – at some point – reach a critical concentration which disturbs lysosomal function and damages the lysosomal integrity. As a consequence, A2E may be liberated from lysosomes and may reach other cell compartments and organelles like mitochondria. The specific and efficient inhibition of mitochondrial COX activity causes mitochondrial dysfunction and triggers the release of AIF and cytochrome c, two apoptosis–inducing proteins. Nuclear accumulation of AIF may trigger the apoptotic cascade leading to death of the cell in the absence of Caspase-3 activation.

Such a model could also explain why AMD is induced mostly in old eyes. Our model requires the accumulation of A2E in an intracellular compartment which might be differentially challenged in different regions of the retina. In addition to rod phagosomes, the phagocytosis of shed cone material might present a greater metabolic load to the lysosomal system of the RPE cells in the region of the macula than in the periphery. Furthermore, the central fundus receives an increased light dose probably favoring the formation of A2E. The highest density of lipofuscin has indeed been found in this region and especially in the perifoveal area[49] correlating also with highest photoreceptor density. Thus, the limits of the lysosomal capacity to 'buffer' A2E might be reached easier in macular reagion of the retina. This could lead to an early release of A2E and to the induction of cell death by the mitochondrial mechanisms described in this chapter.

So far, no efficient AMD therapy or prevention exists. Carotenoids and antioxidants[50], limiting exposure to light or targeting of the precursors of A2E[8] may be useful. Our findings suggest several other promising strategies. One is supplementation with cardiolipin to facilitate retention of cytochrome c in mitochondria. This might be especially promising because the mitochondrial cardiolipin content and function decreases in humans progressively with age to about 60% of the value found in children[51,52]. Another possible strategy is the prevention of A2E formation by, e.g., outcompeting ethanolamine with a secondary amine, or outcompeting retinal with another aldehyde. Since the level of retinal is decreased by retinal dehydrogenase, stimulation of this enzyme may also be useful to counteract AMD.

5. ACKNOWLEDGMENTS

We thank Dr. Paolo Gazzotti (Federal Insitute of Technology, Zurich) for kindly providing rat liver COX, Heike Naumann for expert technical assistance and Prof. T. Seiler for continous support. This work was supported by the Swiss National Science Foundation, the E&B Grimmke Foundation, Germany and the Bruppacher Foundation, Switzerland.

REFERENCES

1. R. Klein, B. E. Klein, K. L. Linton, Prevalence of age-related maculopathy. The Beaver Dam Eye Study, *Ophthalmology* **99**(6), 933-943 (1992).
2. R. Allikmets, N. F. Shroyer, N. Singh, J. M. Seddon, R. A. Lewis, P. S. Bernstein, A. Peiffer, N. A. Zabriskie, Y. Li, A. Hutchinson, M. Dean, J. R. Lupski, M. Leppert, Mutation of the Stargardt disease gene (ABCR) in age-related macular degeneration, *Science* **277**(5333), 1805-1807 (1997).
3. R. Allikmets, Molecular genetics of age-related macular degeneration: current status, *Eur. J. Ophthalmol.* **9**(4), 255-265 (1999).

4. M. B. Gorin, J. C. Breitner, P. T. De Jong, G. S. Hageman, C. C. Klaver, M. H. Kuehn, J. M. Seddon, The genetics of age-related macular degeneration, *Mol. Vis.* **5**, 29 (1999). http://www.molvis.org/molvis/v5/p29/
5. G. Silvestri, Age-related macular degeneration: genetics and implications for detection and treatment, *Mol. Med. Today* **3**(2), 84-91 (1997).
6. K. J. Cruickshanks, R. Klein, B. E. Klein, Sunlight and age-related macular degeneration. The Beaver Dam Eye Study, *Arch. Ophthalmol.* **111**(4), 514-518 (1993).
7. H. M. Leibowitz, D. E. Krueger, L. R. Maunder, R. C. Milton, M. M. Kini, H. A. Kahn, R. J. Nickerson, J. Pool, T. L. Colton, J. P. Ganley, J. I. Loewenstein, T. R. Dawber, The Framingham Eye Study monograph: An ophthalmological and epidemiological study of cataract, glaucoma, diabetic retinopathy, macular degeneration, and visual acuity in a general population of 2631 adults, 1973-1975, *Surv. Ophthalmol.* **24**(Suppl), 335-610 (1980).
8. N. L. Mata, J. Weng, G. H. Travis, Biosynthesis of a major lipofuscin fluorophore in mice and humans with ABCR-mediated retinal and macular degeneration, *Proc. Natl. Acad. Sci. U S A* **97**(13), 7154-7159 (2000).
9. J. Weng, N. L. Mata, S. M. Azarian, R. T. Tzekov, D. G. Birch, G. H. Travis, Insights into the function of Rim protein in photoreceptors and etiology of Stargardt's disease from the phenotype in abcr knockout mice, *Cell* **98**(1), 13-23 (1999).
10. G. E. Eldred, M. R. Lasky, Retinal age pigments generated by self-assembling lysosomotropic detergents, *Nature* **361**(6414), 724-726 (1993).
11. C. A. Parish, M. Hashimoto, K. Nakanishi, J. Dillon, J. Sparrow, Isolation and one-step preparation of A2E and iso-A2E, fluorophores from human retinal pigment epithelium, *Proc. Natl. Acad. Sci. U S A* **95**(25), 14609-14613 (1998).
12. J. R. Sparrow, C. A. Parish, M. Hashimoto, K. Nakanishi, A2E, a lipofuscin fluorophore, in human retinal pigmented epithelial cells in culture, *Invest. Ophthalmol. Vis. Sci.* **40**(12), 2988-2995 (1999).
13. F. G. Holz, F. Schutt, J. Kopitz, H. E. Volcker, [Introduction of the lipofuscin-fluorophor A2E into the lysosomal compartment of human retinal pigment epithelial cells by coupling to LDL particles. An in vitro model of retinal pigment epithelium cell aging], *Ophthalmologe* **96**(12), 781-785 (1999).
14. G. E. Eldred, Lipofuscin fluorophore inhibits lysosomal protein degradation and may cause early stages of macular degeneration, *Gerontology* **41**(Suppl 2), 15-28 (1995).
15. F. G. Holz, F. Schutt, J. Kopitz, G. E. Eldred, F. E. Kruse, H. E. Volcker, M. Cantz, Inhibition of lysosomal degradative functions in RPE cells by a retinoid component of lipofuscin, *Invest. Ophthalmol. Vis. Sci.* **40**(3), 737-743 (1999).
16. J. R. Sparrow, K. Nakanishi, C. A. Parish, The lipofuscin fluorophore A2E mediates blue light-induced damage to retinal pigmented epithelial cells, *Invest. Ophthalmol. Vis. Sci.* **41**(7), 1981-1989 (2000).
17. R. W. Young, D. Bok, Participation of the retinal pigment epithelium in the rod outer segment renewal process, *J. Cell Biol.* **42**(2), 392-403 (1969).
18. R. W. Young, Visual cells and the concept of renewal, *Invest. Ophthalmol. Vis. Sci.* **15**(9), 700-725 (1976).
19. L. Feeney, Lipofuscin and melanin of human retinal pigment epithelium. Fluorescence, enzyme cytochemical, and ultrastructural studies, *Invest. Ophthalmol. Vis. Sci.* **17**(7), 583-600 (1978).
20. T. Ishibashi, R. Patterson, Y. Ohnishi, H. Inomata, S. J. Ryan, Formation of drusen in the human eye, *Am. J. Ophthalmol.* **101**(3), 342-353 (1986).
21. G. Z. Xu, W. W. Li, M. O. Tso, Apoptosis in human retinal degenerations, *Trans. Am. Ophthalmol. Soc.* **94**, 411-430 (1996).
22. J. F. Kerr, A. H. Wyllie, A. R. Currie, Apoptosis: a basic biological phenomenon with wide-ranging implications in tissue kinetics, *Br. J. Cancer* **26**(4), 239-257 (1972).
23. C. B. Thompson, Apoptosis in the pathogenesis and treatment of disease, *Science* **267**(5203), 1456-1462 (1995).
24. I. Wertz, M. R. Hanley, Diverse molecular provocation of programmed cell death, *Trends Biol. Science (TIBS)* **21**, 359-364 (1996).
25. A. Ashkenazi, V. M. Dixit, Death receptors: signaling and modulation, *Science* **281**(5381), 1305-1308 (1998).
26. G. Kroemer, J. C. Reed, Mitochondrial control of cell death, *Nat. Med.* **6**(5), 513-519 (2000).
27. D. C. Wallace, Mitochondrial diseases in man and mouse, *Science* **283**(5407), 1482-1488 (1999).
28. M. Leist, B. Single, H. Naumann, E. Fava, B. Simon, S. Kuhnle, P. Nicotera, Inhibition of mitochondrial ATP generation by nitric oxide switches apoptosis to necrosis, *Exp. Cell Res.* **249**(2), 396-403 (1999).
29. S. A. Susin, H. K. Lorenzo, N. Zamzami, I. Marzo, B. E. Snow, G. M. Brothers, J. Mangion, E. Jacotot, P. Costantini, M. Loeffler, N. Larochette, D. R. Goodlett, R. Aebersold, D. P. Siderovski, J. M. Penninger, G. Kroemer, Molecular characterization of mitochondrial apoptosis-inducing factor, *Nature* **397**(6718), 441-446 (1999).

30. E. Daugas, S. A. Susin, N. Zamzami, K. F. Ferri, T. Irinopoulou, N. Larochette, M. C. Prevost, B. Leber, D. Andrews, J. Penninger, G. Kroemer, Mitochondrio-nuclear translocation of AIF in apoptosis and necrosis, *Faseb J.* **14**(5), 729-739 (2000).

31. S. A. Susin, H. K. Lorenzo, N. Zamzami, I. Marzo, C. Brenner, N. Larochette, M. C. Prevost, P. M. Alzari, G. Kroemer, Mitochondrial release of caspase-2 and -9 during the apoptotic process, *J. Exp. Med.* **189**(2), 381-394 (1999).

32. P. Esser, S. Grisanti, N. Kociok, H. Abts, A. Hueber, K. Unfried, K. Heimann, M. Weller, Expression and upregulation of microtubule-associated protein 1B in cultured retinal pigment epithelial cells, *Invest. Ophthalmol. Vis. Sci.* **38**(13), 2852-2856 (1997).

33. A. Schousboe, E. Meier, J. Dreijer, L. Hertz, Preparation of primary cultures of mouse (rat) cerebellar granule cells, in *A dissection and tissue culture manual of the nervous system*, A. Sahar, *et al.*, Editors. 1989, Liss, A.R.: New York. p. 203-206.

34. M. Leist, C. Volbracht, E. Fava, P. Nicotera, 1-Methyl-4-phenylpyridinium induces autocrine excitotoxicity, protease activation, and neuronal apoptosis, *Mol. Pharmacol.* **54**(5), 789-801 (1998).

35. C. Volbracht, M. Leist, P. Nicotera, ATP controls neuronal apoptosis triggered by microtubule breakdown or potassium deprivation, *Mol. Med.* **5**(7), 477-489 (1999).

36. M. Leist, B. Single, A. F. Castoldi, S. Kuhnle, P. Nicotera, Intracellular adenosine triphosphate (ATP) concentration: a switch in the decision between apoptosis and necrosis, *J. Exp. Med.* **185**(8), 1481-1486 (1997).

37. M. Leist, E. Fava, C. Montecucco, P. Nicotera, Peroxynitrite and nitric oxide donors induce neuronal apoptosis by eliciting autocrine excitotoxicity, *Eur. J. Neurosci.* **9**(7), 1488-1498 (1997).

38. B. Single, M. Leist, P. Nicotera, Simultaneous release of adenylate kinase and cytochrome c in cell death, *Cell Death Differ.* **5**(12), 1001-1003 (1998).

39. M. Latta, G. Kunstle, M. Leist, A. Wendel, Metabolic depletion of ATP by fructose inversely controls CD95- and tumor necrosis factor receptor 1-mediated hepatic apoptosis, *J. Exp. Med.* **191**(11), 1975-1986 (2000).

40. M. Suter, C. E. Reme, C. Grimm, A. Wenzel, M. Jaattela, P. Esser, N. Kociok, M. Leist, C. Richter, Age-related macular degeneration: The lipofuscin component A2E detaches pro-apoptotic proteins from mitochondria and induces apoptosis in mammalian retinal pigment epithelial cells, *J Biol Chem* **275**, 39625-39630 (2000).

41. G. E. Eldred, M. L. Katz, Fluorophores of the human retinal pigment epithelium: separation and spectral characterization, *Exp. Eye Res.* **47**(1), 71-86 (1988).

42. J. Schlegel, M. Schweizer, C. Richter, 'Pore' formation is not required for the hydroperoxide-induced Ca2+ release from rat liver mitochondria, *Biochem. J.* **285**(Pt 1), 65-69 (1992).

43. J. Cai, D. P. Jones, Superoxide in apoptosis. Mitochondrial generation triggered by cytochrome c loss, *J. Biol. Chem.* **273**(19), 11401-11404 (1998).

44. C. Richter, M. Schweizer, A. Cossarizza, C. Franceschi, Control of apoptosis by the cellular ATP level, *FEBS Lett.* **378**(2), 107-110 (1996).

45. J. F. Krebs, R. C. Armstrong, A. Srinivasan, T. Aja, A. M. Wong, A. Aboy, R. Sayers, B. Pham, T. Vu, K. Hoang, D. S. Karanewsky, C. Leist, A. Schmitz, J. C. Wu, K. J. Tomaselli, L. C. Fritz, Activation of membrane-associated procaspase-3 is regulated by Bcl-2, *J Cell Biol* **144**(5), 915-926 (1999).

46. R. J. Carmody, T. G. Cotter, Oxidative stress induces caspase-independent retinal apoptosis in vitro, *Cell Death Differ.* **7**(3), 282-291 (2000).

47. J. Nylandsted, M. Rohde, K. Brand, L. Bastholm, F. Elling, M. Jäättela, Selective depletion of heat shock protein 70 (Hsp70) activates a tumor-specific death program that is independent of caspases and bypasses Bcl-2, *Proc. Natl. Acad. Sci. U S A* **97**(14), 7871-7876 (2000).

48. Z. Salamon, G. Tollin, Surface plasmon resonance studies of complex formation between cytochrome c and bovine cytochrome c oxidase incorporated into a supported planar lipid bilayer. II. Binding of cytochrome c to oxidase-containing cardiolipin/phosphatidylcholine membranes, *Biophys. J.* **71**(2), 858-867 (1996).

49. G. L. Wing, G. C. Blanchard, J. J. Weiter, The topography and age relationship of lipofuscin concentration in the retinal pigment epithelium., *Invest. Ophthalmol. Vis. Sci.* **17**, 601-607 (1978).

50. D. M. Snodderly, Evidence for protection against age-related macular degeneration by carotenoids and antioxidant vitamins, *Am. J. Clin. Nutr.* **62**(6 Suppl), 1448S-1461S (1995).

51. B. N. Ames, M. K. Shigenaga, T. M. Hagen, Mitochondrial decay in aging, *Biochim. Biophys. Acta* **1271**(1), 165-170 (1995).

52. A. Maftah, M. H. Ratinaud, M. Dumas, F. Bonte, A. Meybeck, R. Julien, Human epidermal cells progressively lose their cardiolipins during ageing without change in mitochondrial transmembrane potential, *Mech. Ageing. Dev.* **77**(2), 83-96 (1994).

DHA LEVELS IN ROD OUTER SEGMENTS OF TRANSGENIC MICE EXPRESSING G90D RHODOPSIN MUTATIONS

Robert E. Anderson[1,2,3], Paul A. Sieving[4], Ronald A. Bush[4], Maureen B. Maude[1,3], Ting-Huai Wu[5], Muna I. Naash[2]

1. INTRODUCTION

Previous studies in humans with a variety of inherited retinal degenerations have shown that they have reduced blood levels of docosahexaenoic acid (DHA, 22:6n-3) (ref. 1-10), the major fatty acid in photoreceptor rod outer segment (ROS) membranes[11]. The greatest differences are seen in people with x-linked retinitis pigmentosa[7] and Usher's syndrome type II (ref 10). Since the gene defect in these two types of inherited retinal degeneration is different, it is not likely that the reduction in blood levels of DHA is due directly to the mutation. Until recently, none of the mutations identified in retinitis pigmentosa were found in genes encoding proteins that are involved in lipid metabolism. However, Zhang et al[12] found that two forms of macular degeneration (Stargardt-like macular dystrophy and autosomal dominant macular dystrophy) map to a 0.6 cM interval on chromosome 6q14 that contains a new retinal photoreceptor-specific gene, *ELOVL4*, which has sequence homology to a yeast protein that functions in the synthesis of very long chain fatty acids (also see chapter in this book).

Reduced levels of DHA have been found in plasma from miniature poodles with progressive rod cone degeneration (*prcd*)[13]. ROS from *prcd*-affected animals also have lower DHA levels than controls[14], indicating for the first time a difference in ROS membrane lipid composition in this animal model of human RP. Since studies have shown that DHA is important in maintaining optimal retinal function[15, 16], Aguirre et al.[14] supplemented *prcd*-affected dogs with fish oil, which contains high levels of DHA, to determine the effect on retinal viability in these animals. No beneficial effects of DHA

Departments of [1]Ophthalmology and [2]Cell Biology, University of Oklahoma Health Sciences Center and [3]Dean A. McGee Eye Institute, Oklahoma City, OK; [4]Department of Ophthalmology, Kellogg Eye Center, Ann Arbor, MI; and [5]Department of Ophthalmology, Wilmer Eye Institute JHU, Baltimore, MD.

New Insights Into Retinal Degenerative Diseases
Edited by Anderson *et al.*, Kluwer Academic/Plenum Publishers, 2001

supplementation were observed on either function or structure of the retina of *prcd*-affected dogs. Although supplementation increased ROS DHA levels in both *prcd*-affected and normal animals, the affected animals had less DHA than the normal dogs.

To determine if reduced ROS DHA was only found in *prcd*-affected dogs, we have examined the ROS of a number of mouse and rat lines that contain natural and transgenically mediated, retina-specific mutations that result in a retinal degeneration. Mice with natural *rds*/peripherin or transgenically engineered P216L peripherin mutation have lower DHA levels in ROS compared to normal mice[17]. Rats with P23H or S334ter mutations in the rhodopsin gene also have lower ROS DHA levels than controls (Anderson, et al., unpublished results). In the present study, we examined the DHA levels in the ROS of mice in which the G90D mutation in rhodopsin has been introduced. This mutation, which has been identified as one of the mutations in humans with congenital stationary night blindness (CSNB), a disease in humans that leads to a very slow degeneration of photoreceptor membranes. Two groups of transgenic animals were generated independently (one by Sieving's group, the other by Naash's group) and each had several lines that expressed the transgene at different levels.

2. MATERIALS AND METHODS

2.1. Animals

The animals were born and raised in the Biological Research Laboratory (BRL) at the University of Illinois at Chicago (UIC) or the Department of Ophthalmology, W.K. Kellogg Eye Center at the University of Michigan (UM). Mice at UIC were maintained on a normal mouse diet, whereas mice at UM were fed a high fat breeding chow (Formulab diet 5008, PMI, Nutrition International, Inc., Brentwood, MO.) with a higher n-3 fatty acid content than standard chow (manufacturer's information). Animals were maintained on a 12D/12L cyclic lighting at an intensity that never exceeded 50 lux outside of the cages. The experiments described were in compliance with the Association for Research in Vision and Ophthalmology Statement for the Use of Animals in Ophthalmic and Vision Research and all protocols were reviewed and approved by the animal protocol review committees at all institutions.

2.1.1. Sieving mice

A 15 Kb genomic clone was isolated from an adult Balb/C mouse liver genomic lambda library (Clontech) containing 5 Kb upstream from rhodopsin exon 1, and two site-directed mutagenesis procedures were performed to insert the G90D mutation and to create A337V, which substitutes human V337 opsin for mouse A337 opsin. This allowed the use of the 3A6 antibody (gift of Dr. Robert Molday) to track mutant opsin expressed by the transgene. To avoid problems of rhodopsin overexpression as possibly contributing to rod degeneration[18], and in order to match more closely the genomic constitution of wild type and G90D rhodopsin present in the human condition, the G90D transgenic line was crossed with rhodopsin knockout mice[19]. Two genotypes were then used in this study, (G+/-, R+/+) and (G+/-, R+/-), corresponding to a single transgene allele expressing on either the two wild type allele normal retina or the one wild type allele heterozygous rhoKO retina. These were compared with the wild type (R+/+, two allele) and the heterozygous rhoKO (R+/-, one allele) retina. Animals were housed in controlled lighting measuring 5 lux at cage level on a 12:12 h light:dark cycle.

2.1.2 Naash mice

The G90D transgene was made of 15 Kb of a mouse genomic fragment that contains 5 Kb of all the mouse opsin exons and introns, as well as a 6.0 Kb and 3.5 Kb of the 5' and 3' flanking sequences, respectively. The mutation was introduced in exon 1 by site directed mutagenesis in PCR reaction as described[20]. Several other silent mutations were also introduced to create RFLP and to change the sequence in that region to allow the differentiation between the transgene and the endogenous gene transcripts at the DNA and RNA level. Three transgenic lines were generated that express the transgene at high, medium and low levels. Animals used for this study were generated from mating transgenic mice from these lines to opsin knockout mice (-/-) to express the transgene in opsin hemizygous background.

2.2. Tissue Collection

Animals were euthanized with carbon dioxide asphyxiation and eyes were enucleated. Retinas were removed rapidly and 6-8 from each group were frozen in microfuge tubes and stored at -80°C until they were sent to Oklahoma on dry ice. Tubes were numbered so that all analyses were performed without knowledge of the identity of the samples. Retinas of transgenic mice from the Naash group were removed between postnatal days 20 and 25, while those from the Sieving group were removed at 21 to 31 weeks of age.

2.3. Preparation of Rod Outer Segment Membranes

ROS were prepared from frozen retinas on a discontinuous sucrose gradient as described by Wiegand et al.[21]. Purity of membrane preparations was determined by SDS-polyacrylamide gel electrophoresis.

2.4. Lipid Extraction and Analysis

Lipids were extracted by the procedure of Bligh and Dyer[22]. Aliquots of each lipid extract were used for preparation of methyl esters, which were analyzed by gas liquid chromatography as previously described[21].

2.5. Data Analysis

Results are reported as relative mole percent. Differences between groups were determined by student's t-test.

3. RESULTS

3.1. Relative Expression of the Transgene

The G90D retinas were heterozygous for the transgene and in the wild type background. Three lines from the Naash group had relative levels of expression of the transgene of ~0.5, ~1.7, and ~2.5 compared to the wild type opsin. In the Sieving animals, one group was heterozygous for G90D with an ~1:1 relative expression. Another group had G90D on a wild type background, with a relative expression of ~1:2. A third group of mice was heterozygous for rhodopsin knockout and expressed one normal rhodopsin.

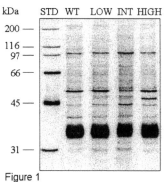

Figure 1

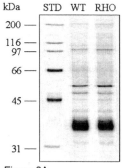

Figure 2A

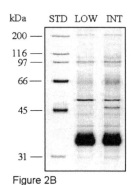

Figure 2B

3.2. Purity of ROS Preparations

Silver-stained polyacrylamide gels of ROS prepared from the Naash mice (Figure 1) and Sieving mice (Figure 2A and 2B) show that the major protein in all preparations is opsin. There is no evidence of contamination of these preparations with membranes from other organelles.

3.3. Fatty Acid Composition of ROS Membranes

The composition of the fatty acids from the total lipids of ROS of G90D transgenic mice from the Naash group is given in Table 1. Results are grouped according to level of saturation and values are presented only for those fatty acids present in amounts greater than 0.2%. Data are compared from mice with low, intermediate, and high levels of expression of the transgene relative to wild type opsin. The retinas from the Naash animals were taken from young animals, while those from the Sieving group were from animals that were older. In the Naash animals, there is no significant difference between the levels in any fatty acid in the low and intermediate expressing groups, compared to the wild type. However, even at 3 weeks of age, there was a significant reduction in the level of DHA in the high expressing animals compared to the wild type. The reduction in n-3 fatty acids in the high expressers was compensated for by increases in saturated, monoenoic, and n-6 fatty acids.

Table 2 contains the fatty acid composition of the total ROS lipids from the Sieving mice, which are older (21-31 weeks) than the Naash mice. The major differences in fatty acids were found in comparing the intermediate-expressing group with the wild type animals, where there was a significant reduction in DHA and total n-3 fatty acids in the intermediate group. Relative increases in the saturated, monoenoic, and n-6 fatty acids compensated for this reduction. The rhodopsin knockout heterozygous animals expressing one normal opsin also have lower ROS DHA than the wild type animals.

Table 1. Fatty Acid Composition of Total Lipids of ROS of G90D Transgenic Mice (Naash)*

FATTY ACID**	WILD TYPE		TRANSGENIC				
			Low***		Intermediate****		High*****
	Mean ± SD		Mean ± SD		Mean ± SD		Mean ± SD
Saturate							
14:0	0.3	0.0	0.4	0.1	0.4	0.1	0.5	0.1
16:0	19.1	0.7	20.0	0.4	19.1	1.0	21.8	0.3[b]
18:0	21.2	1.5	21.1	0.7	22.2	0.4	23.8	0.6
Subtotal	41.0	1.9	41.8	0.6	42.2	1.5	46.7	0.5[c]
Monoenoic							
16:1	0.4	0.0	0.4	0.1	0.5	0.1	0.7	0.3
18:1	6.6	0.2	8.1	1.5	7.5	1.0	10.5	0.3[a]
22:1	0.4	0.2	0.5	0.5	0.5	0.2	0.5	0.1
Subtotal	7.7	0.4	9.3	1.4	8.8	1.0	12.0	0.6[a]
N-6							
18:2	1.0	0.2	1.6	0.5	1.0	0.3	1.3	0.7
20:3	0.5	0.1	0.5	0.1	0.4	0.1	0.5	0.1
20:4	5.7	0.6	5.8	0.2	5.9	0.5	7.3	0.5[c]
22:4	0.9	0.1	1.0	0.1	1.0	0.1	1.1	0.1
22:5	0.3	0.1	0.2	0.0	0.2	0.0	0.3	0.1
24:4	0.3	0.1	0.2	0.1	0.3	0.0	0.2	0.0
Subtotal	8.9	0.8	9.7	0.3	9.3	0.9	11.7	0.5[c]
N-3							
22:5	0.6	0.1	0.7	0.1	0.7	0.0	0.7	0.1
22:6	40.7	2.1	37.5	2.3	37.7	3.5	27.4	0.9[b]
24:6	0.6	0.2	0.6	0.3	0.6	0.2	0.8	0.3
Subtotal	42.5	1.9	39.2	2.0	39.7	3.4	29.7	1.2[a]
N-6/N-3	0.21	0.02	0.25	0.02	0.24	0.04	0.39	0.03[b]

*	Results are expressed as mole % ± S.D. for at least 3 independent ROS preparations.
**	Fatty acids present in less than 0.2% are not listed, but their values are included in the subtotals.
***	0.5:1-Level of expression of the transgene to WT opsin.
****	1.7:1-Level of expression of the transgene to WT opsin.
*****	2.5:1-Level of expression of the transgene to WT opsin.
a -	P<0.001 compared to wild type
b -	P<0.01 compared to wild type
c -	P<0.05 compared to wild type

Table 2. Fatty Acid Composition of Total Lipids of ROS of G90D Transgenic Mice (Sieving)*

FATTY ACID**	WILD TYPE Mean ± SD		TRANSGENIC					
			RhoKo het*** Mean ± SD		Low**** Mean ± SD		Intermediate***** Mean ± SD	
Saturate								
14:0	0.5	0.2	0.7	0.1	0.6	0.2	0.8	0.3
16:0	17.8	1.1	18.6	0.1	20.6	1.5	19.7	1.3
18:0	19.9	1.8	23.5	1.8	19.7	0.6	21.5	0.8
20:0	0.3	0.0	0.3	0.0	0.3	0.0	0.5	0.2^c
Subtotal	38.7	2.1	43.2	1.8^c	41.4	2.4	42.7	2.1^c
Monoenoic								
16:1	0.3	0.0	0.4	0.1	0.4	0.1^c	0.5	0.1^b
18:1	6.9	0.3	8.7	0.9	8.4	0.6^c	10.3	1.8^b
20:1	0.2	0.0	0.2	0.0	0.3	0.0	0.3	0.1^c
22:1	0.4	0.4	0.5	0.0	0.3	0.1	0.6	0.4
24:1	0.2	0.1	0.5	0.1	0.2	0.2	0.5	0.4
Subtotal	8.04	0.8	10.2	0.9^c	9.4	0.9	12.2	2.3^b
N-6								
18:2	0.7	0.1	0.7	0.0	0.8	0.1	0.9	0.1
20:3	0.4	0.1	0.3	0.0	0.4	0.0	0.3	0.0
20:4	2.9	0.5	3.9	0.1	3.4	0.1	3.9	0.3
22:4	0.4	0.1	0.4	0.0	0.4	0.0	0.5	0.1
Subtotal	4.6	0.4	5.5	0.1	5.1	0.2	5.9	0.4^c
N-3								
22:5	0.2	0.0	0.2	0.1	0.2	0.0	0.7	0.2^b
22:6	46.3	2.8	39.4	1.0^c	42.1	3.1	36.5	3.4^b
24:5	0.7	0.1	0.5	0.4	0.6	0.0	0.8	0.2
24:6	1.3	0.1	0.8	0.1^c	1.1	0.1^c	1.0	0.1^a
Subtotal	48.7	2.9	41.1	1.5^c	44.1	3.2	39.2	3.3^b
N-6/N-3	0.09	0.01	0.13	0.01^c	0.12	0.01	0.15	0.02^b

*   Results are expressed as mole % ± S.D. for at least 3 independent ROS preparations.
**   Fatty acids present in less than 0.2% are not listed, but their values are included in the subtotals.
***   Rhodopsin knockout animal expressing one normal opsin.
**** 0.5:1-Level of expression of the transgene to WT opsin.
*****   1:1-Level of expression of the transgene to WT opsin.
a -   P<0.001 compared to wild type
b -   P<0.01 compared to wild type
c -   P<0.05 compared to wild type

There are several noteworthy differences between the data in Table 1 and Table 2. Comparison of the wild type composition shows that the Sieving animals have higher levels of n-3 fatty acids and lower levels of n-6 fatty acid than the Naash animals. The difference in the fatty acid composition is most apparent in the polyunsaturated fatty acids, where the n-6/n-3 ratio of the Naash animals is about twice that of the Sieving animals. This may reflect differences in strain of animals, as well as husbandry conditions and diet. The diet may be the most important variable here because the breeding animals at UM received a high fat diet.

4. DISCUSSION

In this study, we compared the fatty acid compositions of rod outer segments prepared from mice expressing various levels of a G90D mutant opsin, compared to wild type opsin. The two groups of transgenic animal were generated independently by two research groups (Naash and Sieving). Some phenotype differences were observed between the two laboratories, but will not be dealt with here. The Sieving animals showed no evidence of a retinal degeneration, whereas the Naash mice, especially the high expressers, showed loss of photoreceptor cell nuclei at an early age.

Comparison of the wild type compositions from the two laboratories shows significant differences between the animals, which may reflect differences in diet, vivarium environment, and animal strain. It is especially interesting that the Naash animals had significantly greater amounts of n-6 polyunsaturated fatty acids than the Sieving animals and less n-3 acids. These differences are readily apparent in the n-6/n-3 ratios; elevated n-6/n-3 ratios are usually indicative of a dietary deficiency of n-3 fatty acids[21,23,24]. However, in all such instances, the increase in n-6 fatty acids in neuronal membranes is always found in 22:5n-6 in ROS of animals fed n-3 deficient diets, and the reduction in DHA is always accompanied by a one-for-one increase in 22:5n-6. This is definitely not the case in the present study, since 22:5n-6 levels in Table 1 are 0.2-0.3%. However, the levels of 20:4n-6 in the Naash animals are about twice those of the Sieving mice. A possible source of the difference is the breeding chow fed to the Sieving mice, which is higher in n-3 fatty than standard chow. If so, this would be the first time, to our knowledge, that 20:4n-6 was used instead of 22:5n-6 to replace DHA in ROS membranes.

The three different Naash lines express different transgene copy number always on two-allele wild type background, and the high expression animals have putatively five copies of transgene on a background of two wild type alleles. Only these high expressing 2.5:1 mice have reduced DHA at three weeks of age, and these mice show degeneration. It is known that degeneration results from overexpression of even normal rhodopsin[18], although the DHA level in the ROS of these animals has not been determined. Consequently, we cannot conclude that G90D rhodopsin alone causes alteration in ROS DHA. It may be that the majority of DHA alteration can be ascribed to photoreceptor degeneration. However, what is clear from these analyses is that the DHA level in the ROS (Table 1) is dramatically lower than that in ROS from the wild type animals.

In comparing wild type versus heterozygous rho-KO (Table 2), it is apparent that the DHA levels are affected by the quantity of opsin apoprotein expression, since the loss of a single allele of wild type opsin led to a significant decline of about 7% in DHA. In the intermediate expressing Sieving mice (1WT:1G90D), G90D opsin apoprotein is expressed at about 2/3 of the single wild type allele (Sieving et al., in press). Thus, one might expect this group to have DHA levels within 2 or 3% of mice with two wild type alleles, and the low expression group (2WT:1G90D) to actually have higher levels of DHA than wile type, assuming a strict dependence on total ROS opsin. However, the G90D group with two wild type alleles was lower in DHA content (though not significantly, p<0.16) than "wild type", and the group with one wild type and one G90D allele was 10% lower (p<0.01). Of the two groups with one wild type allele, the group with an additional G90D allele had lower levels of DHA (though not significantly, p<0.20). The fact that mice with the same number of wild type alleles tended to have lower DHA levels if they also have a G90D allele suggests that G90D opsin expression may be related to lower DHA levels. However, whether these lower levels are strictly related to the expression of G90D, to total ROS opsin content, or to degeneration cannot be determined from the data in Tables 1 and 2. Further experiments will be required in which we carefully measure the level of opsin expression and degeneration in transgenic mice with different levels of wild type and G90D apoprotein.

The reason for the reduced levels of DHA in ROS membranes of animals expressing mutant proteins is not known. When the low level of DHA in blood lipids of animals and humans with inherited retinal degenerations was first discovered, we thought that the degeneration itself may result from a deficiency of DHA and that, in fact, the mutation may be in a gene encoding proteins that metabolize polyunsaturated fatty acids[13]. However, subsequent studies showed that the retina and retinal pigment epithelium of *prcd*-affected dogs could synthesize DHA from n-3 precursors[25,26]. Also, with the large number of different mutations that are now known, it is apparent that the reduced level of DHA is not the result of an abnormal lipid metabolizing protein. Nevertheless, given the importance of DHA in retinal function, which was established over the last 30 years by studies in numerous laboratories, it seemed reasonable to propose that the loss of DHA could potentially lead to a retinal degeneration. To test this hypothesis, Aguirre et al.[14] carried out a clinical trial *prcd*-affected dogs in which they fed a supplement containing DHA for 5 months. Affected animals received no protective benefit of DHA supplementation, although the level of DHA in their ROS membranes exceeded that of membranes from control animals supplemented with placebo oil. Therefore, it is now clear, at least for the *prcd*-affected dogs, that increasing ROS DHA levels provides no protective benefit.

Why then are the levels of DHA lower in the ROS membranes of animals with inherited retinal degeneration? Several years ago we discovered that albino rats born and raised in bright cyclic light had reduced levels of DHA in the ROS compared to animals raised in dim cyclic light[24,27,28]. These animals also had increased retinal levels of vitamin E and vitamin C, as well as increased activities of glutathione reductase, glutathione peroxidase, and glutathione-S-transferase. We proposed that the animals raised in bright cyclic light were stressed more than those raised in dim cyclic light, and that the response we measured was to an oxidant stress caused by bright cyclic light[29] An interesting feature of these animals was that the reduction in DHA in the ROS was not accompanied by an increase in 22:5n-6 (ref.24,27). Thus, the fatty acid patterns of these animals resemble those that we have found for animals with inherited retinal degenerations, and was completely different from that induced by dietary manipulations. Because we have consistently found lower levels of DHA in the ROS of animals with inherited retinal degeneration, we proposed that the reduced DHA might be a response to a metabolic stress caused by the mutation[30,31].

The results for the G90D mice present something of a paradox, especially for the Sieving animals, because there is no retinal degeneration in these animals (Sieving, personal communication). Therefore, it seems that the mutation alone, in the absence of retinal degeneration, may be sufficient to cause alteration in DHA levels of these membranes. The cause of night blindness in humans with the G90D mutations is thought to be the activation of the rhodopsin molecule in the absence of light[32]. This constitutive activity in G90D mice, though at a level too low to cause "light" damage to the retina, may have an effect similar to bright light rearing on ROS lipid content. It is also interesting, as noted above, that in the rhodopsin knockout heterozygotes, the only difference between them and wild type animals was the level of expression of the normal opsin gene. This result suggests that not only the type, but also the level of expression of integral membrane protein may be sufficient to alter the composition of the fatty acids that makes up the membrane lipid bilayer. This conclusion is reasonable in light of the fact that DHA is co-transported with opsin in membrane synthesis[33], and suggests that G90D opsin may lack this association.

We have now found reduced levels of DHA in the ROS membrane phospholipids of several animals with mutations that cause loss of function or retinal degeneration. Since the clinical trials in dogs discussed above did not show a beneficial affect of DHA supplementation, we propose that the reduced levels of DHA in these animals are an adaptation to an oxidant stress caused by the mutation. To further test this hypothesis, we are currently carrying out clinical trials on transgenic rats expressing P23H or S334ter mutations, both of which have lower DHA levels in their ROS (Anderson et al, unpublished results).

5. ACKNOWLEDGMENTS

This work was supported by grants from the National Institutes of Health/National Eye Institute (EY00871, EY04149, and EY12190 to REA; EY10609 and EY01792 to MIN; and EY6094 to PAS), Research to Prevent Blindness, Inc., New York, NY; and The Foundation Fighting Blindness, Baltimore, MD. Dr. Naash is a recipient of the Research to Prevent Blindness James S. Adams Scholar award. We would like to thank Austra Liepa for overseeing breeding and shipment of the animals at the University of Michigan.

REFERENCES

1. Anderson, R.E., Maude, M.B., Lewis, R.A., Newsome, D.A., and Fishman, G.A. (1987). Abnormal plasma levels of polyunsaturated fatty acid in autosomal dominant retinitis pigmentosa. *Exp. Eye Res.* **44**:155-159.
2. Converse, C.A., Hammer, H.M., Packard, C.J., and Shepherd, J. (1983). Plasma lipid abnormalities in retinitis pigmentosa and related conditions. *Trans. Ophthalmol. Sci. UK* **103**:508-512.
3. Converse, C.A., McLachlan, T., Bow, A.C., Packard, C.J., and Shepherd, J. (1987). Lipid metabolism in retinitis pigmentosa. In Degenerative Retinal Disorders: Clinical and Laboratory Investigations, (Eds. Hollyfield, J.G., Anderson, R.E., and LaVail, M.M.), Alan R. Liss, Inc. (New York), pp. 93-101.
4. Schaefer, E.J., Robins, S.J., Patton, G.M., Sandberg, M.A., Weigel-DiFranco, C.A., Rosner, B., and Berson, E.L. (1995). Red blood cell membrane phosphatidylethanolamine fatty acid content in various forms of retinitis pigmentosa. *J. Lipid Res.* **36**:1427-33.
5. Gong, J., Rosner, B., Rees, D.G., Berson, E.L., Weigel-DiFranco, C.A., and Schaefer, E.J. (1992). Plasma docosahexenoic acid levels in various genetic forms of retinitis pigmentosa. *Invest. Ophthalmol. Vis. Sci.* **33**:2596-2602.
6. Hoffman, D.R., Uauy. R., and Birch, D.G. (1993b). Red blood cell fatty acid levels in patients with autosomal dominant retinitis pigmentosa. *Exp. Eye Res.* **57**:359-368.
7. Hoffman, D.R. and Birch, D.G. (1995). Docosahexaenoic acid in red blood cells of patients with X-linked retinitis pigmentosa. *Invest Ophthalmol. Vis. Sci.* **36**:1009-18.

8. Hoffman, D.R. and Birch, D.G. (1998). Omega 3 fatty acid status in patients with retinitis pigmentosa. *World Rev Nutr Diet* **83**:52-60.

9. Holman, R.T., Bibus, D.M., Jeffrey, G.H., Smethurst, P., and Crofts, J.W. (1994). Abnormal plasma lipids of patients with retinitis pigmentosa. *Lipids.* **29**:61-65.

10. Maude, M.B., Anderson, E.O., and Anderson, R.E. (1998). Polyunsaturated fatty acids are lower in blood lipids of Usher's Type I, but not Usher's Type II. *Invest. Ophthalmol. Vis. Sci.* **39**:2164-2166.

11. Fliesler, S.J. and Anderson, R.E. (1983). Chemistry and metabolism of lipids in the vertebrate retina. In Progress in Lipid Research, (Ed. R.T. Holman), Pergamon Press (London) **22**:79-131.

12. Zhang, K., Kniazeva, M., Han, M., Li, W., Yu, Z., Yang, Z., Li, Y., Metzker, M.L., Allikmets, R., Zack, D.J., Kakuk, L.E., Lagali, P.S., Wong, P.W., MacDonald, I.M., Sieving, P.A., Figueroa, D.J., Austin, C.P., Gould, R.J., Ayyagari, R., Petrukhin, K. (2001). A 5-bp deletion in ELOVL4 is associated with two related forms of autosomal dominant macular dystrophy. *Nat. Genet.* **27**:89-93.

13. Anderson, R.E., Maude, M.B., Alvarez, R.A., Acland, G.M., and Aguirre, G. (1991a). Plasma lipid abnormalities in the miniature poodle with progressive rod-cone degeneration. *Exp. Eye. Res.* **52**:349-355.

14. Aguirre, G.D., Acland, G.M., Maude, M.B., and Anderson, R.E. (1997). Diets enriched in docosahexaenoic acid fail to correct progressive rod-cone degeneration (*prcd*) phenotype. *Invest. Ophthalmol. Vis. Sci.* **38**:2387-2407.

15. Benolken, R.M., Anderson, R.E. and Wheeler, T.G. (1973). Membrane fatty acids associated with the electrical response in visual excitation. *Science* **182**:1253-1254.

16. Wheeler, T.G., Benolken, R.M. and Anderson, R.E. (1975). Visual membrane: Specificity of fatty acid precursors for the electrical response to illumination. *Science* **188**:1312-1314.

17. Anderson, R.E., Maude, M.B., and Bok, D. (2001). Mice with rds and P216L peripherin mutations have lower docosahexaenoic acid levels in rod outer segment membranes than control mice. *Invest. Ophthalmol. Vis. Sci.*, In Press.

18. Tan, E., Wang, Q., Quiambao, A.B., Xu, X., Qtaishat, N.M., Peachey, N.S., Lem, J., Fliesler, S.J., Pepperberg, D.R., Naash, M., and Al-Ubaidi, M.R. (2001). The relationship between opsin overexpression and photoreceptor degeneration. *Invest. Ophthalmol. Vis. Sci.* **42**:589-600.

19. Humphries, M.M., Rancourt, D., Farrar, G.J., Kenna, P., Hazel, M., Bush, R.A., Sieving, P.A., Sheils, D.M., McNally, N., Creighton, P., Erven, A., Boros, A.. (1997). Retinopathy induced in mice by targeted disruption of the rhodopsin gene. *Nat. Genet.* **15**:216-9.

20. Gulya K, Capecchi MR, Humphries P.Naash, M.I., Hollyfield, J.G., Al-Ubaidi, M.R. and Baehr, W. (1993). Simulation of human autosomal dominant retinitis pigmentosa in transgenic mice expressing a mutated murine opsin gene. *Proc. Nat. Acad. Sci. USA* **90**:5499-5503).

21. Wiegand, R.D., Koutz, C.A., Stinson, A.M., and Anderson, R.E. (1991). Conservation of docosahexaenoic acid in rod outer segments of rat retinas during n-3 and n-6 fatty acid deficiency. *J. Neurochem.* **57**:1690-1699.

22. Bligh, E.G. and Dyer, W.J. (1959). A rapid method of total lipid extraction and purification. *Can. J. Biochem. Physiol.* **37**:911-917.

23. Anderson, R.E. and Maude, M.B. (1972). Lipids of ocular tissues: VII. The effects of essential fatty acid deficiency on the phospholipids of the photoreceptor membranes of rat retina. *Arch. Biochem. Biophys.* **151**:270-276.

24. Wiegand, R.D., Koutz, C.A., Chen, H., and Anderson, R.E. (1995). Effect of dietary fat and environmental lighting on the phospholipid molecular species of rat photoreceptor membranes. *Exp. Eye Res.* **60**:291-306.

25. Alvarez, R.A., Aguirre, G.D., Acland, G.M., and Anderson, R.E. (1994). Docosapentaenoic acid is converted to docosahexaenoic acid in the retinas of normal and *prcd*-affected miniature poodle dogs. *Invest. Ophthalmol. Vis. Sci.* **35**:402-408.

26. Chen, H., Ray, J., Scarpino, V., Acland, G.M., Aguirre, G.A., and Anderson, R.E. (1999). Docosahexaenoic acid is synthesized and released by the retinal pigment epithelial cells of *prcd*-affected dogs. *Invest. Ophthalmol. Vis. Sci.* **40**:2418-2422.

27. Penn, J.S. and Anderson, R.E. (1987). Effect of light history on rod outer-segment membrane composition in the rat. *Exp. Eye Res.* **44**:767-778.

28. Penn, J.S., Naash, M.I., and Anderson, R.E. (1987). Effect of light history on retinal antioxidants and light damage susceptibility in the rat. *Exp. Eye Res.* **44**:779-788.

29. Penn, J.S. and Anderson, R.E. (1992). Effects of light history on the rat retina. In *Progress in Retinal Research*, Vol. 11, edited by N. Osborne and G. Chader, Pergamon Press (New York), pp. 75-98.

30. Anderson, R.E., Maude, M.B., Alvarez, R.A., Acland, G., and Aguirre, G.D. (1999a). Inherited retinal Degenerations-Role of Polyunsaturated Fatty Acids. Les Seminaires Ophthalmologiques d'IPSEN, ed Y. Christen, M. Doly, and M.-T. Droy-Lefaix. Irvinn (Paris). pp. 57-65.

31. Anderson, R.E., Maude, M.B., Alvarez, R.A., Acland, G., and Aguirre, G.D. (1999b). A hypothesis to explain the reduced blood levels of docosahexaenoic acid in inherited retinal degenerations caused by mutations in genes encoding retina-specific proteins. *Lipids.* **34**:S235-238.
32. Rodriguez de Turco, E.B., Deretic, D., Bazan, N.G., Papermaster, D.S. (1997). Post-Golgi vesicles cotransport docosahexaenoyl-phospholipids and rhodopsin during frog photoreceptor membrane biogenesis. *J. Biol. Chem.* **272**:10491-70497.

APPLICATION OF POLYMERASE CHAIN REACTION IN DELETION AND INVERSION OF INTERNAL DNA FRAGMENTS

Tong Cheng, Muna I. Naash, and Muayyad R. Al-Ubaidi*

1. INTRODUCTION

Deletion and inversion of an internal DNA fragment has been wildly used in delineating promoter activity and determining the function of certain *cis*-elements in the 5' flanking region of genes (1-5). These are generally achieved by restriction enzyme digestion and ligation. In most cases, such convenient restriction enzymes are not available. In this chapter, we describe a strategy of deleting or inverting a DNA fragment using a PCR based technique. As an example, an 115 bp element within a 931 bp of the immediate 5' flanking region of the mouse peripherin/rds gene (cloned in pBScript vector) was deleted. The created DNA fragment can then be used to test for its promoter activity and the functional importance of the deleted element can be evaluated. We also inverted a 158 bp element within a 3304 bp fragment of the same promoter. Similarly, the functional as well as the directional importance of the manipulated fragment can be studied.

2. MATERIALS AND METHODS

2.1. Polymerase Chain Reaction

Unless otherwise stated, 200 ng template DNA and 10 pmol of each primer were used in a 50 µl PCR amplification. For products of less than 2 kb in size, PCR was performed in Perkin Elmer 480 (Perkin Elmer, Foster City, CA) using 1 cycle at 94°C for 4 min/55°C for 2 min followed by 30 cycles at 94°C for 4 min/55°C for 2 min/72°C for 2 min. For larger size products, PCR was performed with the first 10 cycles at 92/60/68°C (10 sec, denaturation; 30 sec, annealing; 8 min, extension) followed by 20 cycles at 92/60/68°C

Tong Cheng, is presently a member of The Center for Aging and Developmental Biology, University of Rochester, Rochester, New York, 14642, Muna I. Naash, and Muayyad R. Al-Ubaidi are on the Cell Biology faculty of the University of Oklahoma Health Sciences Center, Oklahoma City, Oklahoma, 73104.

(10 sec, denaturation; 30 sec, annealing; 8 min, extension with 20 sec increment each cycle). One unit of a mixture of *Taq* (Boehringer Mannheim, Indianapolis, IN) and *Pfu* polymerases (Stratagene, La Jolla, CA) (20U:1U) was used for each amplification and the PCR products were separated on a 1% TBE agarose gel and detected with ethidium bromide. The products were then cut out of the agarose gel and purified using GeneClean Kit (Clontech, Palo Alto, CA).

2.2. Extension Reaction

Two PCR products were mixed at 1:1 ratio and used in one cycle of extension reaction in presence of all dNTPs and Taq DNA Polymerase (Promega, Madison, WI) at conditions of 94°C for 2 min, 55°C for 2 min and 68°C for 10 min. 5μl of extended product was used as template in a second PCR reaction. Final products were cloned in T-vector (Promega, Madison, WI) and sequenced to confirm that the desired manipulation has been accomplished.

2.3. Primer Design for Deletion and Inversion of an Internal DNA Fragment

Two 50 bp primers (P1 and P2) were synthesized for deletion of internal fragment A (115 bp in size) (Figure 1). Each primer was designed to contain two parts. The 3' portion (25 bp, area 3 for P1 and area 2 for P2) of these two primers were identical to the DNA template directly downstream from the sequences to be deleted (Figure. 1). The 5' part (25 bp, area 4 for P1, area 1 for P2) were designed to be the immediate upstream sequences from the area to be deleted. P3 and P4 (22mers) are designed to be at an PCR amplifiable distance from each other (they represent Reverse and Forward in our case).

For inversion, six different primers were designed, as shown in Figure 3. Both P1 and P2 were composed of two parts with the 3' portion (25 bp, area "a" for P1 and area "g" for P2) identical to the DNA template at the 5' ends of the fragment to be inverted (X). The 5' part (25 bp) of primer P1 (area "h") is the immediately upstream sequence of the fragment X on the complementary strand while the 5' part (25 bp) of the P2 (area "b") is the sequence immediately upstream of the 158 bp fragment X on the complementary strand. Both P3 and P4 (20mers) are located immediately downstream from the 158 bp fragment (fragment "X") on opposite strands and P5 and P6 are complementary to primers P3 and P4, respectively. Primers P7 and P8 are designed to be at PCR amplificable distance outside the working area (they represent Reverse and Forward in our case).

3. RESULTS

3.1. Deletion of an internal 115 bp fragment

The strategy of deleting a 115 bp sequence from a 931 bp fragment is depicted in Figure 1. Two independent PCR amplifications were carried out using P1/P3 and P2/P4 using a 931 bp fragment cloned in pBScript vector as a template. The result is shown in Figure 2 (lanes 2 & 3). The PCR products were gel purified and mixed together for PCR extension and the extended product was used as a template in a third PCR amplification using primers P3/P4. As a result, a 1015 bp fragment including 219 bp vector sequence (Figure 2, lane 4) was obtained. Produced fragment was cloned in pGEM-T vector (Promega, Madison, WI) and sequence analysis revealed that the 115 bp fragment was deleted (data not shown).

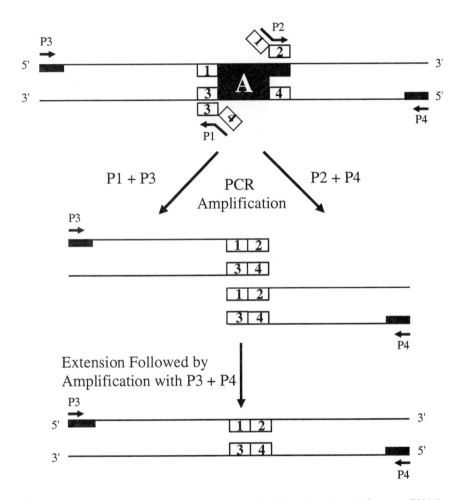

Figure 1. Schematic representation of the strategy of deletion of an internal fragment. DNA is represented by two solid lines as double stranded. The fragment to be deleted (A) is marked with solid box. Primers P1 and P2 are composed of two portions and the corresponding relation of each portion to the original fragment is shown. Primers P3 and P4 are presented as solid boxes and their locations relative to the DNA template are indicated. Primer sets used for each PCR reactions are marked and the PCR products are indicated by arrows.

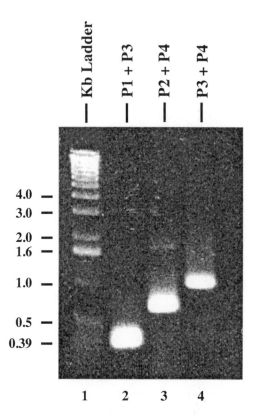

Figure 2.   Deletion of an internal DNA fragment through PCR amplification. 5 µl of each PCR product was loaded on 1% TBE agarose gel stained with ethidium bromide. Primer sets used for each PCR are labeled on top of each lane and the DNA KB ladder was included (Gibco/BRL, Frederick, MD) as a marker.

3.2. Inversion of an internal 158 bp fragment

As shown in Figure 3, the inversion of the 158 bp fragment (fragment X) was accomplished by three independent PCR amplifications using primer sets P1/P2, P3/P7 and P4/P8 and a 3304 bp DNA segment as a template. PCR products A, B and C were obtained as shown in Figure 4 (lanes 2, 3, 4). Products A and C were extended and amplified with P6/P8 to produce product "A+C" (Figure 3 and Figure 4, lane 5) and product "C+B" was produced by the same strategy by extension and amplification with P5/P7 from PCR products C and B (Figure 3, Figure 4, lane 6). Finally, gel purified products "A+C" & "C+B" were extended and amplified using P7/P8 (Figure 3 and Figure 4, lane 7). Sequence analysis of the final PCR product revealed the inversion of the 158 bp fragment (data not shown).

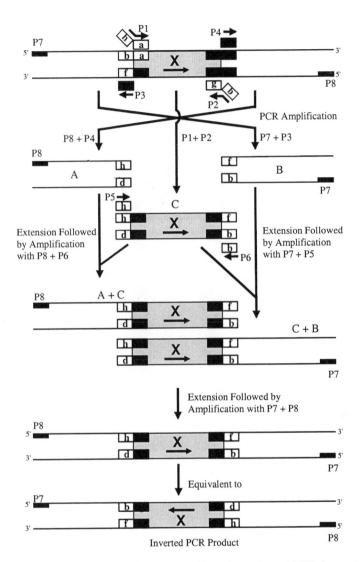

Figure 3. A flow chart depicting strategy of inverting an internal DNA fragment. The fragment to be inverted (X) is marked with solid gray box. Primers P1 and P2 each consisted of two portions and the relative corresponding sequence to each portion within the original fragment are also labeled. Primers P7 and P8 are presented as solid boxes and their locations are also indicated. Primer sets used for each PCR reactions are indicated while the produced PCR product are indicated by arrows.

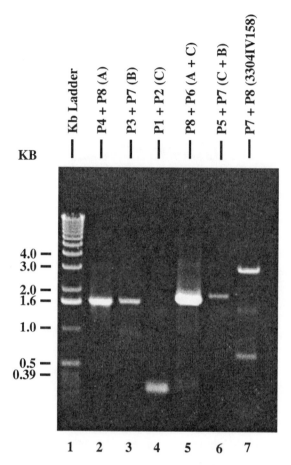

Figure 4. Inversion of an internal DNA fragment through PCR amplification. 5 µl of each PCR product was loaded on 1% TBE agarose gel stained with ethidium bromide. Primer sets used for each PCR are indicated on top of each lane and the DNA KB ladder was included (Gibco/BRL, Frederick, MD) as a marker.

4. DISCUSSION

Deletion and inversion of an internal fragment within the 5' flanking region is an inevitable step to study the regulatory function of certain *cis*-elements. In this manuscript, we described a PCR-based technique to delete or invert an internal segment within the 5' flanking region of the mouse peripherin/rds gene. The conventional means of deletion or inversion of an internal fragment requires the presence of conveniently located restriction enzymes sites. The present strategy overcomes the difficulties encountered by the conventional methods when the restriction enzyme sites are absent or located at inconvenient distances. Another advantage is that the present technique is easy to perform and is less time-consuming.

However, there are also some limitations. For example, although clean and sharp PCR products were obtained in most of the PCR reactions, background were still observed from some of them, as can be seen in Figure 4, lane 7. The purity of the PCR product generally depends on the PCR condition used, the length of primers and their GC-content. All these parameter can be optimized prior to the PCR amplification. The purity of the PCR products can also be assured following gel purification. Since Taq polymerase has lower fidelity, we recommend the use of any of the high fidelity DNA polymerase such as *Pfu* (Stratagene, La Jolla, CA) or DNA polymerase Mix of *Taq* and *Pwo* (Boehringer Mannheim, Indianapolis, IN).

5. ACKNOWLEDGMENT

This research was funded by grants from the National Eye Institute/NIH (EY10609, 11376 and 12190). Dr. Naash is a recipient of the Research to Prevent Blindness James S. Adams Scholar.

REFERENCES

1. Donald J. Zack, Jean Bennett, Yanshu Wang, Carol Davenport, Brenda Klaunberg, John Gearhart, and Jeremy Nathans. (1991). Unusual Topography of Bovine Rhodopsin Promoter-LacZ Fusion Gene Expression in Transgenic Mouse Retinas. *Neuron* **6**: 187-199
2. Higuchi, R. (1990). Recombinant PCR. In: PCR Protocols: a guide to methods and applications, ed. M.A. Innis, D.H. Gelfand, J.J. Sninsky, T.J. White. San Diego: Academic Press, 177-183.
3. Shimin chen and Donald J. Zack. Ret 4, a positive Acting Rhodopsin Regulatory Element Identified Using a Bovine Retina in vitro Transcription System. 1996. *J. Biol. Chem.* **271**, No 45: 28549 - 28557
4. Schanke, J., Quam, L., and Van Ness, B. (1994). Flip PCR as a method to invert DNA sequence motifs. *BioTechniques* **16**:414-416.
5. T. Allen Morris, Wei-Bao Fong, Melissa J. Ward, Hu Hu, and Shao-Ling Fong. Location of Upstream Silencer Elements Involved in the Expression of Cone Transducin α-Subunit (GNAT2). (1997) *Invest. Ophthalmol. Vis. Sci.* **38**: 196 -206

THE cGMP-PHOSPHODIESTERASE β-SUBUNIT GENE

Transcriptional and Post-Transcriptional Regulation

Debora B. Farber, Leonid E. Lerner, Yekaterina E. Gribanova, Mark R. Verardo, Natik I. Piriev, and Barry E. Knox*

1. INTRODUCTION

Mutations in the protein-coding region of the gene encoding the β-subunit of cyclic GMP-phosphodiesterase (β-PDE) cosegregate with retinal degeneration affecting humans[1], mice[2,3] and dogs[4,5]. Moveover, given that the exonic sequences are intact, the lack or the suboptimal expression of the β-PDE gene can cause alterations in phototransduction leading to photoreceptor functional and structural abnormalities. In light of the involvement of the β-PDE gene defects in the development of retinal disease, it is important to understand the events that regulate the expression of this gene in rod photoreceptors of the human retina.

Gene expression in eukaryotic cells is controlled at multiple levels generally grouped in transcriptional and post-transcriptional events. Transcriptional regulation is a major determinant of initiation of protein synthesis and of the level of gene expression. However, there is increasing evidence of the importance of post-transcriptional control mechanisms in determining and particularly in fine-tuning the final amount of the synthesized protein. These regulatory events include modulation of mRNA stability, localization and translation, and of protein stability and modification. Features of the mRNA molecules encoded in their primary as well as secondary structures together with the interactions that mRNAs may have with cellular protein factors play a critical role in translational efficiency. Our interest in understanding the mechanisms regulating β-PDE expression has led us to investigate the transcriptional and, more recently, post-transcriptional regulation of this gene in *in vitro, ex*

---

*Debora B. Farber, Leonid E. Lerner, Yekaterina E. Gribanova, Mark R. Verardo, Natik I. Piriev, Jules Stein Eye Insitute, UCLA School of Medicine, Los Angeles, California, 90095-7000; Barry E. Knox, Departments of Biochemistry & Molecular Biology and Ophthalmology, SUNY HSC at Syracuse, Syracuse, New York 13210.

New Insights Into Retinal Degenerative Diseases
Edited by Anderson *et al.*, Kluwer Academic/Plenum Publishers, 2001

vivo and *in vivo* studies. The results of these studies will ultimately help us to find ways to support and maintain expression of β-PDE in photoreceptor cells at physiological levels in order to prevent retinal degenerations.

2. TRANSCRIPTIONAL STUDIES

In eukaryotes, formation of a preinitiation transcription complex involves the interaction of RNA polymerase II with general transcription factors at the basal promoter region of genes. General transcription factors support basal level of transcription enhanced by the action of activator proteins that interact with specific DNA elements. Most of the gene promoters that have been studied in detail contain either a TATA box or an initiator responsible for selecting the start site of transcription. However, the β-PDE gene does not have either of these basal promoter elements, yet transcription always begins at the same sites[6].

In previous studies, we cloned about 1.4 kb of the 5' flanking region of the human β-PDE gene and identified by sequence comparison several potential elements that may be controlling expression of this gene. These included: an E box, the binding site of the helix-loop-helix family of transcriptional activators[7]; an AP-1 motif, the DNA recognition site for members of the b-Zip family of transcription factors[8]; an NRE-like element, the binding site of the retina-enriched transcription factor Nrl[9,10]; a GC box, which usually binds members of the Sp family of zinc finger transcription factors[11]; a CRE-like element, the binding site for the Crx transcription factor[12]; and a T/A-rich region at -31/-25 bases upstream of the major transcription start site[6].

2.1. Minimal promoter of the β-PDE gene

To search for enhancers and repressors and to determine whether the identified elements were involved in transcriptional activation of the β-PDE gene various length fragments of the 5' flanking region were transiently transfected in human Y-79 retinoblastoma cells, using constructs that contained the luciferase reporter gene (Fig. 1). Y-79 retinoblastoma cells were chosen for these experiments because we had shown before that these cells endogenously express both rod and cone phototransduction mRNAs and, in particular, the β-PDE mRNA[13]. Cells were also co-transfected in these experiments with a plasmid containing the β-galactosidase gene. This was used as an internal control for variations in transfection efficiency. Deletions from -1356 to -93 in the 5'-flanking region of the β-PDE gene did not reduce relative (to β-galactosidase) luciferase activity, but a 5'-end deletion to +4 practically abolished all activity (Fig. 1). This allowed us to infer that the sequence between -93 and +4 is necessary and sufficient for β-PDE expression, constituting the minimal promoter of this gene.

2.2. The β-PDE promoter directs expression to rod photoreceptors in transgenic mice

Transgenic technology was then used to confirm the participation of the 5' flanking sequences in directing the expression of the β-PDE gene to retinal photoreceptors *in vivo*. Several independent transgenic mouse lines carrying either of two fragments of the β-PDE gene upstream sequence, -1118 to +53 or -297 to +53, fused to the LacZ reporter gene were generated. Measurement of β-galactosidase activity in homogenates from retina and other

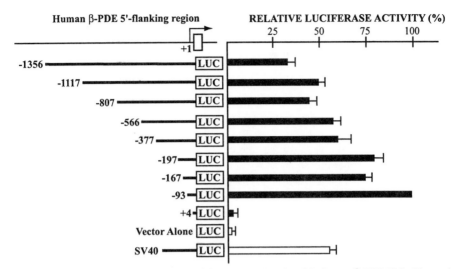

Figure 1. Luciferase activity of constructs containing different lengths of the human β-PDE 5' flanking region. Luciferase activity was normalized to the corresponding β-galactosidase activity for each sample and expressed as percent of the mean activity of the -93 to +53 construct. Values represent the average of at least three transfections and standard deviation bars are shown. (Modified from Di Polo, Lerner, and Farber, 1997[17] with permission from Oxford University Press.)

tissues from these transgenic lines showed the highest levels in the retinas of the -297 to +53 mice, about seven-fold higher than those in the retinas of -1118 to +53 mice. Reporter gene activity was undetectable in heart and muscle of these animals and low in kidney, brain, liver and spleen (Fig. 2). Within the retina, the reporter gene was specifically expressed in rod photoreceptors, as determined by *in situ* labeling of cryosections with 2 mM fluorescein-di-β-galactopyranoside (FDP) and examination by fluorescence microscopy[14]. This localization is consistent with that of endogenous β-PDE[15] and it was further confirmed

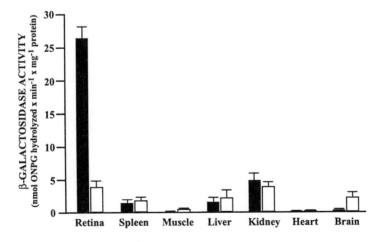

Figure 2. β-galactosidase activity in different tissues of mice carrying the -297 to + 53/LacZ (black bars) or the -1118 to +53/LacZ (white bars) transgenes.

by detecting β-galactosidase with immunocytochemistry. Retinal sections reacted with rabbit anti-β-galactosidase and then with a secondary antibody conjugated to the Cy3 fluorochrome were observed using a confocal microscope[14].

2.3. Studies with transgenic *Xenopus laevis*

Since the mouse transgenic experiments are very time consuming and expensive, in order to find out whether the -93 to +4 sequence of the β-PDE gene 5'-flanking region was indeed an effective promoter, we turned to transgenic experiments using *Xenopus laevis*. Sequences -1356 to +53 and -93 to +53 were cloned in a plasmid containing the GFP gene. The constructs were introduced into *Xenopus* by recombinant enzyme-mediated transgenesis[16]. Multiple transgenic lines carrying either of the two sequences were evaluated for the distribution and the intensity of the GFP-specific green fluorescence in different organs.

Green fluorescence is clearly visible in the eye (Fig. 3). The level of expression was higher in the -93 to +53 animals (Fig. 3b), and the exposure time for the -1356 to +53 image was longer to compensate for this. This accounts for the increased background fluorescence seen in 3a) compared to 3b). GFP fluorescence was restricted to the eye for the -93 to +53 tadpoles. In addition to the eye, faint levels were also detected in the brain of the -1356 to +53 animals. Fig. 3c) shows that the expression of GFP driven by the -93 to +53 promoter is still present after metamorphosis in a 3 cm froglet.

Transgenic *Xenopus* also allowed us to definitively localize the rod-specific expression driven by the -93 to +53 β-PDE promoter. Fig. 3d) shows that cones do not have GFP fluorescence. Expression of GFP was higher in the inner segment and cell body than in the outer segments of rods in all areas of the retina.

2.4. Functional sequences detected in the β-PDE minimal promoter using transfections in *Xenopus* embryos and human Y79 retinoblastoma cells

To determine the sequences in the proximal promoter region important for β-PDE transcription, we transfected Y79 retinoblastoma cells with sequential 5'-end deletion constructs of the -93 to +53 region. Whereas deletion of the segment between -93 and -83 (that possibly contains part of an extended NRE-like sequence)[10] caused a decrease in promoter activity, further deletion from -83 to -72 (that contains the E box) did not change luciferase expression (Fig. 4). Deletion of the sequence between -72 and -61 bp, that entirely eliminated the βAp1/NRE site[17], caused a strong inhibition of gene expression. The significant reduction of luciferase expression observed with the -93 to -61 deletion suggests that the extended βAp1/NRE sequence is implicated in regulation of β-PDE gene transcription.

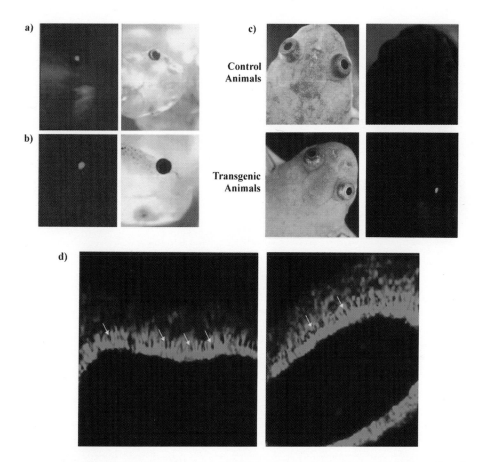

Figure 3. Representative -1356 to +53 (a) and -93 to +53 (b)stage 47 tadpoles photographed using a dissecting microscope under fluorescent (left panels) or bright (right panels) illumination. Animals were positioned laterally with a partial view of the ventral surface. c) A transgenic -93 to +53 froglet is seen at room light and under blue light (the green eye is facing the camera) and it is compared to a non-transgenic sibling. d) Radial cross sections from fixed and frozen eyes of transgenic tadpoles (stage 56) carring the -93 to +53 β-PDE promoter transgene. Cones are indicated by arrows.

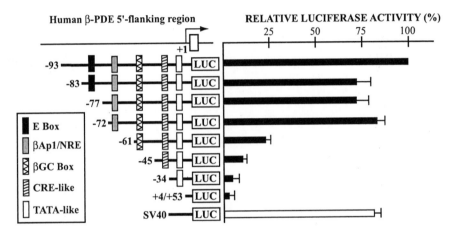

Figure 4. Deletion analysis of the β-PDE promoter. Deletion constructs were transfected in Y79 human retinoblastoma cells. Relative luciferase activity was determined for each transfection as described above. Results are expressed as the mean percent activity produced by the -93 to +53 β-PDE construct and the standard deviation bars are shown. Each transfection was carried out in triplicate and repeated at least 2-4 times.

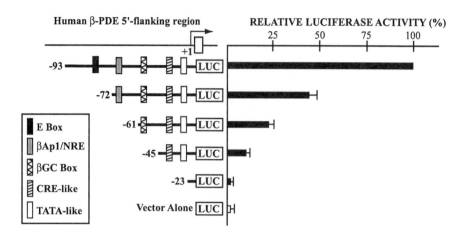

Figure 5. Deletion analysis of the β-PDE promoter with constructs transfected in embryonic *Xenopus leavis*. Relative luciferase activity and expression of results as in Figure 4.

To expand and corroborate our initial findings, we transfected multiple deletion and substitution mutants of the human β-PDE promoter in Y79 retinoblastoma cells and, to confirm these results, in an *ex vivo* system using embryonic *Xenopus laevis*. In the latter method, heads were dissected 48 hours after fertilization (stage 26) and were permeabilized by enzymatic treatment. Groups of 10-12 embryo heads were subjected to cationic lipid (DOTAP)/DNA transfection. Luciferase activity was assayed 72 hours later. The results of the two transfection systems were, in general, quite similar. Deletion of 16 bp from -61 to -45, a region that contains a GC-rich sequence, caused a further significant decrease of luciferase activity (Fig. 5). An extra 3-4 fold reduction was observed with the deletion of the -45 to -23 region where the CRE consensus sequence is located. The level of -23 to +53 promoter activity was not significantly different from that observed with the promoter-less vector.

Fig. 6a shows all the mutations created downstream from -59, using the -72 to +53 sequence as template. The results obtained when the mutant constructs were transfected in Y-79 retinoblastoma cells (Fig. 6b) complemented those from the deletion experiments. Mutating the -56/-53 nucleotides, considerably decreased the activity of the -72 to +53 promoter, and this was further reduced with the -51/-49 mutation; both of these mutations are part of the βGC element. A disruption of the CRE sequence had little effect on the overall activity of the β-PDE promoter.

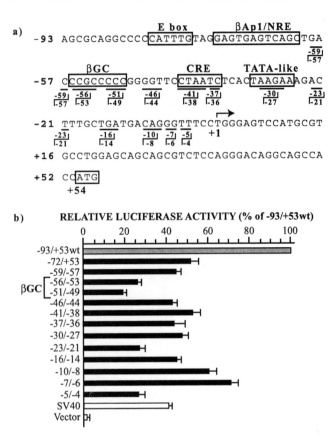

Figure 6. a) Sequence of the human β-PDE proximal promoter region showing the potential regulatory elements (boxed) and the mutations analyzed downstream from -61 (underlined and labeled) using the -72 to +53 construct as template. b) Nucleotide substitution analysis. Constructs were transfected in Y79 human retinoblastoma cells. Relative luciferase activity and expression of results as in Figure 4.

2.5. Transcription factors interacting with the regulatory elements in the β-PDE promoter

2.5.1. Involvement of Nrl

We previously demonstrated the binding of nuclear proteins to the 7 bp AP-1 motif (-69/-63) in the β-PDE promoter using gel shift assays[17]. We established that the AP-1-protein complex was sequence-specific because the intensity of the complexes was reduced upon addition of increasing amounts of unlabeled probe and no significant inhibition of complex formation was observed with excess unlabeled mutated probe. But the nature of the nuclear factor(s) binding to the AP-1 site remained unknown. To determine which of the

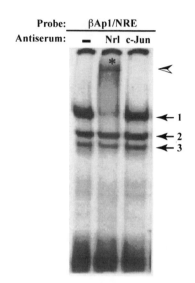

Figure 7. Electrophoretic mobility shift assay using the βAp1/NRE probe and Y79 human retinoblastoma cell extracts incubated without (lane 1) or with antibodies against Nrl (lane 2) and c-Jun (lane 3).

numerous members of the Jun/Fos, CREB or Nrl/Maf families may interact with the β–PDE gene AP-1 site, we used a longer probe than the one tested in previous experiments and investigated the possibility of supershifts with antibodies specific to different transcription factors. A longer probe was important because Nrl in complex with other transcription factors has been shown to bind with stronger affinity to an extended NRE, and the sequences flanking the consensus AP-1 site are involved in this interaction. Three retarded bands were observed when the radiolabeled βAP1/NRE probe was used (Fig. 7). Addition of an antibody against Nrl to the Y79 cell nuclear extract in the gel shift reaction prior to addition of the probe caused a mobility supershift with concomitant disappearance of complex 1. Therefore, complex 1 is likely to contain Nrl or a related protein. In contrast, addition of antibodies against c-Jun or c-Fos (not shown) did not disrupt the mobility of any of the complexes, suggesting that these nuclear factors do not play a significant role in βAP1/NRE-mediated regulation of the β-PDE promoter.

To test whether Nrl is indeed capable of transactivating the β-PDE promoter, we cotransfected 293 embryonic kidney cells that do not express endogenous Nrl with the β-PDE reporter construct and an Nrl expression vector. 3-4-fold transactivation was induced by overexpressed Nrl. Thus, Nrl seems to interact with the β-PDE promoter and upregulate its activity.

2.5.2. Involvement of transcription factors of the Sp family of proteins

To demonstrate that nuclear proteins interact with the βGC element, radiolabeled probes spanning the -66 to -37 sequence without and with mutations in the βGC element were used in gel shift assays with Y79 cell nuclear extracts (Fig. 8a). None of the complexes were observed with the mutated probe (-51/-49m). Furthermore, increasing concentrations of

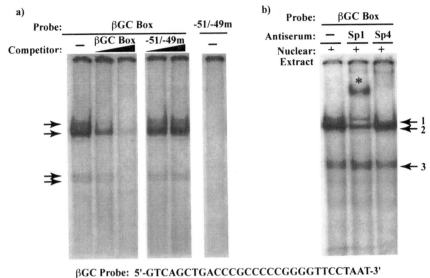

βGC Probe: 5'-GTCAGCTGACCCGCCCCCGGGGTTCCTAAT-3'
-51/-49m: 5'-GTCAGCTGACCCGCCAAAGGGGTTCCTAAT-3'

Figure 8. a) Electrophoretic mobility shift assay using the βGC probe or the -51/-49 mutated probe and Y79 human retinoblastoma cell extracts. Increasing concentrations of unlabeled probe and -51/-49 mutated probe were used as competitors. b) Polyclonal antibodies against Sp1 (lane 2) and Sp4 (lane 3) were incubated with the Y79 retinoblastoma cell extracts before addition of the labeled probe.

unlabeled wild type probe competed efficiently for binding of all the shifted bands but the mutated probes did not, even when added in large excess.

Since GC boxes are known to bind proteins of the Sp family, we tested whether the ubiquitous Sp1 or the CNS-restricted Sp4 were involved in the formation of the retarded complexes observed in the gel mobility shift assays (Fig. 8b). Addition of antibodies against Sp1 altered the mobility of a portion of complex 2 and produced a supershifted band (*)whereas antibodies against Sp4 disrupted all of the binding activity of complex 1. Thus, Sp4 or an antigenically related protein is part of complex 1 and complex 2 is likely to be a mixture of Sp1 and an unidentified protein.

2.5.3. Involvement of Crx

The presence of a consensus CRE sequence (-41/-36) in the minimal promoter of the β-PDE gene prompted us to explore whether Crx, an important transcriptional activator of other retina-specific genes such as rhodopsin, is involved in transcriptional regulation of the β-PDE gene. Co-transfection of overexpressed Crx in 293 human embryonic kidney cells together with the -72 to +53 (Fig. 9) or the -45 to +53 (not shown) promoters increased luciferase activity approximately 4-fold, but it did not cause any effect on the -23 to +53 promoter (not shown). Thus, when overexpressed, Crx modestly transactivates the β-PDE promoter and the -45 to -23 sequence containing CRE is necessary for this effect. We also tested whether Nrl and Crx have functional synergy on the β-PDE promoter. A 10-fold increase in luciferase activity induced by the Crx/Nrl combination indicated that there was

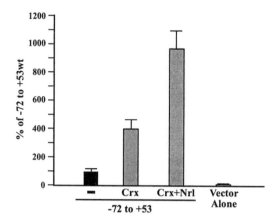

Figure 9. Co-transfection of the -72 to +53 β-PDE promoter and Crx or Crx in combination with Nrl into embryonic kidney cells.

an additive effect. Interestingly, as mentioned before, mutations that completely disrupted the consensus CRE sequence in Y-79 retinoblastoma cells (that have endogenous Crx) had little effect on the overall β-PDE promoter activity.

3. POST-TRANSCRIPTIONAL STUDIES

3.1. Involvement of the mRNA 5' UTR in the regulation of β-PDE expression

Transcription start-sites for the β-PDE gene have been determined in two species, mouse and human[6]. In both cases the β-PDE gene is expressed in the retina as two transcripts that differ by the length of their 5' UTRs. We have analyzed the sequence of the β-PDE mRNAs for potential elements that may be involved in regulation of translation. The longer 5' UTR in both mouse and human mRNAs contain an upstream AUG (Fig 10a). Additional upstream AUGs are known to enhance, in many instances, the efficiency of protein synthesis from the major open reading frame (ORF). Sequence comparison of the 5' UTRs in mouse and human show evolutionary conservation suggesting that they may have a functional significance in β-PDE expression. Other mRNA elements that may affect the initiation step of β-PDE translation – a major determinant of protein synthesis efficiency – include the cap structure, the length, the initiation codon and the sequence surrounding it (the Kozak consensus sequence) and the secondary structure of the 5' UTR.

The secondary structure prediction for both the long and short 5' UTRs of the β-PDE transcripts revealed the presence of hairpin-like structures with free energy values of -15.2 kcal/mol and -4.1 kcal/mol respectively. These values indicate a low probability of involvement of the secondary structure in regulation of protein synthesis (Fig 10b).

Comparison of the nucleotides surrounding the β-PDE translation initiation codon (AACACC<u>AUG</u>A) with the Kozak consensus sequence (GCC$^G/_A$CC<u>AUG</u>G) shows that the β-PDE mRNAs contain an **A** instead of a **G** at the +4 position (Fig. 10a). The importance of this highly conserved **G** at the +4 position has previously been analyzed by mutational studies that showed that its substitution can lead to 'leaky scanning', reducing the efficiency of translation[18].

We have evidence that the Kozak sequence is functionally important for β-PDE gene expression. We are currently performing detailed deletional and mutational analyses of the 5' UTR of the β-PDE mRNA. Results of these experiments will allow us to assess the involvement of the 5' UTR in the β-PDE translational efficiency.

3.2. Involvement of the mRNA 3' UTR in the regulation of β-PDE expression

Regulation of gene expression by 3' UTRs has not been investigated as extensively as by the 5' flanking region. However, we do have examples of the involvement of the 3' UTR in regulation of several diverse processes such as transcription[19], polyadenylation[20], transport from the nucleus to specific regions of the cytoplasm[21,22] stability or instability of message[23] and translation[24]. Ultimately, the 3' UTR may be related to cellular functions such as response to second messengers[25] and stress[26]. Small regions of highly conserved nucleotides in the 3' UTR usually mediate these processes by binding proteins. The vast majority of studies on the 3' UTR shows its involvement in post-transcriptional processes. For example, Liu and Redmond[27] have shown that expression of the RPE65 gene is controlled by a translational inhibitory element (TIE) present in the 3' UTR of the RPE65 mRNA. The recent attention to the role of the 3' UTR in pathology[28] points to this area of β-PDE as an attractive target of investigation in retinal degenerations.

We have created a simple system to assay 3' UTRs for their effects on regulation of gene expression. Full-length 3' UTRs were cloned by 3'RACE (Rapid Amplification of cDNA ends), sequenced, and then subcloned into a modified pGL3-promoter vector (Promega) downstream of the luciferase cDNA under the control of the SV-40 promoter. In addition, fragments of the 3' UTRs were inserted into the pGL3 vector. Each 3' UTR construct

a)

b)
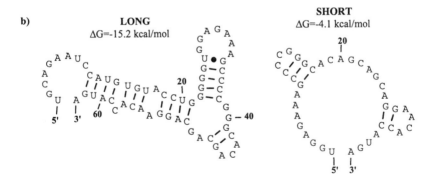

Figure 10. Sequence a) and predicted secondary structures b) of the 5' UTRs of the β-PDE mRNA. The -59 and -34 transcription start sites are indicated. The translation initiation codon is underlined; the upstream AUG is boxed and the nucleotides in the consensus Kozak sequence at -3 and +4 positions are bolded.

was then transiently co-transfected into Y-79 retinoblastoma cells with a plasmid containing the β-galactosidase gene, used as a control for transfection efficiency. Forty-eight hours after transfection, cell lysates were harvested, assayed for both luciferase and β-galactosidase activities, and the relative luciferase activity (to β-galactosidase) was determined.

We found that the β-PDE 3' UTR produces one-tenth of the relative luciferase activity obtained with the SV-40 3'UTR (a gene with a well studied regulation mechanism), the SRB7 subunit of the RNA polymerase II holoenzyme 3' UTR (involved in the eukaryotic transcriptional machinery and expressed in all cell types[29]), and other control 3' UTRs such as those of the cone arrestin and α'-PDE mRNAs. Further, we have determined that certain sequences of the β-PDE 3' UTR upregulate the expression of the luciferase reporter gene, and others decrease or even virtually abolish it. These results indicate that the regulation of β-PDE by its 3' UTR is a complex process involving more than one *cis*-element and possibly multiple *trans*-acting factors. We are currently characterizing and investigating these molecular events.

4. CONCLUSIONS

The studies that we have described in this chapter are a continuation of our long-term research on β-PDE. Many years ago we found that cGMP-phospodiesterase was essential for the integrity and normal function of photoreceptor cells. Those observations were followed by identification of the β-PDE gene as the site of mutations causing retinal degeneration in the *rd* mouse, Irish setter dog, and autosomal recessive retinitis pigmentosa patients. We have now extended our biochemical, genetic and molecular biology research to studies of the molecular events regulating expression in the photoreceptors of the β-PDE gene. We hope that in the near future the knowledge obtained will lead to better diagnostic tools and therapeutic modalities for the retinal degenerations associated with this gene.

5. ACKNOWLEDGEMENTS

We wish to thank Mr. Clyde Yamashita for his invaluable assistance in the preparation of this manuscript. This work was supported by National Institutes of Health grants EY02651 (DBF), EY00367 (LEL), EY11256 and EY12975 (BEK), and Training grant EY07026 (DBF); and grants from The Foundation Fighting Blindness (DBF and BEK). DBF is the recipient of a Research to Prevent Blindness Senior Scientific Investigators Award.

REFERENCES

1. Farber, D. B., and M. Danciger, Identification of genes causing photoreceptor degenerations leading to blindness, *Curr. Opin. Neurobiol.* **7**(5), 666-673. (1997).
2. Bowes, C., T. Li, M. Danciger, L. C. Baxter, M. L. Applebury, and D. B. Farber, Retinal degeneration in the *rd* mouse is caused by a defect in the β subunit of rod cGMP-phosphodiesterase, *Nature* **347**(6294), 677-680 (1990).
3. Pittler, S. J., and W. Baehr, Identification of a nonsense mutation in the rod photoreceptor cGMP phosphodiesterase β subunit gene of the *rd* mouse, *Proc. Natl. Acad. Sci. USA* **88**(19), 8322-8326 (1991).
4. Farber, D. B., J. S. Danciger, and G. Aguirre, The β subunit of cyclic GMP phosphodiesterase mRNA is deficient in canine rod-cone dysplasia 1, *Neuron* **9**(2), 349-356 (1992).
5. Suber, M. L., S. Pittler, N. Qin, G. Wright, V. Holcombe, R. Lee, C. Craft, R. Lolley, W. Baehr, and R. Hurwitz, Irish setter dogs affected with rod-cone dysplasia contain a nonsense mutation in the rod cyclic GMP phosphodiesterase β subunit gene., *Proc. Natl. Acad. Sci. USA* **90**(9), 3968-3972 (1993).

6. Di Polo, A., C. Bowes-Rickman, and D. B. Farber, Isolation and initial characterization of the 5' flanking region of the human and murine cGMP phosphodiesterase β-subunit genes., *Invest. Ophthalmol. Vis. Sci.* **37**(4), 551-560 (1996).

7. Murre, C., G. Bain, M. A. van Dijk, I. Engel, B. A. Furnari, M. E. Massari, J. R. Matthews, M. W. Quong, R. R. Rivera, and M. H. Stuiver, Structure and function of helix-loop-helix proteins, *Biochim. Biophys. Acta* **1218**(2), 129-35 (1994).

8. Angel, P., and M. Karin, The role of Jun, Fos and the AP-1 complex in cell-proliferation and transformation, *Biochim. Biophys. Acta* **1072**(2-3), 129-57 (1991).

9. Kerppola, T. K., and T. Curran, Maf and Nrl can bind to AP-1 sites and form heterodimers with Fos and Jun, *Oncogene* **9**(3), 675-84 (1994).

10. Kerppola, T. K., and T. Curran, A conserved region adjacent to the basic domain is required for recognition of an extended DNA binding site by Maf/Nrl family proteins, *Oncogene* **9**(11), 3149-58 (1994).

11. Briggs, M. R., J. T. Kadonaga, S. P. Bell, and R. Tjian, Purification and biochemical characterization of the promoter-specific transcription factor, Sp1, *Science* **234**(4772), 47-52 (1986).

12. Chen, S., Q. L. Wang, Z. Nie, H. Sun, G. Lennon, N. G. Copeland, D. J. Gilbert, N. A. Jenkins, and D. J. Zack, Crx, a novel Otx-like paired-homeodomain protein, binds to and transactivates photoreceptor cell-specific genes, *Neuron* **19**(5), 1017-30 (1997).

13. Di Polo, A., and D. B. Farber, Rod photoreceptor-specific gene expression in human retinoblastoma cells, *Proc. Natl. Acad. Sci. USA* **92**(9), 4016-4020 (1995).

14. Ogueta, S. B., A. Di Polo, J. G. Flannery, C. K. Yamashita, and D. B. Farber, The human cGMP-PDE beta-subunit promoter region directs expression of the gene to mouse photoreceptors, *Invest. Ophthalmol. Vis. Sci.* **41**(13), 4059-4063. (2000).

15. Farber, D. B., and T. Shuster, in: *The Retina: A Model for Cell Biology Studies*, edited by R. Adler and D. B. Farber (Academic Press Inc., Orlando, Fl., 1986), pp. 239-296.

16. Batni, S., S. S. Mani, C. Schlueter, M. Ji, and B. E. Knox, in: *Methods in Enzymology; Vertebrate phototransduction and the visual cycle, Part B*, edited by K. Palczewski (Academic Press Inc., San Diego, 2000), pp. 50-64.

17. Di Polo, A., L. E. Lerner, and D. B. Farber, Transcriptional activation of the human rod cGMP-phosphodiesterase beta-subunit gene is mediated by an upstream AP-1 element, *Nucleic Acids Res* **25**(19), 3863-7 (1997).

18. Kozak, M., Recognition of AUG and alternative initiator codons is augmented by G in position +4 but is not generally affected by the nucleotides in positions +5 and +6, *Embo Journal* **16**(9), 2482-92 (1997).

19. DePonti-Zilli, L., A. Seiler-Tuyns, and B. M. Paterson, A 40-base-pair sequence in the 3' end of the beta-actin gene regulates beta-actin mRNA transcription during myogenesis, *Proc. Natl. Acad. Sci. USA* **85**(5), 1389-93 (1988).

20. Baker, E. J., in: *mRNA metabolism & post-transcriptional gene regulation*. edited by Harford, J. B., and D. R. Morris (Wiley-Liss, New York 1997), pp. 85-105.

21. Keene, J. D., Why is Hu where? Shuttling of early-response-gene messenger RNA subsets, *Proc. Natl. Acad. Sci. USA* **96**(1), 5-7. (1999).

22. Mayford, M., D. Baranes, K. Podsypanina, and E. R. Kandel, The 3'-untranslated region of CaMKII alpha is a cis-acting signal for the localization and translation of mRNA in dendrites, *Proc. Natl. Acad. Sci. USA* **93**(23), 13250-5 (1996).

23. Wickens, M., P. Anderson, and R. J. Jackson, Life and death in the cytoplasm: messages from the 3' end, *Curr. Opin. in Genet. Devel.* **7**(2), 220-32 (1997).

24. Gray, N. K., and M. Wickens, Control of translation initiation in animals, *Annu. Rev. Cell Dev. Biol.* **14**, 399-458 (1998).

25. Jarzembowski, J. A., and J. S. Malter, Cytoplasmic fate of eukaryotic mRNA: identification and characterization of AU-binding proteins, *Prog. Mol. Subcell. Biol.* **18**, 141-72 (1997).

26. Spicher, A., O. M. Guichert, L. Duret, A. Aslanian, E. M. Sanjines, N. C. Denko, A. J. Giaccia, and H. M. Blau, Highly conserved RNA sequences that are sensors of environmental stress, *Mol. Cell Biol.* **18**(12), 7371-7382. (1998).

27. Liu, S. Y., and T. M. Redmond, Role of the 3'-untranslated region of RPE65 mRNA in the translational regulation of the RPE65 gene: identification of a specific translation inhibitory element, *Arch. of Biochem. Biophys.* **357**(1), 37-44 (1998).

28. Conne, B., A. Stutz, and J. D. Vassalli, The 3' untranslated region of messenger RNA: A molecular 'hotspot' for pathology?, *Nature Medicine* **6**(6), 637-41 (2000).

29. Chao, D. M., E. L. Gadbois, P. J. Murray, S. F. Anderson, M. S. Sonu, J. D. Parvin, and R. A. Young, A mammalian SRB protein associated with an RNA polymerase II holoenzyme, *Nature* **380**(6569), 82-5 (1996).

ORGANIZATION OF THE CHICKEN AND *Xenopus* PERIPHERIN/*rds* GENE

Chibo Li[1], John O'Brien[2], Muayyad R. Al-Ubaidi[3], and Muna I. Naash[3]

1. INTRODUCTION

Peripherin/*rds* is an integral membrane glycoprotein located in the rim region of rod and cone outer segment disk membranes (1). Interest in peripherin/*rds* has increased since the discovery of its association with different forms of human retinal diseases. Over 50 different pathogenic mutations have been identified that are associated with retinitis pigmentosa (RP), pattern dystrophy, retinitis punctata albescens, macular dystrophy (MD), butterfly-shaped MD, or cone-rod dystrophy (2). These mutations include base substitutions causing missense mutations or premature termination, and in-frame insertion/deletion mutations that change the reading frame. The majority of these mutations are located in the large intradiscal loop, emphasizing the important role played by this region in the function of peripherin/*rds*.

Peripherin/*rds* has been isolated and sequenced from human (3), cow (4), dog (5), cat (6), rat (7), mouse (8), chicken (9), *Xenopus* (10), and skate (11). It has been shown to code for proteins containing 346 to 364 amino acids (aa), depending on the species. The predicted polypeptide is composed of four putative transmembrane segments, a relatively small (21 aa) and a large (142 aa) intradiscal loops, and a long C-terminal segment exposed to the cytoplasmic side of the disk membranes [for review, see reference (1)]. *In vitro* biochemical studies suggested a noncovalent association between peripherin/*rds* and rom-1 (12-15), a nonglycosylated transmembrane protein that shares several characteristics with peripherin/*rds*. These characteristics include similar hydropathy profiles, gene organization, localization, and highly conserved amino acid residues (1;12;16). Although these proteins are believed to play a crucial role in normal outer segment structure, the functional activities of peripherin/*rds* and rom-1 at the molecular level are not currently understood.

Peripherin/*rds* belongs to a gene family, which includes rom-1 and several "*rds*-like" genes from non-mammals. The relationship of these *rds*-like genes to peripherin-*rds* and

[1]Ophthalmology, Northwestern University, Tarry building 5727, 300 E Superior St, Chicago, IL 60612; [2]Ophthalmology & Visual Science, University of Texas School of Medicine, 6431 Fannin MSB 7.024, Houston, TX 77030; and [3]Cell Biology, University of Oklahoma Health Sciences Center, 940 Stanton L. Young Blvd., Oklahoma City, OK 73104.

New Insights Into Retinal Degenerative Diseases
Edited by Anderson *et al.*, Kluwer Academic/Plenum Publishers, 2001

rom-1 is unclear, as is their functional significance. An understanding of the organization of these genes may lend some insight into their evolutionary relationships, and hence a better understanding of the genes' functions. At present, the gene structures of only the mammalian peripherin/*rds* (17-19) and rom-1 (20) and the skate peripherin/*rds* (11) have been described. In this study, we analyze the gene structures of the chicken and *Xenopus* peripherin/*rds* and *rds*-like genes.

2. MATERIALS AND METHODS

2.1. Sequence analysis

Sequence data obtained by direct sequencing were analyzed and edited with PC/GENE (Oxford Molecular LTD, Campbell, CA) software. The same program was used to generate a comparison of compiled DNA sequences and generate multiple alignments of predicted amino acid sequences from different species available in the GenBank database. The alignment parameters were as follows: K-tuple value: 1; Gap penalty: 5; Window size: 10; Filtering level: 2.5; Open gap cost: 10; Unit gap cost: 10. The term identity describes amino acids that are completely conserved while the term similarity specifies identical amino acids plus all conservative substitutions.

TABLE 1. Sequence of the primers used in this study.

Primer Name	Species	Sequence 5' to 3'
MIN221	XRDS38	CGACCTCTTAGCCACAATGGCC
MIN222	XRDS38	AAGCCTGATGAATAGACACGAGC
MIN223	XRDS35	AGAGGCGGGTGAAACTTGC
MIN224	XRDS35	GGTCAACTGTAGTCTTCTGGGACG
MIN225	XRDS36	CTTTCCAGAGGCGAGTGAAACTTGC
MIN226	XRDS36	GAAGCAGCATTCCAGTGTGTTCC
MIN239	CRDS1	CTTTGCTTGGAAGGAACCACC
MIN240	CRDS1	GCAGGAAATACCACTGCTTAGC
MIN241	CRDS1	AAATAGTTGTAGGTGCTAATTGCCC
MIN243	CRDS2	AAGACAGAACTACACAGACATCAGG
MIN244	CRDS2	TATTCAGACCAGGACAGGAATGC
MIN245	CRDS2	TGCCTTGGATACATGCTTCTAAGC
MIN246	CRDS2	AGACCAGGACAGGAATGCG
MIN270	CRDS1	CTCGGCTCACTACAGCTACG
MIN271	CRDS1	CGTAGCTGTAGTGAGCCGAG
MIN272	CRDS2	CCTGACCAACAACAGTGCCC
MIN273	CRDS2	GGGCACTGTTGTTGGTCAGG
MIN274	XRDS35	CTGGCTCAGTGTCCTAGTAGGG
MIN275	XRDS35	GTGTCTTCCATCCAGAGAAGGG
MIN276	XRDS35	CTGGGTGAGGGGCTGCAGAG
MIN277	XRDS35	CTCTGCAGCCCCTCACCCAG
MIN278	XRDS36	TAGTGGGGTGCCTGACATTTGG
MIN281	XRDS36	CCCAGATGTTCACGTCATCACCC

2.2. Analysis of gene structure by PCR

Genomic DNA was isolated from muscle tissue of a single chicken or *Xenopus* using published protocols (21). The primers were chosen to amplify potential introns present in the coding region since both mouse and human peripherin/*rds* and rom-1 genes harbor two introns in the coding region (17-20). cDNAs from chicken or *Xenopus* retinas were used for control templates. These cDNAs were obtained by RT-PCR from total RNA isolated from individual chicken or *Xenopus* retina. Sequences of the primers used in this study are shown in table 1. In the first PCR, the primer sets MIN239/MIN240 and MIN243/MIN244 were used to amplify across the coding region of chicken *rds* (CRDS) 1 and 2, respectively. In the second PCR for CRDS1, the primer sets MIN241/MIN271 and MIN270/240 were chosen to amplify across introns I and II, respectively. The second PCR for CRDS2, the primer sets MIN245/MIN273 and MIN272/MIN246 were used to amplify across introns I and II, respectively. The amplifications were performed with genomic and cDNA as templates using Expand long PCR kit (Boehringer Mannheim, Indianapolis, IN). The reactions were performed according to the manufacturer's protocol with 250 ng genomic DNA and 200 ng of each primer in a 50 µl reaction mixture. A negative control (no template) was run with each reaction set. PCR products were cloned into a pGEM-T vector (Promega, Madison, WI) and sequenced from both strands using vector-specific primers.

The same method indicated above for the chicken was used to study the genomic organization of the *Xenopus* homologs. The primer sets MIN225/MIN226, MIN223/MIN224, and MIN221/MIN222 were chosen to amplify across the coding region of *Xenopus* peripherin/*rds* (XRDS) 36, 35, and 38, respectively. The second PCR for XRDS35, the primer sets MIN274/MIN277 and MIN275/MIN276 were used to amplify across introns I and II, respectively. In the second amplifications, the primer sets used to amplify across introns I and II of XRDS36 were MIN278/MIN281. To determine the positions of introns I and II in XRDS38, the PCR product from the initial amplification was subcloned in T-vector and sequenced from both sides.

3. RESULTS

3.1. Genomic organization of CRDS 1 and 2

Analysis of the gene structure of mouse and human peripherin/*rds* and rom-1 revealed the presence of two introns (17-20). Although the sizes of these introns are different, the locations of introns I and II are the same in mouse and human (17;19). The skate homologue, however, consists of two exons and one intron (11) at the same location of intron I in mammalian peripherin/*rds*. To determine if introns were present in the chicken peripherin/*rds* genes, we amplified across the entire coding region using genomic DNA as a template. Exon-intron boundaries were noted by divergence of the genomic sequence from that of the cDNA.

Two peripherin/*rds* homologs from chicken (CRDS1 and 2) have been isolated (9). CRDS1 is the chicken peripherin/*rds* ortholog. To compare the intron/exon organization of chicken peripherin/*rds* to that in other species, we determined the presence of introns in CRDS1 and 2 genes with sequence specific primers that amplify across either the entire coding region and/or separately across introns I and II (Fig. 1A). Two steps of amplifications were performed with chicken genomic DNA and cDNA as a control. In the first amplification, one set of primers was used to amplify across each of the entire coding regions of CRDS1 (MIN239/MIN240) and CRDS2 (MIN243/MIN244). The PCRs yielded the expected fragments from CRDS1 and 2 (1.9 and 1.7 kb, respectively) when the cDNAs were used as a template. However, a fragment of ~9 kb for CRDS1 and a smear for CRDS2 were obtained when genomic DNA was used. In the

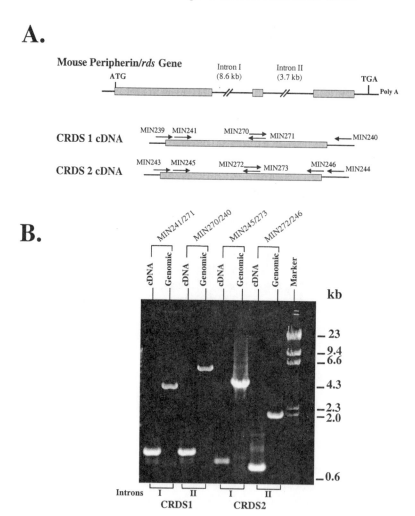

Figure 1. (A) A map showing the genomic organization of peripherin/*rds* from mouse. Arrows represent the positions and directions of the primers used to identify the presence of introns in the Chicken homologs. Filled boxes indicate translated sequences. Horizontal lines represent non-translated sequences. (B) Agarose gel analysis of the PCR products obtained from amplification across the entire coding sequences of CRDS1 and 2 with chicken genomic DNA. Amplification across intron I using genomic DNA revealed the presence of ~3.0 kb-longer fragments for both CRDS1 and 2 than that observed when using cDNA as a template. Amplifications across intron II were ~5.0 and 1.0 kb longer for CRDS1 and 2, respectively.

second amplification, two sets of primers were used for each of CRDS1 and 2 to amplify across the area of putative intron I (MIN241/MIN271 and MIN245/MIN273, respectively) and intron II (MIN270/MIN240 and MIN272/MIN246, respectively). When genomic DNA was used as a template, the PCR products for intron I were ~3.0 kb longer for both CRDS1 and 2 than when the cDNA was used (Fig. 1B). The PCR products for intron II were ~5.0 and 1.0 kb longer for CRDS1 and 2, respectively. The data indicate the presence of both introns in CRDS1 and 2. Sequence analysis showed that the location of both introns in CRDS1 and 2 are the same as in mouse and human.

3.2. Genomic organization of XRDS35, 36, and 38

Three peripherin/*rds* homologs from *Xenopus* (XRDS36, 35 and 38) have been isolated (10). XRDS38 is the *Xenopus* peripherin/*rds* ortholog. A similar method as described for the chicken was used to determine the gene organization of XRDS35, 36 and 38 (Fig. 2A). Two fragments of ~ 10 and 7 kb were obtained from XRDS35 and 38, respectively. Since the coding sequence of the peripherin/*rds* gene is 1.2 kb, this indicates the presence of ~9 and 6 kb of intervening sequences in XRDS35 and 38, respectively. The amplified fragment for XRDS38 was subcloned and sequenced from both directions (Fig. 2B). Two introns were identified at the same locations as the mouse and human peripherin/*rds* introns I and II. The sequence of the exon/intron junction is in agreement with both the 5' and 3' consensus splice site sequences (Fig. 2C).

To identify the presence of introns I and II in XRDS35, two sets of primers were designed to amplify across each intron. The PCR products from genomic DNA indicated the presence of introns I and II with sizes of 0.8 and 7.0 kb, respectively. Although XRDS36 cDNA was amplified by RT-PCR, no amplification was detected in three separate experiments from genomic DNA with the same primers, suggesting the presence of a larger intron(s) that could not be amplified under the conditions used. Therefore, we designed two sets of primers to amplify across each of introns I and II of XRDS36 (data not shown). Again, no PCR products were obtained from genomic DNA while the proper products were obtained from cDNA. Since under the same conditions we were able amplify ~ 10 kb from the same genomic DNA, we concluded that the sizes of introns I and II in XRDS36 are very likely larger than 10 kb each.

4. DISCUSSION

This study was undertaken to determine the exon-intron organization of peripherin/*rds* from chicken and *Xenopus*. Two homologs of peripherin/rds were identified in chicken photoreceptors (9) and three in *Xenopus* (10). CRDS1 and XRDS38 are the orthologs of mammalian peripherin/*rds* while CRDS2, XRDS35 and 36 are more distant relatives and called rds-like proteins (9;10).

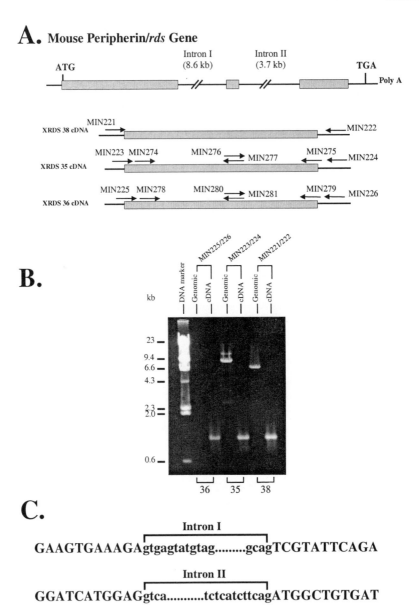

Figure 2. (A) A map showing the genomic organization of peripherin/*rds* from mouse. Arrows represent the positions and directions of the primers used to identify the presence of introns in the *Xenopus* homologs. Filled boxes indicate translated sequences. Horizontal lines represent non-translated sequences. (B) Agarose gel analysis of the PCR products obtained from amplification across the entire coding sequences of XRDS36, 35, and 38 from *Xenopus* genomic DNA. Amplification products from XRDS35 and 38 revealed the presence of ~10 and 7 kb intervening sequences, respectively. All primer pairs amplified a 1.2kb DNA fragment from *Xenopus* peripherin/*rds* cDNAs. No amplification of XRDS36 was detected using genomic DNA. Sequence analysis of the PCR fragments showed that XRDS38 contains two introns, located at the same sites as in the human and the mouse peripherin/*rds* genes. (C) Sequence of the intron-exon junctions of the XRDS38 peripherin/*rds* gene. Exon sequences are in upper case while intron sequences are in lower case.

Studying the organization of the gene may offer some insights into how peripherin/*rds* gene family arose.   The gene structure determined for mammalian peripherin/*rds*, and mammalian rom-1 indicates the presence of three exons and two introns with conserved boundaries (17-20). We have found that the gene organization of CRDS1 and 2 and XRDS38, 36 and 35 also composed of three exons and two introns in the same location as intron I and II in peripherin/*rds*, and mammalian rom-1. Although the sizes of these introns are different, the locations of introns I and II are the same as in mouse and human (17-19). Skate peripherin/*rds*, on the other hand, has only two exons and one intron in the same location as intron I in other species (11). The presence of intron II in the chicken and *Xenopus* and its absence in the skate homologue suggest the presence of two possibilities that could account for the relationship of the skate peripherin/*rds* gene to its counterpart in other species as well as to rds-like or rom-1 genes. First, a gene duplication leading to the evolution of the rds-like and rom-1 proteins may have occurred following the divergence of skate from the ancestral vertebrate line. In addition, intron II must have been acquired between the divergence of the skate and this gene duplication. Alternatively, a gene duplication leading to rom-1 and rds-like genes may have preceded the divergence of the skate. In this case, both introns are presumed to be present in the ancestral gene, with intron II lost from the skate peripherin/*rds* gene. Further work is needed to clarify gene complement in the skate and to resolve the question of when gene duplication may have occurred in this family.

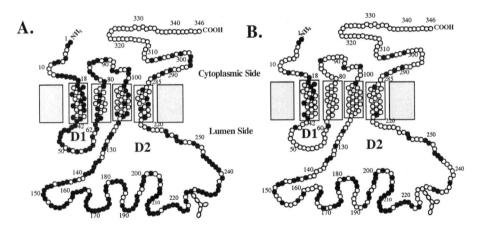

Figure 3. Schematic diagram of peripherin/rds or rom-1 molecule with marked amino acids as filled circle that are identical in all known (A) peripherin/*rds* and (B) peripherin/*rds* and rom-1.

To advance our knowledge of the structure and function of peripherin/*rds*, we aim to identify the highly conserved amino acids throughout species to mark regions of importance to the function of the protein (Fig. 3). Sequence comparisons between nine homologs of peripherin/*rds* at different steps in the evolutionary ladder, including the skate, a species that dates from the early Devonian period nearly 400 million years ago, revealed 54.5% identity on the overall protein level (Fig. 3A). However, different domains share different levels of identity. The transmembrane and C-terminal domains retain about 38% identity, while the large hydrophilic loop between the third and the fourth transmembrane domains shares 73% identity and 93% similarity with peripherin/*rds* from all known species. Interestingly, most of the mutations in the peripherin/*rds* gene associated with human retinal diseases are located in the large intradiscal loop (2), pointing to the greater importance of this loop to the overall function of the protein. The N-terminus and the small intradiscal loop share 61% and 76% identity, respectively. When rom-1 proteins, CRDS2, and XRDS35 and 36 are included in the comparison, the overall identity among the proteins is only 18%, although a high degree of identity is apparent in the large intradiscal loop (Fig. 3B). A high level of identity was also found when the rom-1 proteins from mouse, bovine and human were compared.

5. ACKNOWLEDGMENT

This research was funded by grants from the National Eye Institute/NIH (EY10609, 11376 and 12190) and the Foundation Fighting Blindness. Dr. Naash is a recipient of the Research to Prevent Blindness James S. Adams Scholar.

REFERENCES

1. Molday RS. Peripherin/rds and rom-1: molecular properties and role in photoreceptor cell degeneration. In: Osborne NN, Chader GJ, editors. Progress in retinal and eye research. Great Britain: Pergamon Press Ltd., 1994: 271-299.
2. Kohl S, Giddings I, Besch D, Apfelstedt-Sylla E, Zrenner E, Wissinger B. The role of the peripherin/*RDS* gene in retinal dystrophies. Acta Anatomica 1998; 162:75-78.
3. Travis GH, Christerson L, Danielson PE et al. The human retinal degeneration slow (RDS) gene: chromosome assignment and structure of the mRNA. Genomics 1991; 10:733-739.
4. Connell GJ, Molday RS. Molecular cloning, primary structure, and orientation of the vertebrate photoreceptor cell protein peripherin in the rod outer segment disc membrane. Biochem 1990; 29:4691-4698.
5. Moghrabi WN, Kedzierski W, Travis GH. Canine homolog and exclusion of *retinal degeneration slow* (*rds*) as the gene for *early retinal degeneration* (*erd*) in the dog. Exp Eye Res 1995; 61(5):641-643.
6. Gorin MB, Snyder S, To A, Narfstrom K, Curtis R. The cat RDS transcript: candidate gene analysis and phylogenetic sequence analysis. Mamm Genome 1993; 4:544-548.
7. Begy C, Bridges CD. Nucleotide and predicted protein sequence of rat retinal degeneration slow (rds). Nucl Acids Res 1990; 18:3058-3058.
8. Travis GH, Brennan MB, Danielson PE, Kozak CA, Sutcliffe JG. Identification of a photoreceptor-specific mRNA encoded by the gene responsible for retinal degeneration slow (*rds*). Nature 1989; 338:70-73.
9. Weng J, Belecky-Adams T, Adler R, Travis GH. Identification of two rds/peripherin homologs in the chick retina. Invest Ophthalmol Vis Sci 1998; 39(2):440-443.
10. Kedzierski W, Moghrabi WN, Allen AC et al. Three homologs of rds/peripherin in *Xenopus laevis* photoreceptors that exhibit covalent and non-covalent interactions. J Cell Sci 1996; 109:2551-2560.
11. Li C, Al-Ubaidi MR and Naash MI. Isolation and characterization of the skate peripherin/rds gene. Invest Ophthalmol Vis Sci 1997; 38(4):S219.
12. Moritz OL, Molday RS. Molecular cloning, membrane topology, and localization of bovine rom-1 in rod and cone photoreceptor cells. Invest Ophthalmol Vis Sci 1996; 37(2):352-362.
13. Goldberg AF, Molday RS. Subunit composition of the peripherin/rds-rom-1 disk rim complex from rod photoreceptors: hydrodynamic evidence for a tetrameric quaternary structure. Biochem 1996; 35(19):6144-6149.
14. Kedzierski W, Weng J, Travis GH. Analysis of the rds/peripherin.rom1 complex in transgenic photoreceptors that express a chimeric protein. J Biol Chem 1999; 274(41):29181-29187.

15. Goldberg AFX, Loewen CJR, Molday RS. Cysteine residues of photoreceptor peripherin/rds: role in subunit assembly and autosomal dominant retinitis pigmentosa. Biochem 1998; 37(2):680-685.
16. Bascom RA, Manara S, Collins L, Molday RS, Kalnins VI, McInnes RR. Cloning of the cDNA for a novel photoreceptor membrane protein (rom-1) identifies a disk rim protein family implicated in human retinopathies. Neuron 1992; 8:1171-1184.
17. Ma J, Norton JC, Allen AC et al. *Retinal degeneration slow* (*rds*) in mouse results from simple insertion of a t haplotype-specific element into protein-coding exon II. Genomics 1995; 28:212-219.
18. Cheng T, Al-Ubaidi MR, Naash MI. Structural and developmental analysis of the mouse peripherin/rds gene. Som Cell Mol Genet 1997; 23(2):165-183.
19. Kajiwara K, Hahn LB, Mukai S, Travis GH, Berson EL, Dryja TP. Mutations in the human retinal degeneration slow gene in autosomal dominant retinitis pigmentosa. Nature 1991; 354:480-483.
20. Bascom RA, Schappert K, McInnes RR. Cloning of the human and murine ROM1 genes: genomic organization and sequence conservation. Hum Mol Genet 1993; 2(4):385-391.
21. Sambrook J, Fritsch EF, Maniatis T. Molecular cloning: A laboratory manual. Cold Spring Harbor Laboratory Press, 1989.

CHANGED INTERACTION OF L-TYPE CA^{2+} CHANNELS AND RECEPTOR TYROSINE KINASE IN RPE CELLS FROM RCS RATS

Rita Rosenthal and Olaf Strauß

1. INTRODUCTION

The Royal College of Surgeons (RCS) rat is an animal model for autosomal recessive retinitis pigmentosa.[1-6] The retinal degeneration is caused by a mutation of the MerTK (c-mer) gene which results in a truncated gene product without function.[5] However, it is not fully understood how this genetical defect leads to the retinal degeneration. MerTK is a receptor tyrosine kinase which probably recognizes photoreceptor outer membranes to initiate their phagocytosis. The inability to phagocytose photoreceptor outer membranes has been described as major cause of the photoreceptor degeneration in RCS rats.[1,3,4] In addition, the genetic defect causes a large number of changes in the second-messenger signaling of RPE cells.[7] For example, the tyrosine kinase-dependent regulation of L-type Ca^{2+} channels in RPE cells from RCS rats is different to their regulation in RPE cells from control rats.[8]

Basic fibroblast growth factor (bFGF or FGF2) has been shown to reduce retinal degeneration in RCS rats and to restore the phagocytic capability of RPE cells.[9-12] We found that in RPE cells from control rats bFGF increases the concentration of the cytosolic free Ca^{2+} by activation of L-type Ca^{2+} channels.[8] Purpose of this study is to describe the bFGF-dependent intracellular signaling in RPE cells.

2. METHODS

All experiments were performed using cultured RPE cells from non-dystrophic control rats and from RCS rats. Changes in intracellular free Ca^{2+} were measured using the Ca^{2+}-sensitive fluorescence dye fura-2. Ba^{2+} currents through L-type Ca^{2+} channels were measured in the perforated-patch configuration of the patch-clamp technique. To study interactions of signaling proteins, membrane proteins were isolated and analysed by immunoprecipitation techniques.

New Insights Into Retinal Degenerative Diseases
Edited by Anderson *et al.*, Kluwer Academic/Plenum Publishers, 2001

3. RESULTS

Studies on RPE cells from non-dystrophic control rats revealed that bFGF activates L-type Ca^{2+} channels by close interaction of the bFGF receptor FGFR2 and the α-subunit of L-type channels.[13] Application of bFGF (10 ng/ml) to cells which have been kept under serum-free conditions for 24 h led to an increase in the maximal current amplitude and to a shift of the steady-state activation towards more negative potentials, more close to the resting potential of RPE cells. Inhibition of tyrosine kinases by the broad range blocker lavendustin A (10 μM) prevented the effect of bFGF. In contrast, inhibition of tyrosine kinase of the src-subtype by herbimycin A (10 μM) and inhibition of FGFR1 by SU5402 (20 μM) did not influence the effect of bFGF on L-type channel activity. Thus, FGFR1 and tyrosine kinase of the src-subtype are not involved in the bFGF-induced activation of L-type Ca^{2+} channels. Thus, FGFR2 possibly activates the L-type channel directly. To test this hypothesis, immunoprecipitation experiments were performed. In previous experiments we could show that L-type channels of neuroendocrine subtype (corresponding to α1D-subunit) are predominantly expressed in the RPE. Immunoprecipitation of α1D-subunits from membrane proteins isolated from cells which have been kept in serum-containing cultures led to coprecipitation of FGFR2 but not of FGFR1. This was supported by precipitation of FGFR2 which showed coprecipitation of α1D-subunits whereas precipitation of FGFR1 did not result in the coprecipitation with Ca^{2+} channel α1-subunits. The fact that the coprecipitation has a functional correlation could be seen when comparing cells from serum-containing cultures with cells from serum-free cultures. Precipitation of FGFR2 from membrane proteins of cells from serum-free cultures did not result in the coprecipiation of α1D-subunits. The coprecipitation of FGFR2 and α1D-subunits was observed when cells from serum-free media were incubated with bFGF for 1h. Thus, only in cells stimulated by bFGF coprecipitation of FGFR2 and α1D-subunits could be detected.

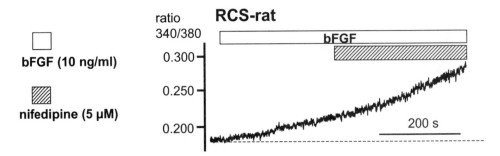

Figure 1. Effect of bFGF on intracellular free Ca^{2+} in RPE cells from RCS rats. Changes in intracellular free Ca^{2+} are depicted as ratio of the fura-2 fluorescence induced by the two excitation wavelengths 340 and 380 nm. Application of bFGF led to an increase in cytosolic free Ca^{2+} which could not be influenced by the L-type Ca^{2+} channel blocker nifedipine.

Application of bFGF to RPE cells from RCS rats led to an increase in intracellular free Ca^{2+}. This increase could not be reduced by additional application of the L-type channel blocker nifedipine (Figure 1). Thus, in RPE cells from RCS rats bFGF seemed to be unable to stimulate L-type channels for generating an increase in cytosolic free Ca^{2+} as second-messenger. The inability of bFGF to stimulate L-type channels was also seen in patch-clamp experiments with cells which have been maintained in serum-free cultures for 24 h.

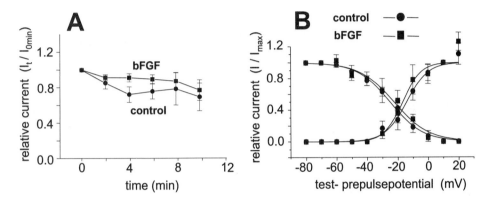

Figure 2. Effect of bFGF on L-type Ca^{2+} channel activity in RPE cells from RCS rats. A.) Plot of the maximal Ba^{2+} current amplitude (voltage-step from –70 to +10 mV) over the experiment time. Current amplitudes were normalized to the amplitude which has been observed after establishing the perforated-patch configuration. No difference was observed between treated and non-treated cells. B.) Effect of bFGF on voltage-dependence: steady-state activation and inactivation curves were measured using standard stimulation protocols. Curves were fitted using the Boltzmann equation. bFGF showed no effect on the voltage-depedence of L-type channels.

In these cells application of bFGF did neither increase the maximal current amplitudes through L-type channels nor led to a shift of the steady-state activation curve (Figure 2).

These data were in contrast to observations made in immunoprecipitation experiments with membrane proteins from cells maintained in serum-containing cultures. Here, precipitation of α1D-subunits showed coprecipitation of FGFR2. Precipitation of FGFR2 led to coprecipitation of α1D-subunits (Figure 3). Thus, in RCS RPE cells FGFR2 can interact with the α1D-subunit of the neuroendocrine L-type channel but fails to increase channel activity.

IP: α_{1D}

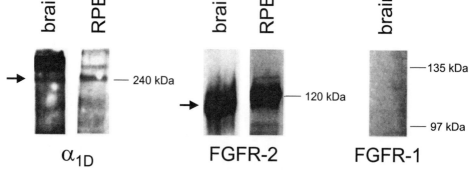

α_{1D} FGFR-2   FGFR-1

— 240 kDa

— 120 kDa

—135 kDa

— 97 kDa

Figure 3. Western blot of immunoprecipitates of membrane proteins from RCS RPE cells and RCS rat brain. Precipitation was performed using antibodies against α1D-subunits of L-type channels. The arrows indicate bands which disappear using the corresponding blocking peptide. The blot was stained from left to right: anti-α1D, anti-FGFR2, and anti-FGFR1. Only FGFR2 can be coprecipitated with Ca^{2+} channel α1D-subunits.

4. DISCUSSION

Investigation of bFGF-dependent signaling in RPE cells revealed the description of a new signaling pathway.[13] We could demonstrate that bFGF led to an increase in the activity of L-type Ca^{2+} channels by close interaction of the bFGF receptor FGFR2 and the α1D-subunit of the Ca^{2+} channel. This effect was dependent on tyrosine kinase activity but independent of FGFR1 activation or tyrosine kinases of the src-subtype.

Voltage-dependent Ca^{2+} channels are composed of at least four subunits.[14, 15] The largest one is the α1-subunit which forms the pore and determines kinetical and pharmacological characteristics. Four different α1-subunits are known to represent L-type channels: the α1S-subunit represents the skeletal muscle subtype, the α1C-subunit the cardiac subtype, the α1F-subunit a retinal subtype[16] and the α1D-subunit the neuroendocrine subtype. The neuroendocrine subtype is expressed for example in neurons, neuroendocrine cells and secreting tissues like islet cells of the pancreas.[15] We found the neuroendocrine subtype expressed in RPE cells.[17] Thus, the L-type channels in RPE cells may be involved in the regulation of secretion.

In RPE cells application of bFGF led to an increase in the concentration of intracellular free Ca^{2+} as a second messenger. This increase could be reduced by inhibition of

L-type channels using the dihydropyridine compound nifedipine. Thus, L-type channels are involved in the intracellular signaling activated by bFGF. Two different bFGF receptors can be responsible for this effect: FGFR1 and FGFR2.[18] For FGFR1 it has been shown that stimulation of this receptor led to the activation of both phospholipase C and a signaling cascade involving src-kinase, grb/Sos, ras/raf and the MAP kinase pathway.[18] However, this signaling cascade has not been confirmed for FGFR2. The involvement of L-type channels in the signaling cascade has not been described for any of these two receptors.

RPE cells express both FGFR1 and FGFR2.[19] Investigating the intracellular signaling induced by bFGF in RPE cells, we found that application of bFGF led to an increase in the activity of L-type channels. This increase was not influenced by blockade of FGFR1 or src-subtype tyrosine kinase. In addition, immunoprecipitation experiments revealed that FGFR2 and the α1D-subunit interact closely to stimulate the L-type channel. Coprecipiation of the receptor with the Ca^{2+} channel was only observed in cells stimulated either by fetal calf serum or by bFGF. Thus, FGFR2 activates a different signaling pathway than FGFR1. This pathway involves direct activation of L-type Ca^{2+} channels by FGFR2. This stimulation of L-type channels can result in two changes in cell function. One may be a change in secretory activity. The other possible effect of bFGF-dependent stimulation of L-type channels could be a change in gene expression. The influence of L-type channels on gene expression has been shown for neurons. Here, stimulation of L-type channels either by repetitive firing or by L-type Ca^{2+} channel openers leads to stimulation of the expression of immediate early genes like c-fos.[14, 20-22] Since bFGF is a rescue factor in the eye, this could be an effect on the RPE to initiate rescue from cell damage in the retina.

We tried to verify a comparable signaling cascade in RPE cells from RCS rats. In these cells, application of bFGF also led to an increase in intracellular free Ca^{2+}. However, application of the L-type channel blocker nifedipine had no effect. Thus, bFGF-dependent intracellular signaling appears to be different in RCS RPE cells. This was confirmed in patch-clamp studies where application of bFGF did not stimulate L-type Ca^{2+} channels. However, immunoprecipitation experiments revealed the same close interaction of FGFR2 and the α1D-subunit as it was observed in RPE cells from control rats. FGFR2 and α1D-subunits can be coprecipitated from membrane proteins of RCS RPE cells but not FGFR1 and α1D-subunits. Thus, in RPE cells from RCS rat, FGFR2 can interact with the L-type Ca^{2+} channel but this interaction does not result in a change of Ca^{2+} channel activity. The reason for this is unclear. However, it has to be a result of the genetic defect in the RCS rat which causes the lack of the receptor tyrosine kinase MerTK.[5] Thus, the presence of MerTK is the prerequisite for a proper regulation of L-type channels. A connection in the MerTK-dependent intracellular signaling with regulation of Ca^{2+} channels has not yet been reported.

In summary, we described a new pathway for bFGF-dependent intracellular signaling. This signaling pathway involves the stimulation of voltage-dependent Ca^{2+} channels by the bFGF receptor FGFR2. In addition, this signaling pathway is disturbed in RPE cells from an animal model for retinitis pigmentosa which lacks the receptor tyrosine kinase MerTK. Future studies on the RPE from RCS rats will reveal new aspects of intracellular signaling of MerTK and the pathophysiology of retinal degeneration in this animal model for retinitis pigmentosa.

5. ACKNOWLEDGEMENTS

This study was supported by the DFG grants Str 480/3-1 and Str 480/8-1, by Pro Retina Germany and The Foundation Fighting Blindness.

REFERENCES

1. M. C. Bourne, D. A. Campbell, and K. Tansley, Heriditary degeneration of rat retina, *Brit. J. Ophthalmol.* **22**, 613-623 (1938).

2. E.-S. El-Hifnawi, Pathomorphology of the retina and its vasculature in hereditary retinal dystrophy in RCS rats, in: *Advances in the Biosciencs*, Vol. 62, E. Zrenner, H. Krastel, H.-H. Goebel, Ed., Pergamon Journals Ltd., Oxford and New York, pp. 417-434 (1987).

3. D. Bok and M. O. Hall, The role of the pigment epithelium in the etiology of inherited retinal dystrophy in the rat, *J. Cell Biol.* **49**, 664-682 (1971).

4. M. M. LaVail and B.-A. Battelle, Influence of eye pigmentation and light deprivation on inherited retinal dystrophy in rat, *Exp. Eye Res.* **21**, 167-192 (1975).

5. P. M. D' Cruz, D. Yasumura, J. Weir, M. T. Matthes, H. Abderrahim, M. M. LaVail, and D. Vollrath, Mutation of the receptor tyrosine kinase gene Mertk in the retinal dystrophic RCS rat, *Hum. Mol. Gen.* **9**, 645-651 (2000).

6. A. Gal, Y. Li, D. A. Thompson, J. Weir, U. Orth, S. G. Jacobson, E. Apfelstedt-Sylla, and D. Vollrath, Mutations in MERTK, the human orthologue of the RCS rat retinal dystrophy gene, cause retinitis pigmentosa, *Nat. Genet.* **26**, 270-271 (2000).

7. O. Strauss, F. Stumpff, S. Mergler, M. Wienrich, and M. Wiederholt, The Royal College of Surgeons rat: an animal model for inherited retinal degeneration with a still unknown genetic defect, *Acta Anat.* **162**, 101-111 (1998).

8. S. Mergler, K. Steinhausen, M. Wiederholt, and O. Strauß, Altered regulation of L-type channels by protein kinase C and protein tyrosine kinases as a pathophysiologic effect in retinal degeneration, *FASEB J.* **12**, 1125-1134 (1998).

9. E. G. Faktorovich, R. H. Steinberg, D. Yasumura, M. T. Matthes, and M. M. LaVail, Photoreceptor degeneration in inherited retinal dystrophy delayed by basic fibroblast growth factor, *Nature* **347**, 83-86 (1990).

10. M. M. LaVail, K. Unoki, D. Yasumura, M. T. Matthes, G. D. Yancopoulos, and R. H. Steinberg, Multiple growth factors, cytokines and neurotrophins rescue photoreceptors from damaging effects of constant light, *Proc. Natl. Acad. Sci.* **89**, 11249-11253 (1992).

11. M. M. LaVail, M. T. Matthes, D. Yasumura, C. Lau, K. Unoki and R. H. Steinberg, Photoreceptor rescue by survival promoting factors in constant light damage and in RCS rats with inherited retinal dystrophy, *Invest. Ophthalmol. Vis. Sci.* **36**, S211 (1995).

12. M. J. McLaren and G. Inana, Inherited retinal degeneration: basic FGF induces phagocytic competence in cultured RPE cells from RCS rats, *FEBS Letters* **412**, 21-29 (1997).

13. R. Rosenthal, H. Thieme, and O. Strauss, Fibroblast growth factor receptor 2 (FGFR2) in brain neurons and retinal pigment epithelial cells act via stimulation of neuroendocrine L-type channels, *FASEB J.* **in press,** (2000).

14. W. A. Catterall, Structure and function of neuronal Ca^{2+} channels and their role in neurotransmitter release, *Cell Calcium* **24**, 307-323 (1998).

15. E. Perez-Reyes and T. Schneider, Molecular biology of calcium channels, *Kid. Int.* **48**, 1111-1124 (1995).

16. T. M. Strom, G. Nyakatura, E. Apfelstedt-Sylla, H. Hellebrand, B. Lorenz, B. H. Weber, K. Wutz, N. Gutwillinger, K. Ruther, B. Drescher, C. Sauer, E. Zrenner, T. Meitinger, A. Rosenthal, and A. Meindl, An L-type calcium-channel gene mutated in incomplete X-linked congenital stationary night blindness, *Nat. Genet.* **19**, 260-263 (1998).

17. O. Strauss, F. Buss, R. Rosenthal, D. Fischer, S. Mergler, F. Stumpff, and H. Thieme, Activation of neuroendocrine L-type channels (alpha1D subunits) in retinal pigment epithelial cells and brain neurons by pp60$^{c\text{-}src}$, *Biochem. Biophys. Res. Commun.* **270**, 806-810 (2000).

18. P. Klint and L. Claesson-Welsh, Signal transduction by fibroblast growth factor receptors, *Front Biosci.* **4**, D165-177 (1999).

19. H. Tanihara, M. Inatani, and Y. Honda, Growth factors and their receptors in the retina and pigment epithelium, *Prog. Retinal Eye Res.* **16**, 271-301 (1997).

20. T. H. Murphy, P. F. Worley, and J. M. Baraban, L-type voltage-sensitive calcium channels mediate synaptic activation of immediate early genes, *Neuron* **7**, 625-635 (1991).

21. H. Bito, K. Deisseroth, and R. W. Tsien, CREB phosphorylation and dephosphorylation: a Ca²⁺- and stimulus duration-dependent switch for hippocampal gene expression, *Cell* **87**, 1203-1214 (1996).

22. K. Deisseroth, E. K. Heist, and R. W. Tsien, Translocation of calmodulin to the nucleus supports CREB phosphorylation in hippocampal neurons, *Nature* **392**, 198-202 (1998).

PERIPHERAL RODS EVADE LIGHT DAMAGE IN ALBINO TROUT

Donald M. Allen, Gordon C. Hendricks and Ted E. Hallows*

1. INTRODUCTION

Albino strains of rats and mice have been the animal models of choice for study of light-induced retinal degeneration. These animals have a nocturnal habit and a rod-rich retina. Exposure to moderate constant illumination or stronger cyclic light (~130 lux) initiates permanent damage to the retina[1]. The initial event appears to require absorption of light by rhodopsin in the rod outer segments (ROS), but this quickly leads to destruction of ROS and culminates in the apoptotic cell death of the rod cell body in the outer nuclear layer (ONL)[2, 3]. Later, the damage extends to surviving cones and retinal pigment epithelium (RPE), a pathology shared with the end stages of other types of inherited retinal degeneration[4]. Albino rodents are more sensitive than normally pigmented strains, primarily due an un-pigmented iris[5-7], but susceptibility also depends on previous light exposure[8], diet[9] and circadian factors[10].

In contrast to nocturnal rodents, surface-dwelling fishes (*teleostei*) are alternative diurnal models for study of retinal photo-degeneration[11]. These species have fully duplex retinas with 3-4 spectral classes of cones, one of which can utilize ultraviolet light [12, 13]. Since they have naked corneas and immobile pupils, they can be accurately dosed with light; moreover, they relatively inexpensive to maintain. This argues that they can be a diurnal model for study of intrinsic resistance to light damage. A melanized iris and well-timed photomechanical movements of both photoreceptors and melanosomes within the apical processes of the RPE normally serve to shield the more sensitive rods from any damage in full daylight[14-16]. Furthermore, as their eye continues to grow, teleosts have the capacity to add new rods from rod precursors in the central retina and from retinal precursors in the circumferential germinal zone (CGZ), so that light damage, once initiated, need not be permanent as it is in the rodent model[17-21]. The specific roles

Donald. M. Allen, Ph.D. and Gordon C. Hendricks, O.D., Department of Science & Mathematics, UTPB, Odessa, TX 79762. Ted E. Hallows, M.S., Utah Division of Wildlife Resources, Kamas State Fish Hatchery, Kamas, UT 84036

played in mitigation of light damage by these intrinsic factors, in addition to those mechanisms of resistance which may inhibit progression to apoptosis, are of great interest. A current goal of research (reviewed elsewhere in this volume) is to contravene or mitigate the degenerative processes leading to apoptosis of the photoreceptors.

Fortunately, the considerable defensive function of melanin in teleosts can be removed by albinism, allowing study of diurnal mechanisms of resistance to light damage possibly not observable in nocturnal rodents. The susceptibility of albino teleosts to light damage has recently been studied in three species: the rainbow trout, *Oncorhyncus mykiss*, the oscar, *Astronotus oscillatus* and the zebrafish, *Danio rerio*. Interestingly, these species differ in the extent to which they sustain light damage. Albino trout exposed to full daylight (100,000 lux) or 3,000 lux constant incandescent light at 11°C for 10 days or longer sustained only shortened or missing ROS in their central retina such that they were essentially night-blind. No losses in photoreceptor nuclei or diminution of cone outer segments (COS) were noted. ROS re-grew and scotopic visual sensitivity returned when fish were placed in reduced cyclic light or experienced the seasonal decline in solar illumination[11, 22]. In albino oscars exposed to 3000 lux constant incandescent light at 28°C for 10 days, the damage progressed beyond truncation of ROS to a loss of the rod cell bodies, but cones were spared[22]. In both trout and oscars normal controls were always undamaged. Finally, in albino zebrafish exposed to 5000 lux of constant halogen light at 28-32°C for 7 days, both rods and cones died apoptotically, but cell proliferation increased, and both rods and cones were subsequently replaced during a 28 day recovery period[23]. That these three species appear to differ in the extent to which the albino strain undergoes light damage is intriguing. Though differences in light exposure and ambient temperature may be partly responsible, species-specific resistance factors are implicated, and invite further investigation.

The degree to which visual pigment losses are correlated with ROS structural damage in an albino fish is important since it has been supposed that resident visual pigment must be present in order for light damage to ensue [3, 4, 7, 24]. The initial purpose of this study was to compare the total amounts of rod visual pigments with the degree of structural damage to rods in the albino rainbow trout model. However, since visual pigment was extracted from entire retinas, we were obliged by initial findings to look at entire retinal slices to correlate histology to amount of visual pigment extracted from the rods.

2. METHODS

2.1. Light treatments and fish

Albino (A) and normally pigmented (N) juvenile rainbow trout were raised separately, first in indoor raceways (late January to late March), then in outdoor raceways (10,000-100,000 lux daylight). Later, albinos and normals were then combined and exposed to constant incandescent light placed directly above an indoor raceway (3000 lux – AS or NS), placed in attenuated natural daylight (10-30 lux – AS or NS), or remained outdoors (AO or NO). Water temperature was 11°C. Light measurements were made at the water surface with a lux-calibrated J-16 digital photometer (Tektronix).

2.2. Visual pigments

Rainbow trout utilize a 503_1-527_2 rhodopsin-porphyropsin " pigment pair" as visual pigments. Analysis of total visual pigment was as reported previously[25]. Entire dark-adapted retinas were frozen in 4% K-Alum and stored in light-tight containers before being thawed and extracted in 2% digitonin. Spectrometric analysis before and after bleaching was performed in 0.2 M hydroxylamine (pH 7.4) to yield a total absorbance per eye. Proportions of rhodopsin and porphyropsin were determined and the abosorbance was converted to nanomoles per eye allowing for the molar contribution of each pigment.

2.3. Histology

Preparation of tissues for thin-sectioning in epoxy resins followed previous methods[22]. After enucleation, a slit was made in the dorsal or nasal sclera of the left or right eye before immersion in ice-cold fixative (2%2% glutaraldehyde-paraformaldehyde in 0.087M phosphate buffer, pH 7.3). After 1 hour, the anterior segment was removed with iris scissors and the eye cup was returned to fixative for 24 hours. After washing, post-fixing in 1% osmium tetroxide and infiltration with an epoxy-araldyte mixture, eye cups were placed in embedding molds and polymerized at 63°C. Sections were then made at 1μm on an RMC ultramicrotome and placed on a drop of 0.5% toluidine blue in 1% sodium borate. Differential staining was effected *en bloc* by heating the slide to dryness on a hot plate. In rodents, which have several layers of nuclei, ONL thickness is used to estimate of cell losses occurring during retinal degenerations. However, since trout have many fewer nuclei, we instead conducted a census of the photoreceptor nuclei in the ONL. Where only a central retina slice was examined, retinas were embedded and sectioned only through the middle of the posterior pole. The number of ONL nuclei, the number of fully transected ROS and the length of at least ten fully transected ROS were averaged for several points along the slice for each individual. Then the values for each individual were averaged to get a group mean. For the estimate of ROS volume, we employed the multiplicand of the group means of ROS length and number of ROS per field[11].

To achieve entire retinal slices in fixed tissue, the dorsal or nasal 1/3 was pared away with iris scissors and the lens removed before returning the eye to fixative for 24 hours. Later, the eye was positioned in flat embedding molds with the pared edge parallel to the intended block face. Depending on which sector had been pared away (nasal or dorsal) and which eye was used (left or right), the block could be sectioned deeper in the block parallel to the dorsal-ventral or nasal-temporal face midway in the eye, often including both cornea and the optic nerve entry. We then examined the retina over a 125 μm field at points 250 μm apart over the entire retinal slice, including the CGZ at each margin. The CGZ in juvenile trout (eye diameter 3.5 to 5.0 mm) was previously measured as about 600 μm, or about 0.6 mm[26]. As before, we measured the number and length of ROS and the number of ONL nuclei in each field in order to reveal differences along the retinal slice.

For comparisons between groups, we employed a one-way analysis of variance with the appropriate *a priori* testing, using the mean square of the overall ANOVA[27]. Results are reported as Means ± 1 standard deviation.

3. RESULTS

3.1. A comparison of amount of visual pigment and central ROS volume

We compared central ROS volumes with total amount of rod visual pigment (rhodopsin and porphyropsin) extracted from entire, dark-adapted retinas in normal and albino trout exposed to full daylight (~10,000 lux - AO or NO) constant light (2,500 lux - AL or NL), or dim daylight (~ 30 lux - AS or NS). Compared to their normal controls, albinos exposed to constant light (AL) or remaining outdoors (AO) had fewer, shorter ROS, but the protected albinos (AS) fared much better (see Table 1). In each case, albinos did not have significantly fewer nuclei in the ONL at the time of sampling, although there was a trend for albinos to have 2-5 per 125 um fewer nuclei than their normal controls. This can be attributed to their slightly large size; overall, albinos grow faster than normals of the same cohort and feeding conditions.

Table 1. Rod visual pigment (VP) and morphology in central retinas of albino (A) and normal (N) rainbow trout held for 60 days in 30 lux daylight (S), 2,500 lux constant light (L) and 10,000 – 100,000 lux daylight (O): Sampled 8/25 at age 220 days. Means $\pm$ 1 SD (n=6). ROS volume is the multiplicand of ROS length and # of ROS.

Group	VP (nmoles/eye)	No. of ROS (per 125μm)	ROS Length (μm)	ROS Volume	ONL nuclei per 125μm	Fish Length (cm)
NO:	4.00 ± 0.25	34.3 ± 1.44	31.0 ± 5.40	1063	63.1 ± 2.62	9.41 ± 0.86
AO:	0.60 ± 0.17	8.62 ± 4.16	3.14 ± 2.19	27	57.8 ± 3.94	9.91 ± 0.86
NL:	3.25 ± 0.31	34.5 ± 1.02	26.4 ± 4.71	910	61.2 ± 1.40	9.45 ± 0.85
AL:	0.65 ± 0.20	12.2 ± 7.34	6.06 ± 3.24	74	59.0 ± 2.22	9.58 ± 0.66
NS:	5.70 ± 0.30	30.5 ± 3.19	37.1 ± 6.56	1131	63.7 ± 3.62	9.66 ± 0.98
AS:	3.78 ± 0.40	29.0 ± 3.40	23.6 ± 4.36	684	61.4 ± 6.69	10.0 ± 1.22

Significance: $p<.05$ (*), $p<.01$(**), $p<.001$(***):
Visual Pigment: NO, NL, AS < NS *; NO, NS, NL,AS >AO,AL***
ROS length: AO or AL < NS, NO, NL or AS***; NS>NL or AS*
No. of ROS per 100 μm: AO or AL < NS,NO,NL or AS ***

Table 2. Central ROS volume and total retinal visual pigment in albino trout, compared to normally pigmented fish from the same treatments. Derived from Table 1.

Group	ROS Volume	Visual Pigment	Difference
AO/NO	0.025	0.15	6.00
AL/NL	0.103	0.20	1.94
AS/NS	0.59	0.66	1.12

In affected albinos, estimated central ROS volume was less than expected in comparison to the amount of visual pigment extracted from entire retinas of fish from the same groups (Tables 1 & 2). The AO/NO ratio was only 2.5% for ROS volume, but 15% for visual pigment, a 6-fold difference. The AL/NL ratio was 10.3% for ROS volume and 20% for visual pigment, a 2-fold difference. In AS retinas, which had shown some recovery and re-growth of ROS over 18 days, the visual pigment was about what one would expect on the basis of central ROS volume (Table 2). Since it seemed unlikely that the specific density of visual pigment within individual ROS could account for these discrepancies[25], we concluded that there must be a non-uniform distribution of damage over the retina. We hypothesized that the additional visual pigment originated from undamaged rods in the peripheral margin which had been discarded prior to histological assay.

3.2. Analysis of ROS length over retinal slices

To test the hypothesis that the peripheral retina was the source of additional visual pigment extracted from entire retinas, we conducted histological analysis over retinal slices on 140 day-old albino and normal trout, using treatments similar to those used before[22]. Fish had remained outdoors 98 days (O), or after 70 days outdoors had been given constant 2,500 lux light for 28 days (L) or after 80 days outdoors had been moved into dim daylight (10-30 lux) for 18 days (S). At this age, the smaller eye (5 mm in average diameter, 7 mm total retinal circumference) enabled us to obtain retinal slices of the entire retina in 1µm sections.

When ROS lengths were examined at various distances along the retinal slices, starting from nasal or dorsal margin, it was clear that the severe truncation of ROS observed in the central retina of affected albinos (AL or AO) did not extend out to the peripheral margins in either plane of section (Fig. 1, Table 3). Instead, within ~ 1 mm from the retinal margin there were near-normal ROS, with a transition zone less than 1mm in width (Fig. 1). In the AS groups, ROS had begun to grow back in the central retina, but not as much as would have been expected if the treatment time had been longer (Table 1). Normal trout (NL, NS or NO) had longer ROS (~35-40µm) throughout the retinal slice (See "N" groups, Table 3; NL-12, Fig 2). The degree of damage to ROS varied somewhat among different individuals.

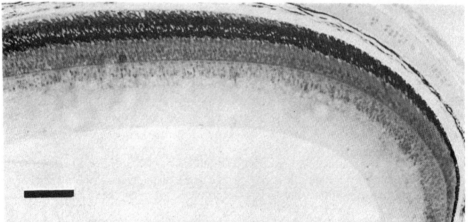

Figure 1. Above: ventral margin of affected (AO) albino, with long ROS only in circumferential zone. Below: nasal margin of normal (NL) with long ROS throughout. Scale bar: 100 μm.

Where shortened ROS remained in the central retina of affected albinos, there were fewer of them (Table 1). Thus, ROS length does not reliably estimate ROS volume over the entire retinal slice. To get a better idea of how ROS volume (and visual pigment extracted from rods) would be affected over the retinal slice, we plotted estimated ROS volume and numbers of photoreceptor nuclei as they occurred over a retinal slice for one normal, two affected and two protected albinos (Fig. 2). ROS volume approached near normal values in the most peripheral 1.0 mm in affected albinos evading the ROS damage sustained by the central retina. The zone of transition in ROS length and volume was typically abrupt, 1 mm or less (Fig. 1; Fig. 2, AO-2 and AO-10). The circumferential band of intact ROS must account for the "extra" visual pigment observed earlier in affected albinos (Table 2). Also, over the retinal slice, there were no apparent losses of ONL nuclei (both rod and cone) in affected albinos in comparison to normally pigmented controls in any of the three treatments. This confirms that light damage does not reduce the number of rods or cones in any part of the retinas in albino trout.

Table 3. Mean ROS length (μm) in normal and albino trout at different points along a retinal slice. Fish placed outdoors at age 42 days (Mar 1) and sampled June 6 at age 140 days. O = outdoors; S =30 lux daylight; L= 2500 lux constant light. ▼= optic nerve.

		Distance along slice (mm)							
Group	Slice	0.5	1.0	2.0	3.0	4.0	5.0	6.0	6.5
AS-2	D→V	37	31	10	6	10	37		
AS-11	D→V	32	37	10	▼	13	40		
AS-1	D→V	25	36	25	▼	7	35		
AS-13	N→T	19	35	25	12 ▼	12	10	37	0
AS-14	N→T	36	37	6	6	▼	7	13	35
AS-15	N→T	25	---	12	---	---	25	35	25
AL-19	D→V	12	0	0	10 ▼	2	18	35	
AL-23	N→T	24	10	0	2 ▼	1	0	4	34
AO-7	D→V	35	19	0	4	5	30		
AO-10	D→V	31	44	7	2 ▼	7	18	38	
AO-24	D→V	25	37	18	9	12	37	35	
AO-5	N→T	24	8	0	▼	2	7		
AO-25	N→T	25	31	4	0 ▼	13	4	32	
NS-6	N→T	25	32	39	36	▼ 25	35	31	
NS-16	N→T	---	---	50	44	▼ 38	41	35	
NL-18	D→V	25	31	35	37	▼	36	25	
NL-20	D→V	25	37	32	37	▼	38	35	
NL-12	N→T	24	32	35	38	▼	37	25	
NL-26	N→T	19	36	37	37	36	32	35	
NO-8	D→V	25	39	35	37	▼ 30	40		
NO-9	D→V	32	36	40	35	34	30	0	

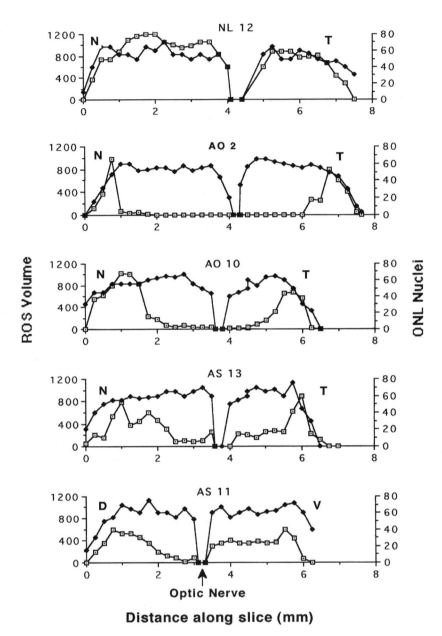

Figure 2. Rod outer segment (ROS) volume and number of ONL nuclei per 125 um at different points along retinal slices.   D=dorsal, V=ventral, N=nasal, T=temporal.

3.3. Screening pigment of the outer iris (argentea)

In teleosts, the reflective guanine crystals of the outer iris (argentea) functions to present a mirror over the normally black uveal tract, making the face of the outer face of the iris indistinguishable from the head scales (anti-predation). It has also been proposed that the argentea actually reflects light into "nooks and crannies" for feeding purposes (Walls, page 236)[14]. Inspection of the iris-angle region of the retinal slices revealed that the argentea remains intact in the albino although there is no melaninization in the inner pigmented layer of the iris epithelium, the pars iridiaca, which is continuous with the RPE (Fig.3 A, PI, see also Walls, page 159)[14]. The argentea may provide sufficient light screening to protect photoreceptors in the PGZ and adjacent retinal margin from light damage even though the inner iris epithelium lacks melanin.

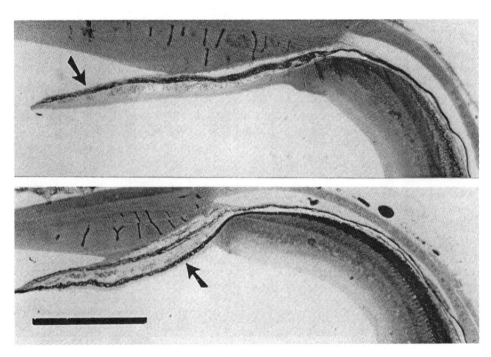

Figure. 3 Above: the iris-angle of albino, arrow ↓ Argentea . Below: arrow ↑ Pars Iridiaca of normal Scale bar = 0.5 mm.

4. DISCUSSION

4.1. Peripheral rods and the argentea

It was clear that in or near to the CGZ, ROS were either protected or were more resistant to light damage. In the former case, peripheral retinal rods in affected albino trout may benefit from a reduction of incoming light below the threshold for partial damage to the ROS. The area protected, including the CGZ, underlies the iris-angle, which contains a reflective argentea (see section 3.2 and Fig. 3). It is important to note that even the incipient stage of light damage is absent in the far periphery. In this "microhabitat", the fact that rods retain long ROS suggests that cell replacement only serves to add new retinal cells in the PGZ, and is not called upon to replace dying rods as may happen in the central retina. If the opacity of the argentea protects the peripheral retina in albinos, then one would expect to find reduced light damage behind the iris-angle even when cell death has occurred in the central retina, as in oscars[11] and zebrafish[23]. The contribution of the argentea to protection in the albino might be tested by surgically removing a small patch of it prior to light exposure.

4.2. Growth factors in the peripheral growth zone

Other aspects of the periphery may be distinct from those of the central retina, not only in terms of local ambient lighting, but also in the titer of growth factors known to rescue rods from the effects of light damage in rodents[28-32]. Among these, insulin-like growth factor (IGF) has been found to stimulate proliferation of rod precursors in the cichlid, *Haplochromis burtoni* while basic fibroblast growth factor (bFGF) stimulated the differentiation of rod precursors into new rods[33]. IGF is produced by cones in this species[34]. In goldfish, *Carrasius aureatus*, the proliferation of new cells which are added at the retinal margin is also stimulated by IGF, which selectively binds to receptors in the CGZ[35, 36]. A second candidate for protection is bFGF, which is produced by photoreceptors and the RPE[32] in rodents. Retinas in bFGF receptor-knockout mice degenerate[37] and in rats, bFGF is up-regulated in response to bright constant light [38, 39], injury[40] or optic nerve section[41]. Furthermore, rod bFGF is asymmetrically distributed in the human retina, with highest concentrations in the far periphery[42]. bFGF is also localized to RPE of goldfish[43] and is widely distributed in the retinas of the frog, *Rana pipiens*[44]. How IGF and bFGF could act as paracrines or autocrines to contravene early stages of light damage to rods in the periphery of the teleost retina is unknown, but their efficacy in promoting survival of rods in light-challenged albino rodents make this an attractive area for further study. If growth factors are concentrated in the CGZ to support ongoing mitotic activity of multi-potent retinal stem cells, a fortuitous but secondary effect may be to maintain peripheral ROS.

4.3. Other resistant factors in teleosts

The presence of functional uv-sensitive cones in juvenile trout, which favor shallow water, suggests that a diurnal habit has selected for resistance to light damage[13]. However, it is plausible that prior exposure to intense light could increase the threshold for light damage as in rodents[8]. The albino trout used in this study were living outdoors in shallow raceways for at least 70 days prior to their exposure to constant light or remaining outdoors. Albino trout lose their central ROS within 10 days after introduction from dim cyclic light into outdoor daylight (AS→AO, Allen et al[11]). Their central retinas then remain ROSless over the summer without apparent losses of ONL nuclei. Perhaps being ROSless prevents the rhodopsin-mediated type of light damage observed previously in rodents [2, 3]. In contrast, normal fish always maintained ROS 25-40 um in length in the central retina although there were minor differences in ROS length (Tables

1, 3). Clearly, the combination of photomechanical movements and melanin confers sufficient protection from diurnal light. Photomechanical movements occur in albinos[45], but whether these are of any benefit is conjectural. Since the cones are positioned in front of the rods at dawn, they may absorb specific wavelengths of light, and it is possible that extending truncated ROS into the un-melanized RPE may also improve exposure to antioxidant capacity and/or permit removal of damaged ROS discs by the RPE or by macrophages. The latter are known to increase in number in outdoor-held albino trout[22].

Some albino teleosts are more susceptible to progressive light damage than trout. As mentioned above, damage to ROS leads to apoptotic death of rods and cones in albinos of zebrafish[23] and losses of rods, but not cones in albino oscars[11]. However, unlike the rodent models, the damage sustained by albinos of teleosts is not permanent, owing to the ability to replace the photoreceptors throughout life[18]. Though some cell death must occur in both normal and albino trout exposed to daylight, the number of nuclei in the ONL is unchanged. In central retinas of trout, rods are replaced by differentiating rod precursors in the ONL which are derived from proliferating stem cells resident in the inner nuclear layer (INL)[21]. If both of these populations show increased mitotic activity in light-challenged albinos as compared to normal trout, this would explain why we have not observed a decline in the number of rod nuclei in albino trout as has been seen in albino oscars and zebrafish [11, 23].

It has been suggested that low temperature (11°C) is protective in albino trout, since oscars and zebrafish were held at 28°C or higher and rat retinas approach 37°C[11]. We have recently completed a study of light damage in channel catfish, *Ictalurus punctatus*, a species known to tolerate a wide range of temperature. Channel catfish are available in partially albinistic or normally pigmented strains, with the former having only a modest melanization of the dorsal retina and almost no melanin in the ventral retina. A modest effect of 3000 lux constant light in shortening ROS did not seem to differ when given at 17 or 30°C (Hendricks and Allen, unpublished). Thus, in this eurythermal species, higher temperature did not overcome a natural resistance to light damage.

Diurnal teleosts have many more cones than nocturnal rodents, and a rod is likely to have a neighboring cone in the typical square mosaic [12, 46, 47]. In albino trout, COS length actually increases during the demise of ROS[48], perhaps because cones are no longer competing with rods for disc precursors. This suggests that COS length may not be auto-correlated to photopic dosage as has been proposed in "photostasis" for ROS of rats, which are functional during the mesopic period and night[24]. According to startle tests, affected albino trout can see well only during the mesopic period[48] and their vision is impaired in full daylight, suggesting excessive light scattering in the absence of melanin. However, during dawn or dusk, cones and rods are both normally functional and therefore in albino retinas lacking ROS, rods may still be electrically coupled to cone activity (for review, see Krizaj[49]). This ongoing activity may signal central rod cell somas to delay undergoing apoptosis even though they have become ROSless.

4.4. Suggestions for further study

In trout, light damage is arrested at an early stage, whereas in zebrafish it proceeds to apoptosis of rods and cones[23]. For the former, a combination of light dose and other extrinsic conditions produces a spatial gradient of light damage between the central retina and the surrounding periphery. Because a maximum state of ROS damage is reached within 14 days in trout[11], an earlier sampling may show *in situ* recapitulation of different stages in the progression of retinal light damage in the transition zone. Where damage proceeds further, as in zebrafish, the same transitional zone may reveal more stages of damage.

To isolate resistance-specific biochemical factors, it would be feasible to ignore the transition zone and compare the central retina with the peripheral retina in the same eye or individual (Fig. 3). The COS-enriched central retina could be separated from the peripheral surround, including the transition zone, by applying a metal coring element of suitable diameter (right cylinder) against the posterior eye-cup. The peripheral retina, minus the transition zone, could be isolated using a larger coring element and taking retinal tissue beyond the transition zone, leaving a larger central core. ROS-enriched control tissue is also available from normal fish or protected albinos. Using this method, differences in lipid compositions of ROS and COS could be analyzed. Since the central core of affected albino trout should contain COS but few ROS, it should be possible to isolate cone-specific proteins or COS-specific lipid compositions from this tissue and determine whether it is a distinct lipid composition which makes COS so much more resistant to light damage than ROS (R.E. Anderson, personal communication). The COS-enriched central core may also offer a comparative method for verifying the importance of cone-specific visual cycle enzymes. For instance, 11-cis retinyl ester hydrolase has been localized to the plasma membrane of the RPE where it could supply chromophore for the rapid regeneration of cone pigments directly from 11-cis retinyl ester in the RPE[50, 51]. If this is true, there should be differences in the amount or activity of the enzyme in the central retina as compared to the periphery or to control retinas which have both ROS and COS intact. A third line of investigation could probe the relative distribution of anti-oxidants in the retina (vitamin E, glutathione, superoxide dismutase (SOD), catalase and ascorbate[52]. The question of whether any of these are adaptively up-regulated in light-challenged diurnal retinas is important, since anti-oxidants are known to be protective when exogenously applied to ocular tissues in rodents[4]. Fortunately, catalase and SOD have now been identified *in situ* by their mRNA transcripts[53].

Finally, we have suggested that a ROSless rod survives during light-challenge in albino trout simply because the absorbing visual pigment has been effectively removed[11]and/or because the rod cell body remains coupled to cone networks. However, in the central retina, which is most damaged and clearly loses almost all rod visual pigment (Table 1), the simultaneous loading of ROS fragments may contribute to lipofuscin in the RPE [11, 54]. This may increase susceptibility of the un-pigmented RPE to light damage in albinos, since lipofuscin enhances photo-oxidation and interferes with phagocytosis in cultured RPE cells [55, 56, 57]. To resolve this issue, one could compare rates of lipofuscin turnover in central and periphery of affected and protected albino trout as done in albino rats[58] (Katz et al., 1999). It would also be of interest to examine whether the increased numbers of macrophages, observed *in situ* in light-challenged albinos[22], actually assist the RPE in preventing accumulation of lipofuscin in the central retina.

5. ACKNOWLEDGEMENTS

One of us (D. M. A.) was supported by an Ashbel Smith Professorship. The Utah Department of Wildlife Resources kindly made space and specimens available.

REFERENCES

1. S. Semple-Rowland and W. W. Dawson, Retinal cyclic light damage threshold for albino rats, *Lab. Animal. Sci.* 37: 289-298 (1987).
2. C. Reme, C. Grimm, F. Hafezi, A. Wenzel and T. P. Williams, Apoptosis in the retina: the silent death of vision., *News Physiol. Sci.* 15: 120-125 (2000).
3. C. Grimm, A. Wenzel, F. Hafezi, S. Yu, T. M. Redmond and C. Reme, Protection of RPE65-deficient mice identifies rhodopsin as a mediator of light-induced retinal degeneration, *Nature Genetics* 25: 63-66 (2000).
4. D. Organisciak and B Winkler, Retinal light damage, practical and theoretical considerations, *Prog. Retinal and Eye Res.* 13: 1-29 (1994).
5. M. Lavail, in: *The Effects of Constant Light on Visual Processes*, edited by T. P Williams and B. N Baker, (Plenum press, NY, 1980), p. 357-387.
6. L. Rapp and T. P. Williams, in: *The Effects of Constant Light on Visual Processes*, edited by T. P. Williams and B. N Baker (Plenum press, NY, 1980), pp. 133-159.
7. T. P. Williams and W. Howell, Action spectrum of retinal light-damage in albino rats. *IOVS* 24: 285-287 (1983).
8. J. S. Penn and R. E. Anderson, Effect of light history on the rod outer segment membrane composition in the rat, *Exp. Eye Res.* 44: 767-778 (1987).
9. W. K. Noell and R. Albrecht, Irreversible effects of visible light on the retina: role of vitamin A, *Science* 172: 76-80 (1966).
10. D. T. Organsiciak, R. Darrow, L. Barsalou, R. Kutty and B. Wiggert, Circadian-dependent retinal light damage in rats, *IOVS* 41: 3694-3701 (2000).
11. D. M. Allen, C. Pipes, K. Deramus and T. Hallows, in: *Retinal Degenerative Diseases and Experimental Therapy* (Plenum Pub., N. Y. (1999), pp 337-350.
12. J. K. Bowmaker and Y. Kunz, Ultraviolet receptors, tetrachromatic colour vision and retinal mosaics in the brown trout (*Salmo trutta*): age –dependent changes, *Vision Res.* 27: 2102-2108 (1987).
13. C. Hawryshyn, M. Arnold, D. Chaisson and P Martin, The ontogeny of ultraviolet photosensitivity in rainbow trout (*Salmo gairdneri*), *Visual Neurosci.* 2: 2247-254 (1989).
14. G. L. Walls, *The Vertebrate Eye* (Cranbrook Inst. Of Science, Bulletin 19, Bloomfield Hills, MI (1942).
15. R. H. Douglas, The function of photomechanical movements in the retina of the rainbow trout (*Salmo gairdneri*), *J. Exp. Biol* 96: 389-403 (1982).
16. B. Burnside and B. Nagle, Retinomotor movements of photoreceptors and retinal pigment epithelium: mechanisms and regulation, *Prog. Retinal Res.* 2: 67-108 (1983).
17. P. R. Johns, Growth of the goldfish eye. III. Source of the new retinal cells, *J. Comp. Neurol.* 176:343-358 (1977).
18. P. R. Johns and R. Fernald, Genesis of rods in the retina of teleost fish, *Nature* 293, 141-142 (1981).
19. R. D. Fernald, Teleost vision: seeing while growing, *J. Exp. Zool. Suppl.* 5:167-180 1991).
20. K. Hoke and R. D. Fernald, Rod photoreceptor neurogenesis, *Prog. Retinal. and Eye Res.* 16: 31-49 (1997).
21. D. Julian, K. Ennis and J. Korenbrot, birth and fate of proliferative cells in the inner nuclear layer of the mature fish retina, *J. Comp. Neurol.* 3394: 271-282 (1998).
22. D. M. Allen and T. Hallows, Solar pruning of retinal rods in albino rainbow trout, *Visual Neurosci.* 14: 589-600 (1997).
23. T. S. Vihtelic and D. R. Hyde, Light-induced rod and cone cell death and regeneration in the adult albino zebrafish (*Danio rerio*) retina, *J. Neurobiol.* 44: 289-307 (2000).
24. J. S. Penn and T. P. Williams, Photostasis: regulation of daily photon-catch by rat retinas in response to various cyclic illuminances, *Exp. Eye Res.* 43: 915-928 (1986).
25. D. M. Allen, E. R. Loew and W. N. McFarland, Seasonal change in the amount of visual pigment in the retinae of fish, *Can. J. Zool.* 60: 281-287 (1982).
26. A. J. Olson, A. Picones, D. Julian and J. I. Korenbrot, A developmental time line in a retinal slice from rainbow trout, *J. Neurosci. Methods* 93: :91-100 (1999).
27. R. Sokal and F. Rholf, *Biometry* (W. H. Freeman, San Fraqncisco, 1969).
28. M. M. LaVail, K. Unoki, D.,Yasumura, M. Matthes and G. Yanucopoulos, Multiple growth factors, cytokines and neurotropins rescue photoreceptors from the damaging effects of constant light, *PNAS* 89:11249-11253 (1992).
29. M. M. LaVail, D. Yasumura, M. Matthes, C. Lau-Villacorta, K. Unoki, C. Sung and R. H. Steinberg, Protection of mouse photoreceptors by survival factors in retinal degenerations, *IOVS* 39: 592-602 (1998).
30. G. Faktorovich, R. H. Steinberg, D. Yasumura, M. Matthes and M. M. LaVail, Basic fibroblast growth factor and local injury protect photoreceptors from light damage in the rat, *Nature* 347: 83-86 (1992).

31. K. Masuda, I. Watanabe, K. Unoki, N. Ohba and T. Marumatsu, Functional rescue of photoreceptors from the damaging effects of constant light by survival-promoting factors in the rat, *IOVS* 36:2142-2146 (1995).
32. P. Campochiaro, S. Hackett and S. Vinores, Growth factors in the retina and retinal pigment epithelium, *Prog. in Retinal and Eye Res.* 15 (2): 547-567 (1996).
33. A. F. Mack, and R. D. Fernald, Regulation of cell division and differentiation in the teleost retina, *Dev. Brain Res.* 76:183-187 (1993).
34. A. F. Mack, S. Balt and R. D. Fernald, Localization and expression of insulin-like growth factor in the teleost retina, *Visual Neurosci.* 12: 457-461 (1995).
35. S. M. Boucher and P. F. Hitchcock, Insulin-related growth factors stimulate proliferation of retinal progenitors in the goldfish, *J. Comp. Neurol..* 394:386-394 (1998).
36. S. M. Boucher and P. F. Hitchcock, Insulin-like growth factor-I binds in the inner plexiform layer and circumferential germinal zone in the retina of the goldfish, *J. Comp. Neurol.* 394:395-401 (1998).
37. P. Campochiaro, M. Chang, S. Vinores, Z. Nie, L. Hjelmeland, A Mansukhani, C. Basilico and D. J. Zack, Photoreceptor degeneration in transgenic mice with photoreceptor-specific expression of a dominant negative fibroblast growth factor receptor, *J. Neurosci.* 16: 1679-1688 (1996).
38. H. Gao and J. Hollyfield, Basic fibroblast growth factor: increased gene expression in inherited and light-induced photoreceptor degeneration, *Exp. Eye Res.* 62: 181-189 (1996).
39. R. Wen, T. Cheng, Y. Song, M. Matthes, D. Yasumura and M. M. LaVail, Continuous exposure to bright light upregulates bFGF and CNTF expression in the rat retina, *Current Eye Res.* 17: 494-500 (1998).
40. W. Cao, R. Wen, F. Li, M. M. LaVail and R. H. Steinberg, Mechanical injury increases bFGF and CNTF mRNA expression in the mouse retina, *Exp. Eye Res.* 65:241-248 (1997).
41. C. Gargini, M.S. Belfiore, S. Bisti, L. Cervetto, K. Valter and J. Stone, The impact of basic fibroblast growth factor on photoreceptor function and morphology, *IOVS* 40: 2088-2099 (1999).
42. Z. Li, J. Chang and A. H. Milam, A gradient of basic fibroblast growth factor in rod photoreceptors in the normal human retina, *Visual Neurosci.* 14: 671-679 (1997).
43. P. A. Raymond, L. Barthel, and M. Rounsifer, Immuno-localization of basic fibroblast growth factor and its receptor in adult goldfish retina, *Exp. Neurol.* 115: 73-78 (1992).
44. R. Blanco, A. Lopez-Roca, J. Soto and J. Blagburn, Basic fibroblast growth factor applied to the optic nerve after injury increases long-term survival in the frog retina, *J. Comp. Neurol.* 423: 646-658 (2000).
45. D. M. Allen, unpublished research.
46. A. H. Lyall, Cone arrangements in teleost retinae, *Quart. J. Microscop. Sci.* 98:189-201 (1957).
47. K. Engstrom, Cone types and cone arrangements in teleost retinae, *Acta Zool.* 44:179-243 (1963).
48. D. M. Allen and T. Hallows, Reversible visual deficits in albino rainbow trout, abstract: Proceedings of the Western Section of the American Fisheries Society, Park City , UT (1995).
49. D. Krizaj, Mesopic state: cellular mechanisms involved in pre- and post-synaptic mixing of rod and cone signals, *Microsc Res. Tech.* 50: 347-359 (2000)
50. N. L. Mata and A. T. C. Tsin, Distribution of 11-cis LRAT, 11-cis RD and 11-cis REH in bovine retinal pigment epithelium membranes, *Biochim. Biophys. Acta.*1394: 16-22 (1998).
51. N. L. Mata, E. Villazana and A. T. C. Tsin, Colocalization of 11-cis retinyl esters and retinyl ester hydrolase activity in retinal pigment epithelium plasma membrane, *IOVS* 39: 11312-1319 (1998).
52. G. J. Handelman and E. A. Dratz, The role of antioxidants in the retina and retinal pigment epithelium and the nature of pro-oxidant-induced damage, *Adv. in Free Radical Biol. and Med.* 2: 1-89 (1986).
53. W. Chen, D. M. Hunt, H. Lu and R. C. Hunt, Expression of antioxidant protective proteins in the rat retina during prenatal and postnatal development, *IOVS* 40: 744-751 (1999).
54. K. Fite, L. Bengston and B. Donaghey, Experimental light damage increases lipofuscin in the retinal pigment epithelium of Japanese quail (*Coturnix coturnix Japonica*), *Exp. Eye Res.* 57: 449-460 (1993).
55. U. T. Brunk, U. Wihlmark, A. Wrigstad, K. Roberg and S-E. Nilsson, Accumulation of lipofuscin within retinal pigment epithelium cells results in enhanced sensitivity to photo-oxidation, *Gerontology* 41 (Suppl 2): 201-211 (1995).
56. U. Wihlmark, A. Wrigstad, K. Roberg, S-E. Nilsson and U. T. Brunk, Lipofuscin accumulation in cultured retinal pigment epithelium cells causes enhanced sensitivity to blue light irradiation, *Free Radical Biol. and Med.* 22 (7): 1229-1234 (1997).
57. S. Sundelin, U. Wihlmark, S.-E. Nilsson and U. Brunk, Lipofuscin accumulation in cultured retinal pigment epithelial cells reduces their phagocytic capacity. *Current Eye Res.* 17: 851-857 (1998).
58. M. L. Katz, C-L. Gao and L. M. Rice, Long-term variations in cyclic light intensity and dietary Vitamin A intake modulate lipofuscin content of the retinal pigment epithelium, *J. Neurosci. Res.* 57:106-116 (1999).

SPACR IN THE IPM: GLYCOPROTEIN IN HUMAN, PROTEOGLYCAN IN MOUSE

Jung W. Lee, Qiuyun Chen, Mary E. Rayborn, Karen G. Shadrach, John W. Crabb, Ignacio R. Rodriguez,* and Joe G. Hollyfield

1. SUMMARY

In this study, we compared the predicted primary structure and glycosaminoglycan (GAG) composition of the human and mouse IPM protein SPACR. Like SPACR in the human IPM, the mouse homologue contains a large central mucin-like domain flanked by consensus sites for N-linked oligosaccharide attachment; both contain one EGF-like domain and at least one HA-binding motif. Unlike human SPACR, which contains no conventional consensus sites for GAG attachment, SPACR in the mouse contains three. Biochemical studies of human and mouse SPACR protein indicate that this novel IPM molecule is a glycoprotein in human and a proteoglycan in the mouse. The presence of consensus sites for GAG attachment in the deduced sequence of SPACR in the mouse and the absence of these sites in human SPACR provide molecular verification of our biochemical results.

2. INTRODUCTION

SPACR is an abundant, interphotoreceptor matrix (IPM), sialoglycoprotein that is retained in the human IPM following saline rinses.[1] Analysis of seven peptides from purified human SPACR[2] revealed 100% identity to the deduced sequence of IMPG1 cDNA (also called GP147[3] and IPM150[4]). The IMPG1/SPACR gene is located on chromosome 6q13-q15.[5,6] A number of macular dystrophies have been linked to this locus in humans.[7-9] Although no specific mutations have yet been identified[5], the gene for this abundant IPM protein is a potential candidate locus for inherited retinal disease.

The gene product of IMPG1 is previously reported to be a chondroitin sulfate proteoglycan core protein in the human IPM (Accession Number NM_001563), but our

Cole Eye Institute, The Cleveland Clinic Foundation, Cleveland, OH 44195, USA, *The National Eye Institute, The National Institutes of Health, Bethesda, MD 20892, USA

carbohydrate analysis indicates that this molecule is not a proteoglycan since it is not a product of chondroitinase ABC digestion and does not react with the highly specific chrondroitin-6-sulfate antibody.[10,11] Additionally, the deduced amino acid sequence of IMPG1 reveals no established glycosaminoglycan (GAG) attachement.[2] Although our biochemical data defining human SPACR as a glycoprotein[2] is unequivocal, analysis of SPACR in the IPM of bovine, rat and mouse, suggest that SPACR in other species may exist as a proteoglycan.[11]

We have cloned the mouse SPACR gene and characterized the mouse SPACR protein as a means of more precisely defining the consensus sites for its secondary modifications. From the deduced amino acid sequence data, we prepared a peptide antibody, which allowed an analysis of the novel distribution of the SPACR polypeptide in the mouse IPM. The presence of GAG attachment sites in the deduced sequence of mouse SPACR and the absence of these sites in human SPACR provide molecular verification of our biochemical results, indicating that differences in post-translational modifications of SPACR may be fundamentally important in SPACR function in foveate and non-foveate retinas.

3. METHODS

The methods employed in these experiments have been described previously in detail.[12] Mouse SPACR cDNA was cloned by screening a retina cDNA library using a human PCR probe derived from the coding region.[12] Human and mouse sequences were aligned using FASTA.

Thirty to fifty mouse retinas were used for each IPM extraction, after two initial rinses with PBS at 4°C for 5 min each to remove the soluble IPM components. SPACR and other insoluble IPM molecules were extracted with 0.1 M TBS, 5 mM DTT at pH 8.0 as previously reported.[11] Extraction was for 2-4 h at 4°C with gentle agitation on a rotating platform. Digestion of the extract with chondroitinase ABC (Seikagaku Corp., Ijamsville, MD, USA) was performed on 50 μl aliquots of 1 mg/ml IPM extract resuspended in 0.1 M Tris-acetate buffer, pH 7.3. Three mU of each enzyme, containing protease inhibitors, was added to the tube. The reaction was allowed to proceed at 37°C for 3 h.

For Western blotting, protein concentrations of 30 μg total IPM protein (obtained as described in mouse IPM extraction) were loaded per lane on 7.5% SDS/PAGE mini-gels. The gels were run at 100 V for 1.5 h at room temperature. Separated IPM samples were electroblotted onto Immobilon-P membranes using a Bio-Rad Semi-Dry Electrophoretic Transfer Cell for 20 min at 18 V. After transfer, the membranes were blocked with 5% Blotto for 2 h at room temperature and the gels were stained with Gel Code Blue (Pierce Chemicals, Rockford, IL) to estimate remaining proteins. The membranes were then incubated in the appropriate antibody (ΔDi6S monoclonal antibody, pre-bleed IgG or anti-SPACR IgG) in 1% BSA/PBS overnight at 4°C. Membranes were washed with PBS-Tween (0.05%) (3 times), and further incubated with alkaline phosphatase-conjugated secondary antibody (1:5000) for 1 h at room temperature. The membranes were washed and the color reaction developed using the substrate 5-bromo-4-chloro-3-indolyl phosphate-nitro blue tetrazolium (BCIP-NBT) or DAB.

The putative SPACR core protein was partially purified from mouse IPM, digested with chondroitinase ABC, separated by SDS/PAGE and detected with ΔDi6S monoclonal antibody. The immunopositive band was excised from the gel, digested with trypsin and the digest analyzed by MALDI mass spectroscopy.

```
Mouse  MKRIFDLPKLRTKRSALFPAA-NICPQESLRQILASLQEYYRLRVCQEVVWEAYRIFLDR 59
       :.:::::: : :::::: ::    .:::::..::: ::: :::::::: :::::::::::
Human  MRRIFDLAKHRTKRSAFFPTGVKVCPQESMKQILDSLQAYYRLRVCQEAVWEAYRIFLDR 116

Mouse  IPDTEEYQDWVSLCQKETFCLFDIGKNFSNSQEHLDLLQQRIKQRSFPGRKDETASMETL 119
       :::: :::::::.::::::::::::::NF::::::::::::::::::: :::: .. .::
Human  IPDTGEYQDWVSICQQETFCLFDIGKNFSNSQEHLDLLQQRIKQRSFPDRKDEISAEKTL 176

Mouse  EAPTEAPVVPTDVSRMSLGPFPLPSDDTDLKEILSVTLKDIQKPTTESITEPIHVSEFSS 179
       :  :  :. :::. .:::::::: ::: : :::  :: : . :::: ::
Human  GEPGETIVISTDVANVSLGPFPLTPDDTLLNEILDNTLNDTKMPTTERETE----FAVLE 232

Mouse  EEKVEFSISLPNHRFKAELTNSGSPYYQELVGQSQLQLQKIFKKLPGFGEIRVLGFRPKK 239
       :..:: :.::.: .:::::: :: ::::::: :.:::..:::::::: .: ::::::::
Human  EQRVELSVSLVNQKFKAELADSQSPYYQELAGKSQLQMQKIFKKLPGFKKIHVLGFRPKK 292

Mouse  EEDGSSSTEIQLMAIFKRDHAEAKSPDSHLLSLDSNKIESERIHHGVIE-DKQPETYLTA 298
       :.:::::::.:: ::::: ::::: : ::: :::::::: ..:: .: ::::: ::::
Human  EKDGSSSTEMQLTAIFKRHSAEAKSPASDLLSFDSNKIESEEVYHGTMEEDKQPEIYLTA 352

Mouse  TDLKKLIIQLLDGDLSLVEGKIPFGDEVTGTL--FRPVTEPDLPKPLADVTEDATLSPEL 356
       ::::.:: . :. . :: :::: : .:: .:     :. .:: : .:::::::::::
Human  TDLKRLISKALEEEQSLDVGTIQFTDEIAGSLPAFGPDTQSELPTSFAVITEDATLSPEL 412

Mouse  PFVEPRLEAVDREGSELPGMSSKDSSWSPPVSASTSRSENLPSF-TPSIFSLDAQSPPPL 415
       : ::::: ::  ::    :.::::: :::: :: :  : : :::::  :
Human  PPVEPQLETVDGAEHGLP-----DTSWSPPAMASTSLSEAPPFFMASSIFSLTDQGTTDT 467

Mouse  MTTGPTALIPKPTLPTIDYSTIRQLPLESSHWPASSSDRELITSSHDTIRDLDGMDVSDT 475
       : :  : :.: :.::.::: :: :   :: :::: :       :  .: :: ::.:::
Human  MATDQTMLVPGLTIPTSDYSAISQLALGISHPPASSDDSRSSAGGEDMVRHLDEMDLSDT 527

Mouse  PALSEISELSGYDSASGQFLEMTTPIPTVRFITTSSETIATKGQELVVFFSLRVANMPFS 535
       :: :: ::: :... :::   .......:: ::.::: ::.::::::::::::::::::
Human  PAPSEVPELSEYVSVPDHFLEDTTPVSALQYITTSSMTIAPKGRELVVFFSLRVANMAFS 587

Mouse  YDLFNKSSLEYQALEQRFTDLLVPYLRSNLTGFKQLEILSFRNGSVIVNSKVRFAKAVPY 595
       ::::::::::::.::::.::.:::::::::NLTGFKQLEIL::NGSVIVNSK :::.::::
Human  NDLFNKSSLEYRALEQQFTQLLVPYLRSNLTGFKQLEILNFRNGSVIVNSKMKFAKSVPY 646

Mouse  NLTQAVRGVLEDLRSTAAQGLNLEIESYSLDIEPADQADPCKLLDCGKFAQCVKNEWTEE 655
       ::::.:::: :::: ::::.::: :::::::: :::::: ::::: .:. :.::::: :::
Human  NLTKAVHGVLEDFRSAAAQQLHLEIDSYSLNIEPADQADPCKFLACGEFAQCVKNERTEE 707

Mouse  AECRCRQGHESHGTLDYQTLNLCPPG-KTCVAGREQATPCRPPDHSTNQAQEPGVKKL-- 714
       :::::::.:.::: :.:::     :: :: :: :  .     ::: :::: :::  . :::
Human  AECRCKPGYDSQGSLDGLEPGLCGPGTKECEVLQGKGAPCRLPDHSENQAYKTSVKKFQN 768

Mouse  RQQNKVVKKRNSKLSAIGFEEFEDQDWEGN 742
       .: :::. :::::. . . .::: ::::::
Human  QQNNKVISKRNSELLTVEYEEFNHQDWEGN 797
```

Fig.1. Optimal global alignment comparison of mouse and human SPACR prepared with the sequence alignment utility and the FASTA algorithm available through the Munich Information Center for Protein sequences. Mouse SPACR predicted amino acid sequence (single letter code). Double dots between aligned residues indicate identity; single dots indicate homology: no dots indicate no homology: dashes indicate interruptions in the sequence allowing for the logical alignment of the two molecules. Asparagine residues in bold (**N**) represent consensus sites for N-linked glycosylation. Serine residues in bold (**S**) represent consensus sites for xylosylation and GAG attachment. The sequences beginning and ending in arginine or lysine that are underlined represent putative hyaluronan-binding motifs. The six conserved cysteine residues (**C** in bold) in the EGF-like motif are involved in interchain disulfide bonding required for stability of the EGF-like motif core proteins present in the mouse IPM.

4. RESULTS AND DISCUSSION

An important first step for the study of the structure-function relationships of this protein was the isolation, cloning and sequencing of the full length cDNA for mouse SPACR. SPACR cDNA in the mouse comprises over 3675 bp containing an open reading frame coding for 742 amino acids. When the deduced amino acid sequences of human and mouse SPACR aligned, several homologous regions were evident (Fig. 1).
(A). Both species contain a large mucin-like domain, with numerous potential O-linked glycosylation sites located in the central part of the sequence. (B). Both species contain clusters of N-linked consensus sites on either side of the mucin domain. Five consensus sites are in perfect alignment and show 100% sequence identity. (C). Both species contain putative HA-binding motifs (three in human, four in mouse). (D). Both species contain an epidermal growth factor (EGF)-like motif, with six conserved cysteine residues in perfect alignment and 65% homology between two species. These conserved cysteine residues participate in the interchain disulfide bonding required for the structural stability of the EGF-like motifs.[15-17]

However, these two species have very distinct structural differences. Unlike human SPACR, which does not contain conventional consensus sites for GAG attachment,[2] three are present in mouse SPACR. GAGs are linked to serine residues in the core protein by way of a trisaccharide bridge, consisting of a xylose residue attached to serine,
followed by two galactose residues.[18] Serine residues functioning as xylosylation sites utilize the consensus sequence SG, SG(acidic) or SGXG.[14,19,20]

Although no other consensus sequence for xylosylation has been documented in any mammalian proteoglycan, two reports implicate the sequence EGSAD in two chicken proteoglycans.[22-24] Such sequences are not present in human or mouse SPACR, although human and mouse SPACR contain a DGS sequence (human D^{295}-S^{297}, mouse D^{242}-S^{244}), and mouse also contains an EGS sequence (E^{369}-S^{371}). It is unlikely that these sites function in GAG attachment since they do not match the five residue sequence thought to be important for xylosylation in chicken.[22,23] The presence of consensus sites for GAG attachment in the deduced sequence of mouse SPACR and the absence of these sites in human SPACR provides molecular verification of our previous biochemical data indicating that the carbohydrate composition of SPACR in these two species is fundamentally different.[25]

Our previous studies demonstrated that human SPACR is a glycoprotein and not a proteoglycan, because the chondroitin-6-sulfate Δdisaccharide (ΔDi6S) monoclonal antibody did not recognize chondroitin 6-sulfate epitope in human extracts digested with chondrotinase ABC.[2] In contrast, mouse SPACR may be represented by one of the prominent chondroitin sulfate proteoglycans in the mouse IPM because of the putative GAG attachment sites identified in the deduced amino acid sequence of mouse SPACR (Fig. 1) and the localization of SPACR expression to photoreceptors.[12] Using the ΔDi6S chondroitin sulfate monoclonal antibody following chondroitinase ABC in the mouse IPM (Fig.2, lanes1-3) several potential bands are evident in the IPM chondroitinase ABC digested IPM samples (Fig. 2, lane 3). To prove that the mouse SPACR is a chondroitin sulfate proteoglycan, highly specific antibody for the chondroitin 6-sulfate was used in the Western blot analysis. We prepared a polyclonal antibody in rabbit to a peptide representing the deduced C-terminal of mouse SPACR open reading frame (Fig. 1). The preimmune IgG fraction does not label the non-digested or the chondroitinase digested mouse IPM. The IgG fraction from the rabbit following the fourth boost does not label the undigested IPM sample (data not shown). But it intensely labels a band at 150 kD in the chondroitinase ABC digested IPM sample (Fig. 2, lane 4), whereas neither the digested nor nondigested IPM sample is labeled when the antibody is preincubated with the peptide to which the antibody was prepared. This suggests that the 150 kD band, which labels with both the ΔDi6S antibody and the antibody prepared to the deduced C-terminal peptide, represents SPACR in the mouse.

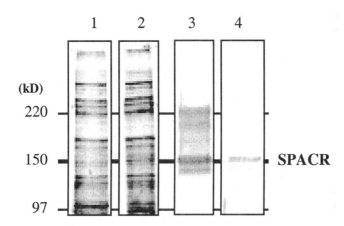

Figure 2. IPM extracts from the mouse retina separated with SDS/PAGE. Lanes 1-2 are Gel Code Blue stained gels. Lanes 2-4 are Western blots from IPM samples transferred from acrylamide gels similar to those shown in lanes 1-2. Numerous protein bands are evident in lanes 1-2. A slight increase in intensity of staining is evident in lane 2 compared to lane 1 above the 220 kD marker, at the 150 kD marker and below the 97 kD marker. Lane 3 shows intense ΔDi6S labeling of the chondroitinase ABC digested sample. Several bands are evident in lane 3 with the most intensely labeled band present at 150 kD. Anti-SPACR IgG fraction was used to label the IPM samples directly (lane 4). The SPACR antibody labels a single band at 150 kD containing the chondroitinase ABC digested IPM.

To further verify the identity of the band that labels in Western blots with these two antibodies, the 150 kD immunoreactive band was isolated, digested with trypsin in situ, and the peptides analyzed by Matrix Assisted Laser Desorption Ionnization Time of Flight (MALDI TOF) mass spectroscopy. The eight peptides (Q^{31}-R^{41}, A^{196}-K^{219}, R^{257}-K^{276}, L^{304}-K^{319}, D^{380}-R^{393}, G^{518}-R^{528}, F^{553}-R^{562}, and E^{689}-K^{710}) from the 150 kD mouse chondroitin sulfate proteoglycan core protein[25] were matched the predicted amino acid sequence of mouse SPACR (Fig. 1). These peptides represent approximately 17 percent of the total deduced sequence. These data demonstrate that the 150 kD chondroitin sulfate core protein present in the IPM that is decorated with both the ΔDi6S antibody and the anti-SPACR antibody is the proteoglycan core protein coded by the SPACR cDNA. These combined data indicate that mouse SPACR is a proteoglycan, but human SPACR is a glycoprotein.

While the functional role of SPACR in the mammalian retina remains to be determined, the presence of this molecule as a glycoprotein in human,[2,3,25] and macaque[26] IPM, and as a proteoglycan in the mouse,[12] rat and bovine,[3,25] suggests a role in foveal specialization. Anthropoid primates are the only mammalian species in which a specialized fovea is present. Foveas in diurnal primates are characterized by the concentration of cone photoreceptors in this region.[27] The extremely high packing density of foveal cones (around 100,000 cones/mm) is in part responsible for the high resolution of visual detail initiated by the cone mosaic in this specialized retinal location.[28] The absence of GAGs on human SPACR will substantially reduce its hydration sphere. This reduction in molecular volume would be permissive for the high packing density of foveal cones in primates. Knowledge of the carbohydrate structure of SPACR in other primate and non-primate species will be important steps in beginning to establish the functional role of this novel IPM molecule.

5. ACKNOWLEDGEMENTS

This work was supported by grants from the National Institutes of Health, National Eye Institute, Bethesda MD and the Foundation Fighting Blindness, Hunt Valley, MD. We thank Dr. Muayyad R. Al-Ubaidi for providing the mouse retina cDNA library; Dr. John R. Hassell for his evaluation on the deduced sequence and his comments on potential GAG attachment sites. We thank Mr. Larry L. Oliver for his expert assistance with the microtomy.

REFERENCES

1. S. Acharya, M. E. Rayborn, and J. G. Hollyfield, Characterization of SPACR, a sialoprotein associated with cones and rods present in the interphotoreceptor matrix of the human retina: immunological and lectin binding analysis, *Glycobiology* **8**, 997-1006 (1998).
2. S. Acharya, I. R. Rodriguez, E. F. Moreira, R. J. Midura, K. Misono, E. Todres, and J. G. Hollyfield, SPACR, a novel interphotoreceptor matrix glycoprotein in human retina that interacts with hyaluronan, *J. Biol. Chem.* **273**, 31599-31606 (1998).
3. L. Tien, M. E. Rayborn, and J. G. Hollyfield, Characterization of the interphotoreceptor matrix surrounding rod photoreceptors in the human retina, *Exp. Eye Res.* **55**, 297-306 (1992).
4. M. Kuehn, and G. Hageman, Expression and characterization of the IPM 150 gene (IMPG1) product, a novel human photoreceptor cell-associated chondroitin-sulfate proteoglycan, *Matrix. Biol.* **18**, 509-518 (1999).
5. A. Gehrig, U. Felbor, R. E. Kelsell, D. M. Hunt, I. H. Maumenee, and B. H. Weber, Assessment of the interphotoreceptor matrix proteoglycan-1 (IMPG1) gene localised to 6q13-q15 in autosomal dominant Stargardt-like disease (ADSTGD), progressive bifocal chorioretinal atrophy (PBCRA), and North Carolina macular dystrophy (MCDR1), *J. Med. Genet.* **35**, 641-645 (1998).
6. U. Felbor, A. Gehrig, C. G. Sauer, A. Marquardt, M. Kohler, M. Schmid, and B. H. Weber, Genomic organization and chromosomal localization of the interphotoreceptor matrix proteoglycan-1 (IMPG1) gene: a candidate for 6q-linked retinopathies, *Cytogenet. Cell Genet.* **81**, 12-17 (1998).
7. R. Kelsell, K. Gregory-Evans, C. Gregory-Evans, G. Holder, M. Jay, B. Weber, A. Moore, A. Bird, and D. Hunt, Localization of a gene (CORD7) for a dominant cone-rod dystrophy to chromosome 6q, *Am. J. Humm, Genet,* **63**, 274-279 (1998).

8. C. Sauer, H. Schworm, M. Ulbig, A. Blakenagel, K. Rohrschneider, D. Pauleikhoff, T. Grimm, and B. Weber, An ancestral core haplotype defines the critical region harbouring the North Carolina macular dystrophy gene (MCDR1), *J. Med. Genet.* **34**, 961-966 (1997).

9. E. Stone, B. Nichols, A. Kimura, T. Weingeist, A. Drack, and V. Sheffield, Clinical features of a Stargardt-like dominant progressive macular dystrophy with genetic linkage to chromosome 6q, *Arch. Ophthalmol.* **112**, 765-772 (1994).

10. S. Acharya, I. R. Rodriguez, and J. G. Hollyfield, The gene product of IMPG1 is the glycoprotein SPACR, not an IPM proteoglycan. In *Retinal Degenerative Diseases and Experimental Therapy.* edited by J. G. Hollyfield, R. E. Anderson and M. M. LaVail (Kluwer Academic/Plenum Publishers, NY, 1999), pp. 183-187.

11. J. G. Hollyfield, M. E. Rayborn, R. J. Midura, K. G. Shadrach, and S. Acharya, Chondroitin sulfate proteoglycan core protein in the interphotoreceptor matrix: a comparative study using biochemical and immunocytochemical analysis, *Exp. Eye Res.* **69**, 311-322 (1999).

12. J. W. Lee, Q. Chen, M. E. Rayborn, K. G. Shadrach, J. W. Crabb, I. R. Rodriguez, and J. G. Hollyfield, SPACR in the interphotoreceptor matrix of the mouse retina: molecular, biochemical and immunohistochemical characterization, *Exp Eye Res.* **71**, 341-52 (2000).

13. D. M. Mann, Y. Yamaguchi, M. A. Bourdon, and E. Ruoslahti, Analysis of glycosaminoglycan substitution in decorin by site-directed mutagenesis, *J. Biol. Chem.* **265**, 5317-5323 (1990).

14. B. Yang, B. L. Yang, R. C. Savani, and E. T. Turley, Identification of a common hyaluronan binding motif in the hyaluronan binding proteins RHAMM, CD44 and link protein. *EMBO J.* **13**, 286-296 (1994).

15. A. Gray, T. J. Dull, and A. Ulrich, Nucleotide sequence of epidermal growth factor cDNA predicts a 128,000-molecular weight protein precursor, *Nature* **303**, 722-725 (1983).

16. U. Novenberg, H. Wille, J. M. Wolff, R. Frank, and F. G. Rathjen, The chicken neural extracellular matrix molecule restrictin: similarity with EGF-, fibronectin type III-, and fibrinogen-like motifs, *Neuron* **8**, 849-863 (1992).

17. J. Scott, M. Ufdea, M. Quiroga, R. Sanchez-Pescador, N. Fong, M. Selby, W. Rutter, and G. I. Bell, Structure of a mouse submaxillary messenger RNA in coding epidermal growth factor and seven related proteins, *Science* **221**, 236-240 (1983).

18. L. Roden, In *The biochemistry of glycoproteins and proteoglycans.* edited by W. Lennarz (Plenum Publishing Co., NY, 1980). pp. 267-371.

19. M. A. Bourdon, A. Oldberg, M. Pierschbacher, and E. Ruoslahti, Molecular cloning and sequence analysis of a chondroitin sulfate proteoglycan cDNA, *Proc. Nat. Acad. Sci. USA* **82**, 1321-1325 (1985).

20. R. K. Chopra, C. H. Pearson, G. A. Pringle, D. S. Fackre, and P. G. Scott, Dermatan sulphate is located on serine-4 of bovine skin proteodermatan sulphate: Demonstration that most molecules possess only one glycosaminoglycan chain and comparison of amino acid sequences around glycosylation sites in different proteoglycans, *Biochem. J.* **232**, 277-279 (1985).

21. W. Li, J.-P. Vergnes, P. K. Cornuet, and J. R. Hassell, cDNA clone to chick corneal chondroitin/dermatan sulfate proteoglycan reveals identity to decorin, *Arch. Biochem. Biophys.* **296**, 190-197 (1992).

22. D. McCormick, M. van der Rest, J. Goodship, G. Lozano, Y. Ninomiya, and B. R. Olsen, Structure of the glycosaminoglycan domain in the type IX collagen-proteoglycan, *Proc. Natl. Acad. Sci. USA,* **84**, 4044-4048 (1987).

23. M. Iwasaki, M. E. Rayborn, A. Tawara, and J. G. Hollyfield, Proteoglycans in the mouse interphotoreceptor matrix. V. Distribution at the apical surface of the pigment epithelium before and after retinal separation, *Exp. Eye Res.* **54**, 415-432 (1992).

24. H. Lazarus, W. Sly, J. Kyle, and G. Hageman, Photoreceptor degeneration and altered distribution of interphotoreceptor matrix proteoglycans in the mucopolysaccharidosis VII mouse, *Exp. Eye Res.* **56**, 531-541 (1993).

25. J. G. Hollyfield, Hyaluronan and the functional organization of the interphotoreceptor matrix, *Invest. Ophthalmol. Vis. Sci.* **40**, 2767-2769 (1999).

26. J. G. Hollyfield, M. E. Rayborn, K. Nishiyama, K. G. Shadrach, M. Miyagi, J. W. Crabb, and I. R. Rodriguez, Interphotoreceptor matrix in the fovea and peripheral retina of the primate macaca mulatta: distribution and glycoforms of SPACR and SPACRCAN, *Exp Eye Res.* **72**, 49-61 (2001).

27. H. Gao, and J. G. Hollyfield, Aging of the human retina: differential loss of neurons and retinal pigment epithelial cells, *Invest. Ophthalmol. Vis. Sci.* **33**, 1-17 (1992).

28. R. Rodieck, *The First Steps in Seeing.* (Sinauer Associates, Inc., Sunderland, MA, USA, 1998).

CONE ARRESTIN EXPRESSION AND INDUCTION IN RETINOBLASTOMA CELLS

Yushun Zhang,[#] Aimin Li,[#] Xuemei Zhu,[#] Ching H. Wong, Bruce Brown, and Cheryl M. Craft[*]

1. INTRODUCTION

Arrestins are regulatory proteins that down-regulate phosphorylated G-protein coupled receptors (GPCRs). The arrestin superfamily includes visual arrestins, beta-arrestins (βarrestins) and insect chemosensory arrestins. Two members of visual arrestins have been identified in vertebrate photoreceptors: rod arrestin (also known as S-antigen or Arrestin1) and cone arrestin (X-arrestin or Arrestin4). Rod arrestin was the first member to be molecularly characterized,[1] and its structure and function have been extensively studied in recent years. In vertebrate rod photoreceptors, rod arrestin quenches the light-induced phototransduction cascade by binding preferentially to the light-activated, phosphorylated rhodopsin.[2, 3] Subsequently, βarrestin1[4] and βarrestin2[5] were identified and shown to have a ubiquitous expression pattern. βarrestins have an analogous function to rod arrestin in the termination of GPCR signaling by sterically inhibiting the coupling of phosphorylated receptors to their respective G proteins.[4, 5] Recent evidence suggests additional roles for βarrestins in both termination of receptor signaling and signaling to downstream effectors by interacting with clathrin-coated pits and the tyrosine kinase c-*Src*.[6-11]

Cone arrestin was first cloned from a rat pineal library and discovered to be highly expressed in retinal cone photoreceptors and pinealocytes.[12] Based on this observation, a human retina cDNA was characterized and the gene was mapped to the X chromosome.[12, 13] Its high sequence homology to other arrestins and its cone photoreceptor localization suggest that cone arrestin may play an important role in the

---

[#] These authors contributed equally to this work.

[*] Yushun Zhang, Aimin Li, Xuemei Zhu, Ching H. Wong, Bruce Brown, and Cheryl M. Craft, The Mary D. Allen Laboratory for Vision Research, Doheny Eye Institute and Department of Cell and Neurobiology, the Keck School of Medicine of at University of Southern California, BMT401, 1333 San Pablo Street, Los Angeles, California 90089-9112.

New Insights Into Retinal Degenerative Diseases
Edited by Anderson *et al.*, Kluwer Academic/Plenum Publishers, 2001

modulation of cone phototransduction as rod arrestin does in rod phototransduction; however, the actual function of cone arrestin is still unknown.

In this study, an affinity-purified anti-peptide polyclonal antibody against the carboxyl terminus of human cone arrestin was prepared and characterized. Direct evidence for the localization of cone arrestin in all three classes of cone photoreceptors in human retina was shown by immunohistochemistry. The expression of cone arrestin was also examined in two retinoblastoma cell lines treated with differentiating agents. Its expression in these cell lines provides potential *in vitro* models for future studies in cone arrestin gene regulation and function in relation to cell growth and differentiation.

2. METHODS

2.1. Preparation of Anti-Human Cone Arrestin Polyclonal Antibody

Human cone arrestin antiserum was prepared by immunization of two New Zealand white rabbits with a synthesized peptide corresponding to the carboxyl terminus of human cone arrestin (GENBANK accession number U03626, aa376-387, QKAVEAEGDEGS).[12] The antiserum was purified by affinity chromatography against the peptide and named Luminaire Founder (LUMIf) (Research Genetics, Huntsville, AL).

2.2. Cell Culture and Treatment

Y79 and Weri-Rb-1 retinoblastoma cells (American Type Tissue Culture, Rockville, MD) were maintained in suspension culture in RPMI 1640 medium as previously described.[14] For monolayer culture, cells were plated at 3×10^5 cells/ml into 60- or 100-mm poly-D-lysine pre-coated tissue culture dishes (Becton Dickson, San Jose, CA). Cells were grown for 48 hrs prior to the addition of dibutyryl cyclic AMP (dbcAMP, 2 mM), sodium butyrate (2 mM), N,N'-hexamethylene-bis-acetamide (HMBA, 1mg/ml), all-*trans* retinoic acid (RA, 10 μM) or the drug vehicle, dimethylsulphoxide (DMSO, 0.1%). The cells were then cultured for an additional period of time, as stated in the results. Media were changed and chemicals were re-administered every 2 days.

2.3. Immunoblot Analysis

Human tissues used in this study were generously provided by the Alzheimer's Disease Center, Neuropathology Core, the Keck School of Medicine at University of Southern California (IRB # 845011). Y79 and Weri-Rb-1 cells were treated as described above. At the end of the incubation period, the cells were washed 3 times with phosphate buffered saline (PBS), scraped in 500 μl of PBS, and sonicated 1 sec x 10 on ice. Equal amounts (50 μg) of soluble proteins from various human tissues or proteins from whole cell homogenates of Y79 and Weri-Rb-1 were electrophoresed on 11.5% sodium dodecal sulfate-polyacrylamide gel electrophoresis (SDS-PAGE) and were electrophoretically transferred to Immobilon-P membranes (Millipore, Bedford, MA) as previously described.[15]

The immobilized proteins were detected with either LUMIf (1:200,000) or the anti-S-antigen (rod arrestin) monoclonal antibody (Mab C10C10, 1:5,000)[16] followed with anti-rabbit and anti-mouse secondary antibodies, respectively, using an Enhanced Chemiluminescence Kit (Amersham).

2.4. Northern Blot Analysis

Weri-Rb-1 cells, treated with different chemicals, were processed for total RNA isolation by removing the medium and immediately lysing the cells in 1 ml of the RNA STAT-60[TM] reagents (TEL-TEST, INC.) per 100-mm dish. Total cellular RNA was isolated by following the manufacturer's instruction. 20 µg of total RNA was blotted and probed with an $[\alpha\text{-}^{32}P]$ dCTP-labeled, random primed hCAR full-length cDNA (#5B) probe[12] as described previously[17] The membrane was then stripped and re-hybridized with a ^{32}P-labeled β-actin probe to determine loading efficiency.

2.5. Immunohistochemistry and Confocal Microscopy

Human eyes with a post-mortem time of 3 to 8 hrs were fixed in 4% paraformaldehyde. The eyes were processed overnight using an automated Shandon Citadel 2000 dehydrator. Dissected parts of the eye were embedded in paraffin and sectioned with a JUNG RM2035 ultratome (Leica, Foster City, CA).

Retinal sections of approximately 6 µm though the optic nerves were fixed and blocked as described previously.[18] The sections were incubated overnight at 4°C with a mixture of primary antibodies containing LUMIf (1:200,000) and one of the following antibodies: anti-red/green opsin (Mab COS-1, 1:10,000),[19] anti-blue opsin (Mab OS-2, 1:10,000),[19] anti-S-antigen (Mab C10C10, 1:1,000), and a universal arrestin antibody (Mab 5C6.47, 1:2,000).[12] Following the washing steps, the sections were incubated with a secondary antibody mixture containing a Cy3-conjugated goat anti-rabbit secondary antibody and a Dichloroforo Triazinyl Amino Fluorescein (DTAF)-conjugated goat anti-mouse secondary antibody for 1 hr at room temperature. After 3 washes with PBS, the slides were mounted and photographed using a Zeiss confocal microscope (Zeiss, Inc., Thornwood, NY).[18]

3. RESULTS

3.1. Cone Arrestin Is Expressed in All Cone Photoreceptors in Human Retina

Immunoblot analysis of various human tissues was used to characterize the affinity-purified polyclonal antibody LUMIf. A single protein of approximately 46 kDa was identified in retina and pineal gland but not in other tissues examined (Figure 1), which is compatible with the published data on the tissue distribution of cone arrestin mRNA in rat.[12] The immunoreactive band in retina and pineal was completely blocked by 10X (mol: mol) of the specific peptide used to generate the antibody (data not shown), confirming the specificity of the antibody. In separate experiments, the human cone arrestin cDNA was subcloned into a mammalian expression vector and transfected into

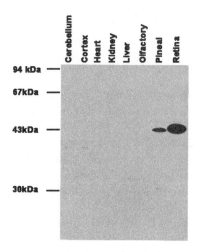

Figure 1. Immunoblot analysis of human cone arrestin expression in various human tissues. Molecular weight markers are indicated on the left.

COS-7 cells, which were then subjected to the same immunoblot assay. The recombinant cone arrestin protein was also recognized by LUMIf as a single band of similar molecular weight (data not shown).

After confirming its antibody activity and specificity against cone arrestin, LUMIf was used with either COS-1 or OS-2 for an indirect immunofluorescent dual localization study. COS-1 and OS-2 are well-characterized monoclonal antibodies that recognize red/green and blue opsins, respectively.[19, 20] Although both COS-1 (Figure 2Aa) and LUMIf (Figure 2Ab) staining were detected in identical cells in the photoreceptor layer, the cellular localization of red/green opsin and cone arrestin does not overlap (Figure 2Ac). COS-1 staining was found in the outer and inner segments, as well as the cytoplasmic membrane of the cone photoreceptor cells (Figure 2Aa), while LUMIf immunoreactivity was seen diffusely distributed throughout the entire cytoplasm of the cell body, including the synaptic processes (Figure 2Ab). The same staining pattern was seen with LUMIf and OS-2 double staining (Figure 2B), confirming that immunoreactive cone arrestin is also expressed in blue cones. It should be noted that some of the LUMIf-positive staining photoreceptors, supposedly red and green cones, were not stained with the blue opsin antibody; however, no blue opsin-positive cells were observed without the LUMIf antibody staining (Figure 2B). We did not detect any LUMIf-positive staining cells that are negative in red/green opsin immunoreactivity (blue cones) in figure 2A, probably because blue cones are relatively scarce compared with red/green cones in human retina.[19, 21]

In retina sections with LUMIf and C10C10 double staining (Figure 2C), all rod photoreceptors were labeled green by C10C10-DTAF-conjugated-anti-mouse secondary antibody (Figure 2Ca), while all cone photoreceptors were labeled red by LUMIf-Cy-3-conjugated-anti-rabbit secondary antibody (Figure 2Cb). No overlap was seen in any

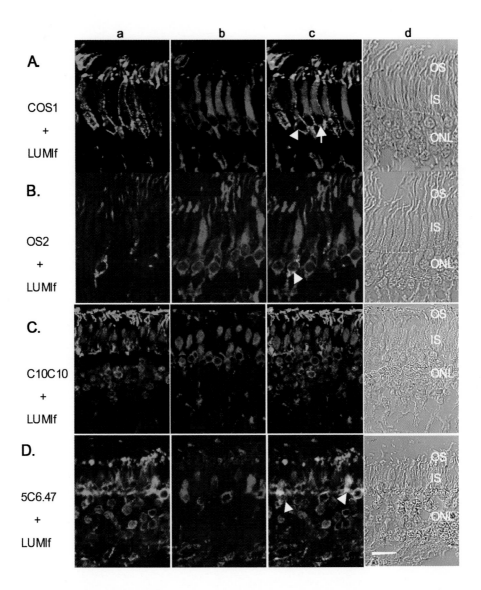

Figure 2. Immunofluorescent double-labeling of human retinal sections with the anti-cone arrestin antibody LUMIf (b) and one of the following antibodies (a): A, COS-1 (anti-red/green opsin); B, OS-2 (anti-blue opsin); C, C10C10 (anti-rod arrestin); D, 5C6.47 (universal arrestin antibody). Human eyes were processed for immunohistochemistry as described in Methods. In each set of images, both Images a and b were taken at the same area on the sample using the same lens and zoom factor. Image a shows the DTAF-conjugated anti-mouse secondary antibody using the Argon 488 nm-barrier filter. Image b shows the Cy3-conjugated anti-rabbit secondary antibody using the HeNe 543 nm barrier filter. Image c is the result of images a and b superimposition, showing the final result of immunofluorescent double-labeling. Arrows identify the cells with both antibody staining. Image d is a phase-contrast picture of the same area. OS, outer segment. IS, inner segment. ONL, outer nuclear layer. Scale bar: 20 μm.

photoreceptors (Figure 2Cc), although C10C10 has been reported to label blue cones.[22] Again, this is because blue cones are rare in human retina.[19, 21] In contrast, in retinal sections with LUMIf and 5C6.47 double staining, all photoreceptors, including rods and cones, were labeled green by 5C6.47-DTAF-conjugated anti-mouse secondary antibody (Figure 2Da), while LUMIf stained only a subpopulation of the photoreceptors. In the morphologically distinct cone photoreceptors, LUMIf staining overlapped with 5C6.47 staining (Figure 2Db & c), suggesting that both antibodies may recognize the same arrestin protein.

The staining with LUMIf as the primary antibody was completely blocked by the specific human cone arrestin peptide (10X) used to generate the antisera (data not shown). No detectable immunoreactive staining was observed in control sections with no primary antibody incubation (data not shown).

3.2. Cone Arrestin Is Expressed in Weri-Rb-1 and Y79 Retinoblastoma cells

Since both rod and cone photoreceptor enriched proteins and mRNAs have been observed in Weri-Rb-1 and Y79 retinoblastoma cell lines, it is interesting to compare the expression pattern of cone verses rod arrestin in these cell lines. We examined cone and rod arrestin expression by immunoblot analyses of whole cell homogenates of Weri-Rb-1 and Y79 cells treated with various agents. Although cone arrestin is expressed in untreated Y79 and Weri-Rb-1 cells at low levels, butyrate and RA treatment for 6 days increased cone arrestin protein levels in both cell lines, while dbcAMP and HMBA did not affect its expression in either of the cell lines (Figure 3A). Northern blot analysis verified that the mRNA for cone arrestin changed in parallel to its protein level in both cell lines following the above treatments (Figure 3B). Treatment with 2 mM of butyrate for 6 days induced cell death in about 70% of Weri-Rb-1 and Y79 cells, while RA caused 20-30% of the cells to die. The limited cells that survived developed morphological changes including extension of neurite-like processes previously shown[23-26] (data not shown). Rod arrestin was below detection with Mab C10C10 in both of the cell lines either in the control cells or following different treatments under our experimental conditions (Figure 3A). Our immunoblot analysis also shows that both βarrestin1 and βarrestin2 are expressed in the cell lines as well as human retina at comparable low levels and do not change after the treatments (data not shown). We also detected glial fibrillary acidic protein (GFAP) expression in both cell lines by immunoblot analysis with no obvious change following the above treatments (data not shown).

4. DISCUSSION

Previously, X-arrestin immunoreactive staining was observed in red and green cones by double-labeling experiments with an antibody (anti-X-arrestin) developed against the identical epitope to LUMIf; however, X-arrestin expression in blue cones was only implicated by the fact that a minor subset of cone photoreceptors were double-labeled with the anti-X-arrestin antibody and the anti-S-antigen monoclonal antibody A9C6, which recognizes not only rods but also blue cone photoreceptors.[27] In the current study, red/green and blue cones were immunofluorescently labeled selectively with cone-specific monoclonal antibodies COS-1 and OS-2, respectively, and our newly generated

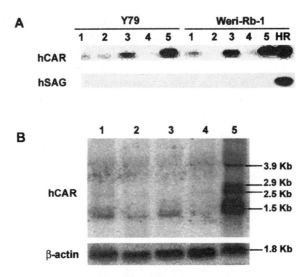

Figure 3. A. Immunoblot analysis of human cone arrestin (hCAR) and human S-antigen (hSAG) expression in Y79 and Weri-Rb-1 retinoblastoma cells treated with various agents. 1, control; 2, dbcAMP; 3, butyrate; 4, HMBA; 5, RA; HR, human retina. B. Northern blot analysis of human cone arrestin mRNA expression in Weri-Rb-1 cells treated with various reagents. β-actin was used as a control. 1-5, the same as A. Sizes of the hybridized mRNAs are indicated on the right.

antibody LUMIf recognized cone arrestin. Our double-labeling experiments clearly show that cone arrestin is specifically localized in all three classes of cone photoreceptors in human retina. This is the first direct evidence for the localization of cone arrestin to blue cones.

In an attempt to search for potential *in vitro* models for studying the gene expression regulation and potential function of cone arrestin, we tested two retinoblastoma cell lines, Y79 and Weri-Rb-1. Y79 cells are capable of differentiating into cells with morphological characteristics of photoreceptors, conventional neurons, glia and pigment epithelial cells when treated with appropriate combinations of substrate and differentiating agents.[23] Bogenmann and coworkers[28] observed cone cell-specific genes expressed in Y79 cells, suggesting that properly cultured retinoblastoma cells can be used to investigate cone photoreceptor cell development. In contrast, rod-specific genes are either absent or expressed at lower levels in Y79 cells.[23, 24, 29] These observations suggest that Y79 cells are predominantly of cone neuronal origin. Weri-Rb-1 is an independently derived retinoblastoma cell line that is from the same type of childhood retinal tumor as Y79. However, they either originated from functionally distinct retinal neuronal populations or have phenotypically diverged during propagation in culture because they are biochemically different in lipid content and metabolism.[30]

Although expression of a functional cone phototransduction cascade has been observed in human retinoblastoma tumors,[29] whether or not cone arrestin is expressed in either retinoblastoma tumors or cell lines derived from them has not been documented. In this study, we establish that cone arrestin is expressed in both Y79 and Weri-Rb-1 cells, while rod arrestin is below detection under our experimental conditions, consistent with the established concept that retinoblastoma cell lines are of cone neuronal origin.

To explore the potential to use the cell lines for studying cone arrestin's gene expression regulation, we tested four known differentiating agents, dbcAMP, butyrate, HMBA, and RA in both cell lines to determine their effect on cone arrestin mRNA and protein expression. We found that butyrate and RA stimulate cone arrestin expression at both mRNA and protein levels in both cell lines, along with cell death and morphological differentiation, while dbcAMP and HMBA did not alter cone arrestin expression in the cells, although extensive morphological differentiation was induced by HMBA.

The differential effects of these chemicals on gene expression and morphological differentiation in Y79 cells have been reported. Butyrate and RA, both naturally occurring agents, induce diverse effects, including growth inhibition, differentiation, and apoptosis in many cell types including retinoblastoma cells.[25, 31-33] They also modulate gene expression of N-acetyltransferase (AA-NAT),[34] hydroxyindole-O-methyltransferase (HIOMT),[34] p53[26] and Bcl-2[26] in these cells. Treatment of Y79 cells with 2 mM of butyrate resulted in a dramatic increase of interphotoreceptor retinoid-binding protein (IRBP).[35] Recoverin, another photoreceptor-specific protein that is expressed in both rods and cones, is also greatly up regulated by 2 mM butyrate treatment of Y79 cells, but only moderately increased by 4 mM dbcAMP treatment.[36] Based on these observations, it was suggested that butyrate induces biochemical differentiation of Y79 cells, possibly into photoreceptor-like cells. Our results further suggest that butyrate induces a subpopulation of Y79 cells to differentiate toward cone cell lineage while inducing another subpopulation into cell death, probably through apoptosis.

RA has been shown to have only a minor effect in inducing neuronal differentiation in Y79 cells and therefore may not be an appropriate differentiating agent in early retinal development.[37] However, RA up-regulates cell adhesion and immune recognition molecules associated with reduced growth and tumor cell differentiation in Y79 cells,[25] suggesting that RA may have a potential role in regulating growth and development of retinoblastoma tumors. We show in this study that RA has a dramatic effect on up-regulating cone arrestin expression in both Y79 and Weri-Rb-1 cells but only a minor effect on their morphology. Since RA is a known potent regulator of many genes, it may have physiological significance in regulating cone arrestin gene expression *in vivo*.

The failure to stimulate cone arrestin expression by treatment with dbcAMP, a cAMP analogue, suggests that cone arrestin is neither diurnally regulated by the cyclic change of intracellular cAMP level, nor by PKA-related pathways. Indeed, cone arrestin expression in mouse retina is not affected either by circadian rhythms or by light/dark cycle (unpublished data). HMBA, a potent differentiating agent with clinical applications, induces significant morphological differentiation of Y79 cells; however, it did not have a major effect on either photoreceptor, neuronal or glial cell markers.[24] It is thus not surprising that it did not affect cone arrestin expression in either Y79 or Weri-Rb-1 cells, although extensive morphological differentiation was observed.

There are at least three possible mechanisms for the up-regulation of cone arrestin by butyrate and RA. First, each agent acts at the gene level to increase the transcription and/or translation of cone arrestin. Second, each stimulates more cells to differentiate toward cone cell lineage. Third, each induces apoptosis in undifferentiated cells so that the percentage of viable cells expressing cone arrestin increases. Further studies are ongoing in our laboratory to address the molecular mechanisms of cone arrestin upregulation in Y79 and Weri-Rb-1 cells and their correlation with its gene regulation *in vivo.*

5. ACKNOWLEDGEMENTS

This work is dedicated to the memory and scientific contributions of Richard N. Lolley. We would also like to acknowledge Mary D. Allen for her generous continued support of vision research. We wish to gratefully acknowledge Dr. Larry A. Donoso for the S-antigen monoclonal antibodies C10C10 and A9C6; Dr. Laura Smith Lang for the universal arrestin monoclonal antibody 5C6.47; Dr. Agoston Szel for the anti-red/green opsin monoclonal antibody COS-1 and anti-blue opsin monoclonal antibody OS-2; and Dr. Carol Miller for the human tissues.

These studies were supported, in part, from NIH grants EY00395 (CMC and RNL), EY03042 Core Vision Research Center grant (Doheny Eye Institute), L. K. Whittier Foundation (CMC), L.K. Whittier Fellowship support for Y. Zhan, and P59-AG051542 from the National Institute of Aging (Carol Miller, M.D.). CMC is the Mary D. Allen Professor for Vision Research, Doheny Eye Institute.

REFERENCES

1.   T. Shinohara, B. Dietzschold, C. M. Craft, G. Wistow, J. J. Early, L. A. Donoso, J. Horwitz, and R. Tao, Primary and secondary structure of bovine retinal S antigen (48-kDa protein), Proc Natl Acad Sci U S A **84**(20), 6975-6979 (1987).

2.   H. Kuhn, S. W. Hall, and U. Wilden, Light-induced binding of 48-kDa protein to photoreceptor membranes is highly enhanced by phosphorylation of rhodopsin, FEBS Lett **176**(2), 473-478 (1984).

3.   H. Kuhn, and U. Wilden, Deactivation of photoactivated rhodopsin by rhodopsin-kinase and arrestin, J Recept Res 7(1-4), 283-298 (1987).

4.   M. J. Lohse, J. L. Benovic, J. Codina, M. G. Caron, and R. J. Lefkowitz, beta-Arrestin: a protein that regulates beta-adrenergic receptor function, Science **248**(4962), 1547-1550 (1990).

5.   H. Attramadal, J. L. Arriza, C. Aoki, T. M. Dawson, J. Codina, M. M. Kwatra, S. H. Snyder, M. G. Caron, and R. J. Lefkowitz, Beta-arrestin2, a novel member of the arrestin/beta-arrestin gene family, J Biol Chem **267**(25), 17882-17890 (1992).

6.   R. J. Lefkowitz, G protein-coupled receptors. III. New roles for receptor kinases and beta-arrestins in receptor signaling and desensitization, J Biol Chem **273**(30), 18677-18680 (1998).

7.   R. A. Hall, R. T. Premont, and R. J. Lefkowitz, Heptahelical receptor signaling: beyond the G protein paradigm, J Cell Biol **145**(5), 927-932 (1999).

8.   O. B. Goodman, Jr., J. G. Krupnick, F. Santini, V. V. Gurevich, R. B. Penn, A. W. Gagnon, J. H. Keen, and J. L. Benovic, Beta-arrestin acts as a clathrin adaptor in endocytosis of the beta2- adrenergic receptor, Nature **383**(6599), 447-450 (1996).

9.   O. B. Goodman, Jr., J. G. Krupnick, V. V. Gurevich, J. L. Benovic, and J. H. Keen, Arrestin/clathrin interaction. Localization of the arrestin binding locus to the clathrin terminal domain, J Biol Chem **272**(23), 15017-15022 (1997).

10.   L. M. Luttrell, S. S. Ferguson, Y. Daaka, W. E. Miller, S. Maudsley, G. J. Della Rocca, F. Lin, H. Kawakatsu, K. Owada, D. K. Luttrell, M. G. Caron, and R. J. Lefkowitz, Beta-arrestin-dependent formation of beta2 adrenergic receptor-Src protein kinase complexes [see comments], Science **283**(5402), 655-661 (1999).

11.   W. E. Miller, S. Maudsley, S. Ahn, K. D. Khan, L. M. Luttrell, and R. J. Lefkowitz, beta-arrestin1 interacts with the catalytic domain of the tyrosine kinase c-SRC. Role of beta-arrestin1-dependent targeting of c-SRC in receptor endocytosis, J Biol Chem **275**(15), 11312-11319 (2000).

12.   C. M. Craft, D. H. Whitmore, and A. F. Wiechmann, Cone arrestin identified by targeting expression of a functional family [published erratum appears in J Biol Chem 1994 Jul 1;269(26):17756], J Biol Chem **269**(6), 4613-4619 (1994).

13. A. Murakami, T. Yajima, H. Sakuma, M. J. McLaren, and G. Inana, X-arrestin: a new retinal arrestin mapping to the X chromosome, FEBS Lett **334**(2), 203-209 (1993).
14. N. Bobola, P. Briata, C. Ilengo, N. Rosatto, C. Craft, G. Corte, and R. Ravazzolo, OTX2 homeodomain protein binds a DNA element necessary for interphotoreceptor retinoid binding protein gene expression, Mech Dev **82**(1-2), 165-169 (1999).
15. C. M. Craft, J. Xu, V. Z. Slepak, X. Zhan-Poe, X. Zhu, B. Brown, and R. N. Lolley, PhLPs and PhLOPs in the phosducin family of G beta gamma binding proteins, Biochemistry **37**(45), 15758-15772 (1998).
16. C. M. Craft, D. H. Whitmore, and L. A. Donoso, Differential expression of mRNA and protein encoding retinal and pineal S-antigen during the light/dark cycle, J Neurochem **55**(5), 1461-1473 (1990).
17. C. M. Craft, J. Murage, B. Brown, and X. Zhan-Poe, Bovine Arylalkylamine N-Acetyltransferase Activity Correlated with mRNA Expression in Pineal and Retina, Brain Research. Molecular Brain Research **65**(1), 44-51 (1999).
18. X. Zhu, and C. M. Craft, Modulation of CRX transactivation activity by phosducin isoforms, Mol Cell Biol **20**(14), 5216-5226 (2000).
19. A. Szel, T. Diamantstein, and P. Rohlich, Identification of the blue-sensitive cones in the mammalian retina by anti-visual pigment antibody, J Comp Neurol **273**(4), 593-602 (1988).
20. A. Szel, L. Takacs, E. Monostori, T. Diamantstein, I. Vigh-Teichmann, and P. Rohlich, Monoclonal antibody-recognizing cone visual pigment, Exp Eye Res **43**(6), 871-883 (1986).
21. F. M. DeMonasterio, S. J. Schein, and E. P. McCrane, Staining of blue-sensitive cones of the macaque retina by a fluorescent dye, Science **213**(4513), 1278-1281 (1981).
22. I. Nir, and N. Ransom, S-antigen in rods and cones of the primate retina: different labeling patterns are revealed with antibodies directed against specific domains in the molecule, J Histochem Cytochem **40**(3), 343-352 (1992).
23. G. J. Chader, Multipotential differentiation of human Y-79 retinoblastoma cells in attachment culture, Cell Differ **20**(2-3), 209-216 (1987).
24. S. Rajagopalan, M. Rodrigues, T. Polk, D. Wilson, G. J. Chader, and B. J. Hayden, Modulation of retinoblastoma cell characteristics by hexamethylene bis- acetamide and other differentiating agents in culture, J Histochem Cytochem **41**(9), 1331-1337 (1993).
25. R. M. Conway, M. C. Madigan, N. J. King, F. A. Billson, and P. L. Penfold, Human retinoblastoma: in vitro differentiation and immunoglobulin superfamily antigen modulation by retinoic acid, Cancer Immunol Immunother **44**(4), 189-196 (1997).
26. M. C. Madigan, G. Chaudhri, P. L. Penfold, and R. M. Conway, Sodium butyrate modulates p53 and Bcl-2 expression in human retinoblastoma cell lines, Oncol Res **11**(7), 331-337 (1999).
27. H. Sakuma, G. Inana, A. Murakami, T. Higashide, and M. J. McLaren, Immunolocalization of X-arrestin in human cone photoreceptors, FEBS Lett **382**(1-2), 105-110 (1996).
28. E. Bogenmann, M. A. Lochrie, and M. I. Simon, Cone cell-specific genes expressed in retinoblastoma, Science **240**(4848), 76-78 (1988).
29. R. L. Hurwitz, E. Bogenmann, R. L. Font, V. Holcombe, and D. Clark, Expression of the functional cone phototransduction cascade in retinoblastoma., J.Clin.Invest. **85**,1872-1878 (1990).
30. M. A. Yorek, P. H. Figard, T. L. Kaduce, and A. A. Spector, A comparison of lipid metabolism in two human retinoblastoma cell lines, Invest Ophthalmol Vis Sci **26**(8), 1148-1154 (1985).
31. M. Giuliano, M. Lauricella, G. Calvaruso, M. Carabillo, S. Emanuele, R. Vento, and G. Tesoriere, The apoptotic effects and synergistic interaction of sodium butyrate and MG132 in human retinoblastoma Y79 cells, Cancer Res **59**(21), 5586-5595 (1999).
32. M. Lauricella, M. Giuliano, S. Emanuele, M. Carabillo, R. Vento, and G. Tesoriere, Increased cyclin E level in retinoblastoma cells during programmed cell death, Cell Mol Biol (Noisy-le-grand) **44**(8), 1229-1235 (1998).
33. M. Lauricella, M. Giuliano, S. Emanuele, R. Vento, and G. Tesoriere, Apoptotic effects of different drugs on cultured retinoblastoma Y79 cells, Tumour Biol **19**(5), 356-363 (1998).
34. A. F. Wiechmann, and M. A. Burden, Regulation of AA-NAT and HIOMT gene expression by butyrate and cyclic AMP in Y79 human retinoblastoma cells, J Pineal Res **27**(2), 116-121 (1999).
35. A. P. Kyritsis, B. Wiggert, L. Lee, and G. J. Chader, Butyrate enhances the synthesis of interphotoreceptor retinoid-binding protein (IRBP) by Y-79 human retinoblastoma cells, J Cell Physiol **124**(2), 233-239 (1985).
36. A. F. Wiechmann, Recoverin in cultured human retinoblastoma cells: enhanced expression during morphological differentiation, J Neurochem **67**(1), 105-110 (1996).
37. M. Campbell, and G. J. Chader, Retinoblastoma cells in tissue culture, Ophthalmic Paediatr Genet **9**(3), 171-199 (1988).

nob: A MOUSE MODEL OF CSNB1

Machelle T. Pardue, Sherry L. Ball, Sophie I. Candille, Maureen A. McCall, Ronald G. Gregg, and Neal S. Peachey[*]

1. INTRODUCTION

Congenital stationary night blindness (CSNB) refers to a group of disorders in which patients have normal to near-normal vision under photopic conditions, reduced sensitivity under scotopic conditions, but no evidence of photoreceptor degeneration. Several subtypes of CSNB can be distinguished by fundus appearance, the inheritance pattern, electroretinogram (ERG) recording, the extent of cone involvement, and other measures of ocular function (Ripps, 1982; Miyake et al., 1986). CSNB can also be classified according to the underlying defect, which may involve the phototransduction cascade (Sieving et al., 1995; Dryja et al., 1996) or the communication between photoreceptor and bipolar cells (Ripps, 1982; Miyake, et al., 1986). The latter form of CSNB, often referred to as the Schubert-Bornschein (1952) type, is characterized by a 'negative' ERG under dark-adapted conditions, in which the amplitude of the ERG b-wave is drastically reduced. Because the a- and b-waves of the ERG reflect the mass response of the photoreceptors and the rod depolarizing bipolar cells (DBCs), respectively, (Robson & Frishman, 1998) the negative ERG waveform indicates a defect in communication between normally functioning rod photoreceptors and second order neurons. Miyake et al. (1986) subdivided this form of CSNB into complete (CSNB1) and incomplete (CSNB2) based on ERG waveforms, refractive error, and the extent of rod and cone pathway involvement.

Figure 1 shows ERGs obtained from a normal subject and from representative patients with complete or incomplete CSNB. In the normal subject, under dark-adapted

[*] MTP, Research Service (151), Department of Ophthalmology, Emory University, Atlanta VA Medical Center, 1670 Clairmont Rd., Decatur, GA, 30033. SLB and NSP, Cole Eye Institute, Cleveland Clinic Foundation, Cleveland VA Medical Center, Cleveland, OH. SIC, MAM and RGG, Departments of Biochemistry & Molecular Biology, Ophthalmology & Visual Science, Psychological & Brain Sciences, University of Louisville, Louisville, KY.

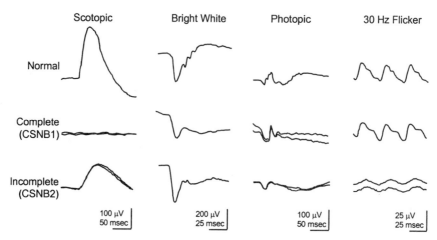

Figure 1. ERG recording from normal subjects and CSNB patients under different adaptation conditions, illustrating the differences in the ERG waveform between complete and incomplete CSNB. Adapted with permission from Miyake et al., 1994.

conditions (scotopic), a relatively dim flash normally elicits an ERG dominated by the b-wave. In patients with CSNB1, however, the b-wave is virtually missing. In contrast patients with CSNB2 retain a b-wave response of reduced amplitude. When a bright flash is presented under the same conditions (Fig. 1; bright white), the normal ERG waveform has a negative polarity a-wave followed by a large amplitude b-wave upon which oscillatory potentials (OPs) are superimposed. OPs are high frequency wavelets on the ascending portion of the b-wave thought to be generated by cells in the inner retina (Wachtmeister, 1998). Both CSNB1 and CSNB2 patients have a normal amplitude a-wave, however in CSNB1 patients the b-wave and OPs are absent. In CSNB2 patients, the b-wave amplitude is small but some OPs are present. Under cone mediated conditions (photopic and 30 Hz flicker), CSNB1 patients generate well-defined responses, although their waveform is abnormal. In comparison, the cone ERGs of CSNB2 patients is greatly reduced in amplitude. Taken together, these results suggest that CSNB1 involves a defect that completely blocks the rod-to-DBC pathway (thus no b-wave or OPs), while communication from cones to bipolar cells is somewhat preserved. In contrast, the defect underlying the incomplete form of CSNB (CSNB2) appears to compromise all forms of post-receptor neurotransmission.

We have discovered a naturally-occurring mouse mutant that closely resembles CSNB1 (Pardue et al., 1998). Like affected patients, the *nob* mouse has no ERG b-wave, a decrease in visual sensitivity, and normal retinal architecture. The *nob* gene is located on the X chromosome (Candille et al., 1999), in a location near the recently cloned CSNB1 human gene (Bech-Hansen et al., 2000; Pusch et al., 2000). This evidence supports the hypothesis that *nob* is a mouse model of CSNB1. As reviewed below, the *nob* retina has been probed using immunocytochemical markers and the synaptic region, the outer plexiform layer (OPL), has been examined at the EM level to begin to elucidate the underlying mechanism.

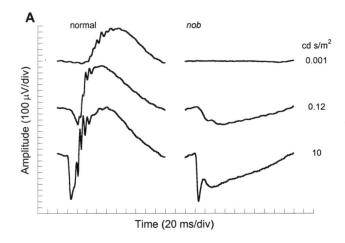

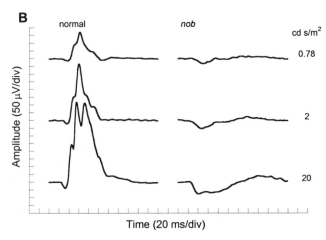

Figure 2. ERG recordings from 2 month old *nob* mice and littermate normals. (A) *nob* ERG rod mediated responses showing the characteristic absence of the b-wave and OPs. (B) Under cone mediated conditions, the *nob* ERG b-wave is absent.

2. *nob* ERG PHENOTYPE

Because the *nob* gene has not yet been cloned, affected animals are identified by ERG recordings (Pardue et al., 1998). The responses shown in Figure 2 were obtained from anesthetized mice using a Diagnosys Espion signal averaging system with an amplifier band-pass of 0.1 – 1000 Hz.

Figure 2A (top graph) shows ERGs obtained to flash stimuli presented to the dark-adapted mouse eye. In response to a dim flash (0.001 cd s/m^2), the normal mouse generates a distinct b-wave with OPs, while the *nob* mouse ERG is flat. At higher flash intensities, the normal b-wave grows in amplitude and the a-wave becomes increasingly apparent. In contrast, the *nob* waveforms are characterized by the absence of a b-wave and OPs, while the a-wave has normal amplitude and timing (cf., Pardue et al., 1998).

Cone ERGs obtained after 10 minutes of light adaptation at 20 cd/m^2 are shown in Figure 2B. At all flash intensities, the normal response is dominated by the positive polarity b-wave, while the *nob* ERG has a response of negative polarity.

These waveforms are similar to those obtained in CSNB1 patients. Note the flat response elicited in the dark-adapted condition to a dim flash in both cases (Fig. 1 and 2A) and the absence of the b-wave and OPs with a high intensity flash. In both the *nob* mouse and CSNB1 patients, the cone pathway is affected. In CSNB1 patients, the abnormality is seen as a decrease in amplitude while in the *nob* mice, the cone ERG has a negative polarity. Thus, the *nob* mouse appears to have normally functioning photoreceptors but a blockage of transmission to the secondary neurons producing the b-wave and OPs. Consistent with this hypothesis, similar dark-adapted ERG waveforms are obtained when the retina is treated with the glutamate agonist aspartate (Reiser et al., 1996), which blocks post-synaptic transmission.

3. VISUAL SENSITIVITY

A negative ERG is not pathognomonic for CSNB since other diseases also are associated with this waveform. However, unlike CSNB, visual sensitivity may be near-normal in disorders presenting with negative ERGs, including X-linked juvenile retinoschisis (Peachey et al., 1987) and Duchenne muscular dystrophy (Tremblay et al., 1994). To determine whether the *nob* defect disrupts transmission to the visual cortex, visually evoked potentials (VEP) were recorded and visually mediated behavior was tested in *nob* mice (Pardue et al., 2000). As shown in Figure 3, the VEP of a normal mouse has a negative amplitude wave correlated with the onset of the flash and a positive wave with the offset. In contrast, the *nob* mouse has only a slow positive wave at the

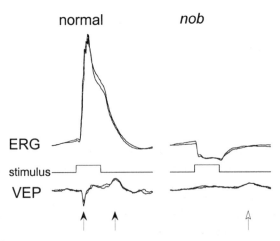

Figure 3. ERG and VEP recordings from a normal and a *nob* mouse to a long duration flash. Note the presence of a slow positive VEP response in the *nob* mouse (arrows). Stimulus tracing represents 250 msec along the x-axis and 20 µV along the y-axis. Used with permission from Pardue et al., 1998.

offset of the flash, with no detectable response correlated with flash onset. These results suggest that the *nob* mouse retina is transmitting visual information to the visual cortex, perhaps through the hyperpolarizing bipolar cell (HBC) or "OFF" pathway. In CSNB1 patients, this hypothesis has gained support from ERG recordings using long duration flashes, where the response to flash onset is greatly impaired while the response to flash offset is generally spared (Miyake et al., 1987; Quigley et al., 1996; Alexander et al., 1992).

To determine whether this visual response is adequate to produce functional vision, normal and *nob* mice were tested in an active avoidance behavioral paradigm using a visual cue (Pardue et al., 2000). Within 7 days, both *nob* mice and littermate controls could be trained to avoid a shock preceded by a bright light stimulus. When the intensity of the light stimulus was dimmed, normal littermates continued to avoid the shock while the performance of *nob* mice decreased dramatically (data not shown). These results show that the *nob* mice can learn a visual task, but have decreased visual sensitivity.

4. RETINAL ARCHITECTURE

Many hereditary retinal disorders, like retinitis pigmentosa, are associated with a defect in the rod photoreceptor pathway (Phelan and Bok, 2000). However, unlike CSNB, these disorders involve a progressive degeneration of rod and cone photoreceptors along with a decline in visual function. In contrast, the retinal structure of a CSNB patient was found to be normal (Kato et al., 1987).

Figure 4 shows representative micrographs from an 8 month old *nob* mouse and a control littermate. At this age, and at all others examined, the laminar structure of the retina was not different between *nob* mice and normal littermates and there were no signs of degeneration of any cell type (cf., Pardue et al., 1998).

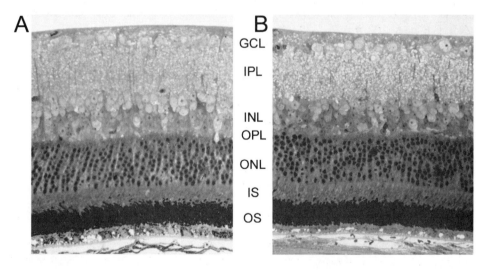

Figure 4. Light micrographs of an 8 month old control (A) and *nob* (B) retina show no signs of retinal degeneration. Original magnification 100X.

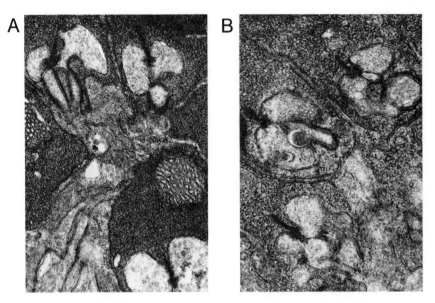

Figure 5: Electron micrograph of a normal littermate (A) and a *nob* mouse (B) showing the synaptic region between the photoreceptors and bipolar cells. Ribbon and conventional synapses were present in the *nob* mice and appeared to have the same morphology as the unaffected littermates.

The ERG results suggest a defect at the synapse between the photoreceptors and rod DBCs. To confirm that the bipolar cells were morphologically intact, *nob* mice were mated with transgenic mice that express an L7-β-galactosidase fusion gene in rod DBCs as well as in cerebellar Purkinje cells (Oberdick et al., 1990). Retinal sections from the offspring of these matings were reacted with X-gal to produce a blue by-product. The bipolar cell morphology and density was similar between *nob* mice and littermate controls (data not shown).

Using transmission electron microcopy, the synaptic region within the OPL was examined in five retinas from *nob* and littermate controls. As shown in Figure 5, no qualitative differences in the ribbon and conventional synapses were detected between *nob* and control retinas. Bipolar and horizontal cell processes could be identified in the *nob* retinas in association with a ribbon synapse. Therefore, the *nob* defect does not appear to cause a gross morphological change at the photoreceptor-to-bipolar cell synapse.

5. INHERITANCE PATTERN AND LINKAGE ANALYSIS

The inheritance pattern of the *nob* trait was determined by breeding *nob* to wild-type mice (Pardue et al., 1998). When a *nob* male was mated to a wild-type female, all offspring had a normal ERG. When females from this generation were then mated to wild-type males, 50% of male offspring were affected and all females appeared normal. This is a typical X-linked recessive inheritance pattern.

Once the inheritance pattern was established, linkage analysis was undertaken to determine the exact location of the *nob* gene using intra- and interspecific breeding strategies. Mice were phenotyped with the ERG and DNA extracted from tail biopsies was analyzed using PCR for published markers on the X chromosome (Candille et al., 1999). Using two-point LOD scores (*Z*) calculated between each marker and the *nob* phenotype (Candille et al., 1999), the *nob* gene has been localized to a centromeric region of the X-chromosome. As shown in Figure 6, this is homologous to the region of the X-chromosome where the gene for human CSNB1 (*NYX*) has been found (Bech-Hansen, 2000; Pusch et al., 2000). Note that this is distinct from the locus for the CACNA1F gene that is responsible for CSNB2 (Bech-Hansen et al., 1998; Strom et al., 1998).

6. MECHANISM UNDERLYING THE *nob* DEFECT

Within the OPL, rod photoreceptors synapse with DBCs and cone photoreceptors synapse with both DBCs and HBCs. Therefore, a possible location of the *nob* defect could be at the synapse between the photoreceptors and DBCs. Disruption of visual transmission along this pathway would explain the absence of the rod ERG b-wave and the VEP "ON" response. In this scenario the cone to HBC pathway would be preserved and drive a visual signal in the ERG of opposite polarity, as well as produce the "OFF" response in the VEP. This visual information could account for the normal vision that we measure behaviorally to relatively bright stimuli.

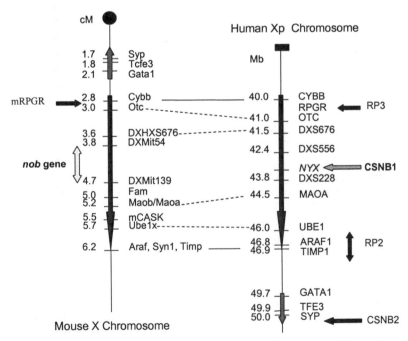

Figure 6. Homologous region of the mouse (left) and human (right) X chromosomes, in the region containing the mouse *nob* gene.

Knock-out models of mGluR6 (Masu et al., 1995) and $G\alpha_0$ (Dhingra et al., 2000), which are proteins involved in the transmission of the visual signal between rod photoreceptors and DBCs, also have a selective reduction in the rod ERG b-wave. Neither of these proteins is the *nob* gene product. mGluR6 is not located on the X chromosome (Masu et al., 1995) and normal mGluR6 labeling is seen in the *nob* retina (data not shown). In addition, $G\alpha_0$ knock-out mice are phenotypically different from *nob*. $G\alpha_0$ knock-out mice are hyperactive, exhibit a turning behavior and do not survive past the first postnatal month (Jian et al., 1998).

The gene responsible for human CSNB1 is a novel leucine-rich proteoglycan called nyctalopin, the function of which is not yet known (Bech-Hanson et al., 2000; Pusch et al., 2000). However, if *nob* involves the homologous mouse gene, then the *nob* mouse may prove useful in determining the function of nyctalopin in the retina and could serve as a model in which to study therapeutic strategies.

7. CONCLUSIONS

The data presented here provide evidence that the *nob* mouse is a model of CSNB1. In every aspect investigated, the *nob* phenotype matches that of patients with CSNB1. Identification of the *nob* gene will confirm this. Further study of the *nob* mouse should provide further insight into the cascade of events that results in transmission of the visual signal from photoreceptors to bipolar cells.

8. ACKNOWLEDGEMENTS

This work was supported by the Veterans Administration's Merit Review Entry Program and National Institutes of Health RO1 grant EY12354.

REFERENCES

Alexander, K.R., Fishman, G.A., Peachey, N.S., Marchese, A.L., and Tso, M.O.M., 1992, 'On' response defect in paraneoplastic night blindness with cutaneous malignant melanoma. *Invest. Ophthalmol. Vis. Sci.* **33**:477-483.

Bech-Hansen, N., Naylor, M.J., Maybaum, T.A., Sparkes, R.L., Koop, B., Birch, D.G., Bergen A.A.B., Prinsen, C.F.M., Polomeno, R.C., Gall, A., Drack, A.V., Musarella, M.A., Jacobson, S.G., Young, R.S.L., and Weleber, R.G., 2000, Mutations in NYX, encoding the leucine-rich proteoglycan nyctalopin, cause X-linked complete congenital stationary night blindness, *Nature Genetics* **26**: 319-323.

Bech-Hansen, N., Naylor, M.J., Maybaum, T.A., Pearce, W.G., Koop, B., Fishman, G.A., Mets, M., Musarella, M.A. & Boycott, K.M., 1998, Loss-of-function mutations in a calcium-channel α_1-subunit gene in Xp11.23 cause incomplete X-linked congenital stationary night blindness, *Nature Genetics* **19**:264-267.

Candille, S.I., Pardue, M.T., McCall, M.A., Peachey, N.S., and Gregg, R.G., 1999, Localization of the mouse *nob* (*no b-wave*) gene to the centromeric region of the X chromosome, *Invest. Ophthalmol. Vis. Sci.* **40**:2748-2751.

Dhingra, A., Lyubarsky, A., Jiang, M., Pugh, E.N., Jr., Birnbaumer, L., Sterling, P., and Vardi, N., 2000, The light response of ON bipolar neurons requires $G\alpha_0$, *J. Neurosci.* **20**:9053-9058.

Dryja, T.P., Berson, E.L., Rao, V.R., and Oprian, D.D., 1993, Heterozygous missense mutation in the rhodopsin gene as a cause of congenital stationary night blindness, *Nature Genetics* **4**:280-283.

Dryja, T.P., Hahn, L.B., Reboul, T., and Arnaud, B.,1996, Missense mutation in the gene encoding the α subunit of rod transducin in the Nougaret form of congenital stationary night blindness, *Nature Genetics* 13:358-360.

Gray, E.G. and Pease, H.L., 1971, On understanding the organization of the retinal synapses, *Brain Res.* 35:1-15.

Gregor P., Reeves, R.H., Jabs, E.W., Yang, X., Dackowski, W., Rochelle, J.M., Brown, R.H.J., Haines, J.L., O'Hara, B.F., and Uhl, G.R., 1993, Chromosomal localization of glutamate receptor genes: relationship to familial amyotrophic lateral sclerosis and other neurological disorders of mice and humans, *Proc. Natl. Acad. Sci., U.S.A.* 90:3053-3057.

Huang, S.H., Pittler, S.J., Huang, X., Oliveira, L., Berson, E.L., and Dryja, T.P., 1995, Autosomal recessive retinitis pigmentosa caused by mutations in the α subunit of rod cGMP phosphodiesterase, *Nature Genetics* 11:468-471.

Jiang, M., Gold, M.S., Boulay, G., Spicher, K., Peyton, M., Brabet, P., Srinivasan, Y., Rudolph, U., Ellison, G., and Birnbaumer, L., 1998, Multiple neurological abnormalities in mice deficient in the G protein G$_o$, *Proc. Natl. Acad. Sci.* 95:3269-3274.

Kato, M., Aonuma, H., Kawamura, H., Miura, Y., and Watanabe, I., 1987, Possible pathogenesis of congenital stationary night blindness, *Jap. J. Ophthalmol.* 31:88-101.

Masu M., Iwakabe H., Tagawa Y., Miyoshi T., Yamashita M., Fukuda Y., Sasaki H., Hiroi K., Nakamura Y., Shigemoto R., Takada, M., Nakamura, K., Nakao, K., Katsuki, M. and Nakanishi, S., 1995, Specific deficit of the On response in visual transmission by targeted disruption of the mGluR6 gene, *Cell* 80:757-65.

Miyake, Y. Horiguchi, M., Terasaki, H. and Kondo, M., 1994, Scotopic threshold response in complete and incomplete types of congenital stationary night blindness, *Invest. Ophthalmol. Vis. Sci* 35:3770-3775.

Miyake, Y., Yagasaki, K., Horiguchi, M., and Kawase, Y., 1987, On- and off-response in photopic electroretinogram in complete and incomplete types of congenital stationary night blindness, *Jpn. J. Ophthalmol.* 31:81-87.

Miyake, Y., Yagasaki, K., Horiguchi, M., Kawase, Y., and Kanda, T., 1986, Congenital stationary night blindness with negative electroretinogram: A new classification, *Arch. Ophthalmol.* 104:1013-1016.

Molday, R.S., 1994, Peripherin/rds and rom-1: Molecular properties and role in photoreceptor cell degeneration, *Progr. Ret. Eye Res.* 13:271-299.

Nawy, S. and Jahr, C.E., 1991, cGMP-gated conductance in retinal bipolar cells is suppressed by the photoreceptor transmitter, *Neuron* 7:677-683.

Oberdick, J., Smeyne, R.J., Mann, J.R., Zackson, S. and Morgan, J.I., 1990, A promoter that drives transgene expression in cerebellar purkinje and retinal bipolar neurons, *Science* 248:223-226.

Pardue, M.T., Ball, S.L., Blake, T.D., McCall, M.A., and Peachey, N.S., 2000, Functional and anatomical features of *nob* mice, ARVO abstracts, *Invest. Ophthalmol. Vis. Sci.* 41:S202

Pardue, M.T., Candille, S., LaVail, M.M., Gregg, R.G., McCall, M.A. and Peachey, N.S., 1998, A potential mouse model of X-linked congenital stationary night blindness, *Invest. Ophthalmol. Vis. Sci.* 39:2443-2449.

Peachey, N.S., Fishman, G.A., Derlacki, D.J. and Brigell, M.G., 1987, Psychophysical and electroretinographic findings in X-linked juvenile retinoschisis, *Arch. Ophthalmol.* 105:513-516.

Phelan , J.K. and Bok, D., 2000, A brief review of retinitis pigmentosa and the identified retinitis pigmentosa genes. *Mol. Vis.* 6:116-124.

Pusch, C.M., Zeitz, C., Brandu, O., Pesch, K., Achatz, H., Feil, S., Scharfe, C., Maurer, J., Jacobi, F.K., Pinckers, A., Andreasson, S., Hardcastle, A., Wissinger, B., Berger, W., and Meindl, A., 2000, The complete form of X-linked congenital stationary night blindness is caused by mutations in a gene encoding a leucine-rich repeat protein, *Nature Genetics* 26:324-327.

Quigley, M., Roy, M.S., Barsoum-Homsy, M., Chevretter, L., Jacob, J.L., and Milot, J., 1996, On- and off-responses in the photopic electroretinogram in complete-type congenital stationary night blindness. *Docum. Ophthamologica* 92:159-165.

Reiser, M.A, Williams, T. P. and Pugh, E.N. Jr., 1996, The effect of light history on the aspartate-isolated fast PIII responses of the albino rat retina, *Invest. Ophthalmol. Vis. Sci.* 37:221-229.

Ripps, H., 1982, Night blindness revisited; from man to molecules, *Invest. Ophthalmol. Vis. Sci.* 23:588-609.

Robson, J.G. and Frishman, L.J., 1998, Dissecting the dark-adapted electroretinogram. *Docum. Ophthalmologica* 95:187-215.

Satoh, H., Aoki, K., Watanabe, S. and Kaneko, A., 1998, L-type calcium channels in the axon terminal of mouse bipolar cells, *NeuroReport* 9:2161-2165.

Schubert, G. and Bornschein, H., 1952, Beitrag zur analyse des menschlichen elektrretinogramms. *Ophthalmologica* 123:396-413.

Sieving, P.A., Richards, J.E., Naarendorp, F., Bingham, E.L., Scott, K. and Alpern, M.A., 1995, Dark-light: model of nightblindness from the human rhodopsin gly-90-asp mutation. *Proc. Natl. Acad. Sci., USA* **92**:880-884.

Strom, T.M., Nyakatura, G., Apfelstedt-Sylla, E., Hellebrand, H., Lorenz, B., Weber, B.H.F., Wutz, K., Gutwillinger, N., Ruther, K., Drescher, B., Sauer, C., Zrenner, E., Meitinger, T., Rosenthal, A., and Meindl, A., 1998, An L-type calcium-channel gene mutated in the incomplete X-linked congenital stationary night blindness, *Nature Genetics* **19**:260-263.

Tremblay, F., DeBecker, I., Riddell, D.C., and Dooley, J.M., 1994, Duchenne muscular dystrophy: negative scotopic bright-flash electroretinogram and normal dark adaptation, *Can. J. Ophthalmol.* **29**:280-283.

Wachtmeister, L., 1998, Oscillatory potentials in the retina, What do they reveal, *Prog. Retin. Eye Res.* **17**:485-521.

UNUSUAL FREQUENCIES OF RHODOPSIN MUTATIONS AND POLYMORPHISMS IN SOUTHERN AFRICAN PATIENTS WITH RETINITIS PIGMENTOSA

Jacquie Greenberg, Lisa Roberts and Rajkumar Ramesar[1]

1. INTRODUCTION

The average adult eye contains approximately 92 000 000 rods and 5 000 000 cones distributed throughout the retina. Cone cells function in colour perception, while rod cells are required for vision in low light. Degeneration of primarily the rod photoreceptor cells is the cause of Retinitis Pigmentosa (RP). This is a group of hereditary disorders characterised by night-blindness and gradual constriction of the visual fields, resulting in partial or total blindness. RP can be inherited in dominant, recessive, X-linked, mitochondrial or digenic modes. It may also present as part of a syndrome.[1]

The first autosomal dominant retinitis pigmentosa (ADRP) locus was mapped to the long arm of chromosome 3 in 1989.[2] Shortly thereafter, a disease-causing mutation in the rhodopsin (RHO) gene, which also maps to the long arm of chromosome 3, was identified in a patient with ADRP.[3] RHO encodes a G-protein coupled receptor (GPCR) which acts in the first stage of the signal transduction cascade in rods, mediating vision in dim light. Mutations in RHO have been reported as the most common cause of RP and account for 20-30% of ADRP cases,[4] although this figure could be as high as 50%, as has been reported in the UK.[5] About 100 different mutations in RHO have been identified in RP patients around the world.[6] The majority of these are in ADRP families, but also include two mutations known to cause recessive RP.[7]

Two mutation hotspots have been described in the RHO gene. Pro-347-Leu has been seen in patients from the US, UK, German and Japan and Thr-58-Arg has been reported in ADRP families from the US and UK. It has been proposed that screening these hotspots using a simple restriction enzyme-based assay (the rapid screen) in cases of sporadic RP and ADRP would be a worthwhile approach.[5]

---

[1] Department of Human Genetics, UCT Medical School, Observatory, South Africa.

New Insights Into Retinal Degenerative Diseases
Edited by Anderson *et al.*, Kluwer Academic/Plenum Publishers, 2001

2. METHOD

For a mutation screening programme, six pairs of primers were designed to cover the 5 exons of RHO, including the intron/exon boundaries[8] (table 1).

Table 1. Primer sequences used for mutation screening of the RHO gene.

Primer	Primer Sequence (5' → 3')
Exon 1a (fwd)	ttc gca gca ttc ttg ggt gg
Exon1a (rev)	agc agg atg tag ttg aga gg
Exon1b (fwd)	caa ctt cct cac gct cta cg
Exon 1b (rev)	cat tga cag gac agg aga ag
Exon 2(fwd)	ccg cct gct gac tgc ctt gca g
Exon 2(rev)	gct tct tcc ctt ctg ctc agt g
Exon 3 (fwd)	ttg gct gtt ccc aag tcc ct
Exon 3 (rev)	tcc aga cca tgg ctc ctc ca
Exon 4 (fwd)	tca cgg ctc tga ggg tcc ag
Exon 4 (rev)	gag tag ctt gtc ctt ggc ag
Exon 5 (fwd)	act caa gcc tct tgc ctt cc
Exon 5 (rev)	gcc aca gag tcc tag gca gg

2.1. The Rapid Screen

Primers for exons 1a and 5 were used for the 2 hotspot mutations:

The Pro-347-Leu mutation (CCG-CTG) in exon 5 destroys a *Msp1* restriction site, and the Thr-58-Arg mutation (ACG-AGG) in exon 1a creates a *Dde1* restriction site. (Boehringer Mannheim restriction enzymes were used.) The amplification of exons 1a and 5 was followed by restriction enzyme digestion, and electrophoresis on 4% agarose gels visualised with ethidium bromide staining.

The rapid screen was used to test 152 patients for the 2 hotspot mutations. Eighty of these patients had a family history indicating ADRP, while 72 were sporadic cases.

2.2. SSCP Analysis

Mutation detection screening was undertaken using all 5 exons of the Rhodopsin gene and single stranded conformational polymorphism analysis (SSCP). Two SSCP gel conditions were used for each exon to ensure reproducibility and high confidence levels. Polymorphisms or disease-causing mutations detected in this manner were sequenced either manually or on an ABI 377 Automated Sequencer (using the PE BIOSYSTEMS BigDye™ Terminator Cycle Sequencing kit), and sequence changes were confirmed with restriction enzyme digestion.

The SSCP method was used to test 60 South African individuals with a family history indicating ADRP.

3. RESULTS

A total of four disease-causing mutations have been identified in the cohort of SA ADRP patients screened to date (Table 2).

Table 2. Disease-causing mutations in the RHO gene in South African ADRP patients.

Exon	Mutation	Base change	RFLP
1a	Thr58Arg	C to G	+*DdeI*
1b	Gly109Arg	G to A	-*FokI*
3	Asp190Asn	G to A	-*RsaI*
5	Pro347Leu	C to T	-*MspI*

Two of these mutations were detected by the rapid screen; one codon 347 mutation, and one codon 58 mutation.

In addition, a donor splice site mutation in intron 4 was detected in an individual with a family history indicating recessive RP. This G to T transversion at position 4335 was present in the homozygous state as confirmed by the loss of a *BanI* restriction site.

Furthermore, an unusually high frequency of a polymorphism in the 5' non-coding region of RHO was detected. This single nucleotide substitution of a G for an A at position 269 of RHO can be confirmed by the creation of a *KspI* site.

A total of 71 patients with RP, as well as 54 known ADRP patients, were investigated for the presence of the A269G polymorphism (table 3). In addition, 53 unaffected unrelated Caucasian controls were studied, as well as 54 indigenous black African controls (Figure1).

Table 3. Frequency of A269G in affected and unaffected population groups

Phenotype	No. Individuals	No. Black individuals	Heterozygotes (%)	Homozygotes (%)
RP	71	3	37	10
ADRP	54	10	26	20
Caucasian Controls	53	-	30	4
Black Controls	54	-	15	79

Figure 1. Digital image of banding pattern of A269G polymorphism detected on SSCP.

4. DISCUSSION

The rapid screen resulted in the detection of only 2 rhodopsin mutations in 80 ADRP individuals (2.5%). A similar screen of 120 patients in the UK revealed that 6% of ADRP patients carried the codon 347 or codon 58 mutations.[5] The one SA ADRP patient found to carry the codon 347 mutation is of black African origin. This was the first report of a disease-causing RHO mutation in an indigenous black African family with RP.[9,10]

Only 4 mutations (including the novel Gly109Arg mutation[11]) have been identified in a cohort of 60 ADRP patients in whom the full RHO gene was screened, using SSCP analysis and the rapid screen. This reflects a low frequency of RHO mutations in SA ADRP patients (6.6%) when compared to 50% in the UK[5] and 20-30% globally.[4]

In addition, a homozygous donor splice site mutation was found in intron 4 of a patient with a family history indicating recessive RP. There has been some discussion as to whether this mutation results in a dominant or recessive allele,[12] but to our knowledge, it has never been reported in the homozygous state. Further investigation of extended family members is being undertaken.

Finally, an unusually high frequency of the A269G polymorphism in the 5' non-coding region of the RHO gene was identified. This polymorphism was detected in 46% of SA ADRP patients, as opposed to 14% of a similar cohort reported in the US.[13] Interestingly, not only is the frequency high in SA Caucasian controls (34%), but it is present in 94% of indigenous black African unaffected, unrelated individuals. The significance of this high frequency should be investigated, as polymorphisms have been found to be extremely useful in studies of protein structure and function. RHO is thought to constitute up to 85% of the total protein present in the rod outer segment,[6] implying a structural role which may be functionally affected by this polymorphism.

These unusual findings together with the two novel ADRP genes that were localised in SA (RP13 and RP17 in 1994 and 1995 respectively)[14,15] clearly emphasise the genetic heterogeneity of Retinal Degenerative Diseases (RDD). In addition, the 'molecular demography' for RDDs in SA may be different from other international study sites, and future genetic analyses could yield further interesting findings.

ACKNOWLEDGEMENT

This research was supported by grants from the Retinal Preservation Foundation of South Africa and the Foundation Fighting Blindness (USA). We are most grateful to Sister Lecia Bartmann for her efforts in tracing family members and the Genetic Sisters of the Department of National Health for visiting family members throughout the country and collecting blood. We are indebted to the SA RP family members for their participation in the study. We are grateful to Professor T. Dryja for assistance and advice with the exon 5 rhodopsin primers.

REFERENCES

1. G. Clarke, E. Heon, and R. R. McInnes, Recent Advances in the molecular basis of inherited photoreceptor degeneration, *Clin. Genet.* **57**: 313-329 (2000).
2. P. McWilliam, J. Farrar, P. Kenna, D. G. Bradley, M. M. Humphries, E. M. Sharp et al., Autosomal Dominant Retinitis Pigmentosa (ADRP): localisation of an ADRP gene to the long arm of chromosome 3, *Genomics* **5**: 619-622 (1989).
3. T. P. Dryja, T. L. McGee, E. Reichel, L. B. Hahn, G. S. Cowley, D. W. Yandell et al., A point mutation of the rhodopsin gene in one form of retinitis pigmentosa, *Nature* **343**:364-366 (1990).
4. J. Phenan, and D. Bok, A brief review of retinitis pigmentosa and the identified retinitis pigmentosa genes, Mol. Vis. **6**: 116-124 (2000). http://www.molvis.org/molvis/v6/216/.
5. E. Tarttelin, M. Al-Maghtheh, J. Keen, S. Bhattacharya, and C. Ingelhearn, Simple tests for rhodopsin involvement in retinitis pigmentosa, *J. Med. Genet.* **33**: 262-263 (1996).
6. S. Van Soenst, A. Westerveld, P. T. V. M. de Jong , E. M. Bleeker-Wagemakers, and A. A. B. Bergen, Retinitis Pigmentosa: Defined from a molecular point of view, *Surv. Ophth.* **43**(4): 321-359 (Jan-Feb 1999).
7. C. Inglehearn, Molecular Genetics of human retinal dystrophies, *Eye* **12**: 571-579 (1998).
8. C. Inglehearn, T. Keen, R. Bashir, M. Jay, F. Fitzke, A. C. Bird et al., A completed screen for mutations of the rhodopsin gene in a panel of patients with autosomal dominant retinitis pigmentosa, *Hum. Mol. Genet.* **1**(1): 41-45 (1992).
9. J. Greenberg, T. Franz, R. Goliath, and R. Ramesar, A photoreceptor gene mutation in an indigenous black African family with retinitis pigmentosa using a rapid screening approach for common rhodopsin mutations, *South Afr. Med. J.* **89**(8): 877-878 (1999).
10. L. Roberts, R. Ramesar, and J. Greenberg, Low frequency of rhodopsin mutations in South African patients with autosomal dominant retinitis pigmentosa, *Clin. Genet.* **58**: 77-78 (2000).
11. R. Goliath, S. Bardien, A. September, R. Martin, R. Ramesar, and J. Greenberg, Rhodopsin mutation Gly109Arg in a family with autosomal dominant retinitis pigmentosa, *Hum. Mut.* Supplement 1: 40-41 (1998).
12. P. Rosenfeld, L. Hahn, M. Sandberg, T. Dryja, and E. Berson, Low incidence of retinitis pigmentosa among heterozygous carriers of a specific rhodopsin splice site mutation, *I. O. V. S* **36** (11): 2186-2192 (October 1995).
13. C. Sung, C. Davenport, J. Hennessey, I. Maumenee, S. Jacobson, J. Heckenlively et al., Rhodopsin mutations in ADRP, *Proc. Natl. Acad. Sci. USA* **88**: 6481-6485 (August 1991).
14. J. Greenberg, R. Goliath, P. Beighton, and R. Ramesar, A new locus for autosomal dominant retinitis pigmentosa on the short arm of chromosome 17, *Hum. Mol. Genet.* **3**: 913-918 (1994).
15. S. Bardien, N. Ebenezar, J. Greenberg, C. F. Inglehearn, L. Bartmann, R. Goliath et al., An eighth locus for autosomal dominant retinitis pigmentosa is genetically linked to chromosome 17q, *Hum. Mol. Genet.* **4**: 1459-1462 (1995).

MIGRATORY HISTORY OF POPULATIONS AND ITS USE IN DETERMINING RESEARCH DIRECTION FOR RETINAL DEGENERATIVE DISORDERS

Raj Ramesar*, Alison September, George Rebello, Jacquie Greenberg and Rene Goliath

1. ABSTRACT

South Africa represents an anthropological and geographical *cul de sac* to which several ancient African populations have migrated from Central, East and West Africa. This subcontinent may thus represent access to some of the most ancient *homo sapiens* genetic material that is still in the process of being passed from one generation to the next. The indigenous African subpopulations may also carry some of the most ancient (non-lethal) disease-causing mutations. Apart from this interesting niche for the study of inherited disorders, this subcontinent has also attracted waves of immigrations over the past 500 years. The first major European settlement was by Dutch seamen in the region of the Cape of Good Hope. This was followed by colonists from Germany and France and other parts of Europe, including a large contingent from Britain. From the East, immigrants also hailed from India and Malaysia. With regard to the potential of genetic investigations, there are significant numbers of communities that have remained as true as is possible to their ancestral lineages from their lands of origin. This has been reinforced by the sociopolitical practise of apartheid. The high degree of clinical and genetic heterogeneity in our cohort of research subjects with retinal degenerative disorders, coupled with the observation of certain disorders segregating in a manner which fits a 'land of origin' scenario has led us to use the information of ancestral lineages to assist in deterring the strategy to defining the underlying genetic causes in RDDs.

---

* Department of Human Genetics, University of Cape Town, South Africa

New Insights Into Retinal Degenerative Diseases
Edited by Anderson *et al.*, Kluwer Academic/Plenum Publishers, 2001

2. INTRODUCTION

The largely untapped 'old world' populations and identifiable isolated immigrant populations in Africa are regarded as very worthy reagents for 'new' disease-gene identification. Identification of such new genes is likely to lead to fresh insights into disease mechanisms underlying retinal degenerative disorders (RDDs).

Migrant populations may resemble their parental populations genetically, and their origins may also be valuable in identifying predominant mutations that are likely to have been introduced into their adopted countries. In addition, studies of migrant groups are useful in attempting to distinguish host and environmental influences.

2.1. Migratory History to Southern Africa

2.1.1. Indigenous Populations

There has been a strong migratory history of populations to the Southern tip of the African continent. The original reported inhabitants of Southern Africa were the Khoi-San (Bushman/Hottentot) peoples. The current Bantu-speaking inhabitants of the country originally migrated to South Africa from central and west Africa. This grouping currently constitute the majority of the country's population, numbering some 30 (of a total of 40) million individuals.

2.1.2. Immigrant populations

Caucasian migrations to Southern Africa have emerged largely from England, Holland, Germany, France, and to a lesser extent from Italy, Portugal, Greece and other European countries. Asian immigrants emigrated mostly from three different parts of the Indian subcontinent.

2.1.3 Status of material to be mapped/genetically characterised

Internationally, as disease-causing genes have been localised and mutations identified in the majority of familial RDDs, a large proportion of RDD subjects have remained recalcitrant to genetic dissection because of their categorisation as 'sporadic', i.e. with no family history (due to problems in ascertainment). Various strategies have been used to determine the genetic defect in such subjects with limited success. A rationale is proposed for a supposedly more productive route for identification of defective genes.

Table 1. Population Groups in South Africa

Indig African	Mixed Ancestry	Caucasian	Asian/Indian
30 845 000	3 596 500	4 287 200	1 031 820

Table 2. Status of the Biological Material Archive

Form of RDD	number of subjects	families
Autosomal dominant RP (ADRP)	624	63
Recessive RP (ARRP)	194	51
Macular degeneration (MD)	116	68
Stargardt Disease (SD)	173	90
Usher Syndrome (US)	125	40
X-linked RP (XLRP)	77	18
RDD Indeterminate (RDDI)	284	133

Table 3. Biological material by population group

	Caucasian	Mixed Ancestry	Indigenous African	Asian Indian	Unknown
ADRP	444	89	52	5	28
ARRP	132	0?	17	12	23
MD	91	9	2	0	5
SD	233	6	10	1	20
US	98	0	11	11	4
XLRP	40	4	11	5	16
RDDI	201	12	30	3	32

3. RESEARCH FINDINGS AND THE WAY AHEAD

Screening material from familial RDDs has resulted in our localizing two novel loci, RP13 and RP17. Furthermore, several families have been shown to carry mutations in *rhodopsin, peripherin/rds, RPE65 and TIMP3*. Overall findings indicate a significantly lower level of mutations in *rhodopsin* and *peripherin/rds* in our cohort of RDD subjects than that reported elsewhere (see poster: Greenberg et al).

Interesting population-specific observations include: a higher incidence of recessive RP in Caucasians of English descent in South Africa, than in the Afrikaner (Dutch, German and French-derived) population group. On the other hand, Stargardt Disease occurs at a higher incidence in the Afrikaner population than in any of the other population groups. Usher Syndrome seems to have a relatively higher incidence within RDDs in the Indian subpopulation.

It would make sense to screen the disorders according not only to the obvious identified candidate genes, but specifically, and quickly for the mutations which have been described in parental populations. In the molecularly uncharacterised subjects with 'sporadic' RDD, also, the most appropriate strategy, across a spread of archived biological material, is the use of information on mutations in geographic and ethnic parental populations. Obviously, there will be a proportion of individuals in whom the genetic changes will not be resolved, because of new mutations and differential effects of the environment. This strategy will eventually identify a group of subjects in whom the disease is due to genuinely 'new mutations'.

4. CONCLUSION

In light of the imminent molecular definition of most genetic retinal disorders, considerable new effort needs to be placed into recruiting new indigenous African subjects and families, and towards ascertaining whether retinal and other inherited disorders occur at a lower incidence in African populations than in other population groups. In our search for individuals and populations with RDDs we may have neglected populations which appear resistant to acquiring such instigatory mutations. The mechanisms which govern this process of protection may well have an impact in our search for a therapy for RDDs.

THE SOUTHWEST EYE REGISTRY
Distribution of disease types and mutations

Dianna H. Wheaton, Stephen P. Daiger, and David G. Birch[*]

1. INTRODUCTION

The Southwest Eye Registry (SER) is a regional database encompassing Texas and the neighboring states of New Mexico, Oklahoma, and Louisiana, with the Dallas metropolitan area at the core. Patients with inherited retinal degenerations or allied disorders are referred for inclusion in the database by ophthalmologists specializing in retinal diseases. The SER was established to serve as a formal mechanism to maintain clinical, functional, and molecular genetic data generated by the Retina Foundation of the Southwest (RFSW; Dallas, TX), the Human Genetics Center (Univ. Texas; Houston, TX) and collaborating laboratories. Reported here is an initial analysis of this database.

2. METHODS

The database population (n=2259) includes patients (n=1727), as well as unaffected and at-risk relatives (n=532). Individuals are categorized according to disease status and genetic subtype. Blood samples are obtained from RFSW patients diagnosed with hereditary retinal disease. DNA is extracted from these samples, catalogued and stored on-site in -80° freezers. A portion of the archived DNA is sent to collaborating genetic researchers for mutation analysis.

Data maintained in the registry include visual function measures [visual acuity, electroretinogram (ERG), visual field, dark adaptation], pedigree, blood sample/DNA tracking, and results of genetic analysis. ERG data are available for 85% of SER patients for subsequent genotype/phenotype characterization. Longitudinal visual function data spanning up to 18 years are available for a number of individuals, a portion of these

[*] Dianna H. Wheaton and David G. Birch, Retina Foundation of the Southwest, Dallas, Texas, 75231. Stephen P. Daiger, Human Genetics Center and Dept. of Ophthalmology and Visual Science, University of Texas, Houston, Texas, 77030.

New Insights Into Retinal Degenerative Diseases
Edited by Anderson *et al.*, Kluwer Academic/Plenum Publishers, 2001

patients participated in a prospective natural history study of degenerative retinal disease[1].

2.1. Database Population Affected With Retinal Disease

The database population was first evaluated by disease status and individuals were subsequently grouped into two categories, affected and normal/at-risk relatives. The normal/at-risk relatives group was not subjected to further evaluation. The database population affected with retinal disease was further subdivided by specific presenting diagnosis. The total number of individuals with the same diagnosis was tabulated and a percent distribution by disease calculated for the SER database.

The database was then evaluated according to family index. Each family was restricted to the initial presenting subject i.e. the proband, thus allowing only one affected subject per family. Demographic information obtained at the time of ascertainment was reviewed to determine the geographic location of the proband's residence (city, state, and county). Individuals derived from outside the central and north Texas region were excluded from further evaluation. Refining the SER population in this manner allowed a rudimentary evaluation of disease prevalence.

3. RESULTS

Evaluation of the total database population affected with retinal disease (n=1727) led to the following disease distributions: 49% retinitis pigmentosa (RP, n=844), 20% macular degeneration (MD, n=352), 9% cone/cone-rod dystrophy (C/CR, n=159), and the remaining 22% distributed across 12 additional disease classifications (Figure 1).

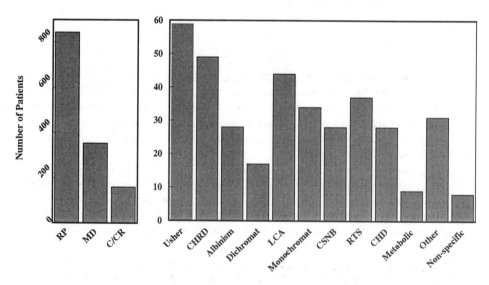

Figure 1. SER distribution by disease. RP, retinitis pigmentosa; MD, macular degeneration; CRD, cone-rod dystrophy; CHRD, chorioretinal degeneration; LCA, Leber congenital amaurosis; CSNB, congenital stationary night blindness; RTS, retinoschisis; CHD, chorioderemia

The major disease categories listed in Figure 1 include a variety of genetic subtypes and disease sub-categories. For example, the macular degeneration category encompassed age-related (AMD), juvenile (JMD), autosomal dominant (adMD), and other clinically defined macular dystrophies (e.g. vitelliform, white dot). Metabolic diseases included ceroid lipofuscinosis and long-chain 3-hydroxyacyl-CoA dehydrogenase (LCHAD). The "Other" category included syndromes (e.g. Bardet-Biedl, Prader-Willi, Stickler) and dominant drusen. The "Non-specific" category covered familial retinal degenerations of undefined nature.

3.1. Genetic Analysis

Genetic analysis identified causative mutations for 180 affected individuals (64 probands), attributable to the *rhodopsin (RHO)*, *RDS/peripherin (RDS)*, *RP1*, *CRX*, *AIPL1*, *RPGR*, and *ABCR* genes. Distributions of these mutations are shown in Figure 2.

Mutations in *RHO*, *RDS*, and *RP1* are commonly associated with autosomal dominant retinal degeneration. *CRX* mutations are associated with cone/cone-rod disease. *AIPL1* mutations are reported to account for up to 20% of Leber congenital amaurosis[2]. *RPGR* mutations are associated with x-linked retinitis pigmentosa, and *ABCR* mutations are thought to account for juvenile macular degeneration.

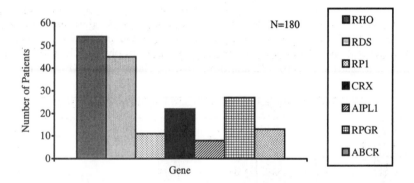

Figure 2. Distribution of gene mutations for all affected individuals

Distribution of gene mutations by proband: *RHO* (n=25), *RDS* (n=14), *RP1* (n=2), *CRX* (n=4), *AIPL1* (n=3), *RPGR* (n=3), *ABCR* (n=13). Among all unrelated patients with adRP, the frequencies of disease-causing mutations detected in this population were 22% *RHO*, 5% *RDS*, and 2% *RP1*.

3.2. Population Perspective

The proband population (n=1261) was further refined by geographic location such that those with residency outside central and north Texas were excluded (n=131).

Additional probands were excluded due to incomplete demographic information (n=11). The AMD/other MD subcategories (n=167) were excluded from the macular degeneration group because of indeterminate genetic origins.

The resulting group of probands (n=952) was evaluated by county of residence to determine each county's contribution to the database. Eighty counties contributed one or more probands to the database. However, four counties (Collin, Dallas, Denton, Tarrant) were found to contribute the majority (69%, n=656) of the subjects. These four counties constitute a region often referred to as the Dallas metroplex. The number of probands per county population was evaluated to provide an indication as to which county(s) had the highest percentage proband inclusion in the SER. Collin and Dallas counties were found to have the highest percentage of probands, 0.022% and 0.018%, respectively (Table 1).

Table 1. Percentage Probands Per County Population

	County Population[3]	# of Probands	% Probands in each County
Collin	456,612	98	0.0215%
Dallas	2,062,100	364	0.0177%
Denton	404,074	53	0.0131%
Tarrant	1,382,442	141	0.0102%

Collin and Dallas county probands (n=462) were utilized for further evaluation due to their similar level of percent probands/population referred to the SER for testing. Prevalences of the clinically defined disease categories were estimated based on the combined population of Collin and Dallas counties (Table 2).

Table 2. Distribution by Disease and Estimated Prevalence

	Probands	Prevalence
Retinitis Pigmentosa	242	1/10,000
Cone/Cone-rod Dystrophy	58	1/43,000
Hereditary Macular Degeneration (adMD, JMD)	53	1/48,000
Usher Syndrome	26	1/97,000
Chorioretinal Degeneration	16	1/157,000
Albinism	11	1/230,000
Dichromat	10	1/252,000
Leber Congenital Amaurosis	8	1/315,000
Monochromat	8	1/315,000
Congenital Stationary Night Blindness	6	1/420,000
Retinoschisis	6	1/420,000
Choroideremia	5	1/504,000
Metabolic (Ceroid lipofuscinosis)	2	1/1,259,000
Other	8	n.d.
Non-specific Retinal Degeneration	3	n.d.

adMD, autosomal dominant macular degeneration; JMD, juvenile macular degeneration; n.d. not done

The three most prevalent disease categories, namely, retinitis pigmentosa, hereditary macular degeneration, and cone/cone-rod dystrophy were reviewed and classified according to genetic subtype. The resulting distributions are illustrated in Figure 3.

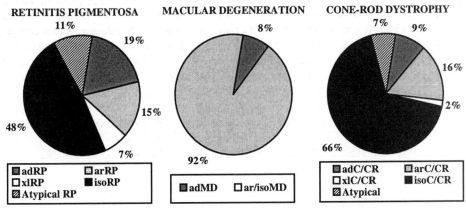

Figure 3. Distribution of genetic subtypes for major disease category

4. DISCUSSION

The SER was established to organize and track the vast amount of clinical, functional, and molecular genetic data that the RFSW accumulates. Additional SER objectives included the ability to perform detailed genotype/phenotype analysis, and the expeditious distribution of DNA samples for genetic analysis. Archived DNA could readily be mobilized from a clinically defined patient group to screen for new genes as they are cloned.

After several years of operation, an interim analysis was performed to characterize this database population. The total database population was evaluated to determine the number of affected individuals and the distribution of disease types. It was determined that 76% of the SER population was affected with a retinal disease, the remainder were unaffected or at-risk relatives.

The affected population was then evaluated to identify only probands, such that each family contributed only one subject. Finally, the geographic location of residence was evaluated for each proband, those with residency outside central and north Texas were excluded. This stepwise restriction of the database population was undertaken to refine the patient base for population statistics.

The resulting cohort of central/north Texas probands were grouped according to county of residence. Eighty counties from the aforementioned region contributed one or more probands. It was determined that four neighboring counties contributed the majority (69%) of the subjects. The proband population derived from this four county region became the focus of a population-based comparision.

The number of probands per county population was evaluated to determine percent probands derived from each county's populus. This population-based percentage provided an indication as to which county(s) had the highest rate of referral to the SER.

Collin and Dallas counties (0.022%; 0.018%, respectively) had the highest referral rates. Therefore, the combined population for these two counties was used to calculate an estimated rate of prevalence for each of the disease categories in the SER.

The prevalence estimates listed in Table 2 are based on our population of patients referred through local ophthalmologists. Obviously, not all recently diagnosed patients are referred to our research foundation, nor do all those who are referred actually choose to participate. In addition, individuals affected with a vision-specific disease (e.g. RP, cone-rod dystrophy) are more likely to be referred than individuals in which the vision disorder is but one manifestation of the disease (e.g. albinism, ceroid lipofuscinosis). Thus the listed disease prevalences should be viewed as minimal estimations.

The estimated prevalence of retinitis pigmentosa probands from our refined database population is 1 in 10,000. On average n=1.3 affected individuals per family were registered with the SER. Thus our prevalence estimate is reasonably consistent with the previously reported general population prevalence of retinitis pigmentosa of 1 in 4,000[4-6]. The estimated proband prevalence for Usher syndrome was 1/97,000 compared to published population estimates of 3/100,000[7] to 4.4/100,000[8].

5. CONCLUSIONS

The SER has the potential to serve as a valuable resource to genetic research involving inherited, degenerative retinal disease. Data available in the registry has allowed detailed genotype/phenotype analysis, and will continue to be a resource for future studies. This well-characterized patient registry also provides an ideal population for clinical treatment trials.

6. ACKNOWLEDGEMENTS

The authors would like to thank Dennis Hoffman, Cecille Doan, and Kathryn Boyd for their assistance collecting and preparing the vast amount of demographic data, and Kirsten Locke for providing valuable assistance with data entry and file maintenance of SER records. Key collaborators for gene mutation analysis included Lori Sullivan, Melanie Sohocki, Anand Swaroop, and Gabriel Travis. The authors would also like to thank patients and family members who participate in the SER and the numerous physicians who refer patients for our research. This work is supported in part by NIH grants EY05235 and EY07142, and by grants from the Foundation Fighting Blindness and Hermann Eye Fund.

REFERENCES

1. DG Birch, JL Anderson, and GE Fish, Yearly rates of rod and cone functional loss in retinitis pigmentosa and cone-rod dystrophy, *Ophthalmol.* **106**:258-268 (1999).
2. MM Sohocki, SJ Bowne, LS Sullivan, S Blackshaw, CL Cepko, AM Payne, SS Battacharya, S Khaliq, S Qasim Mehdi, DG Birch, WR Harrison, FFB Elder, JR Heckenlively, and SP Daiger, Mutations in a new photoreceptor-pineal gene on 17p cause Leber congenital amaurosis, *Nat. Genet.* **24**:79-83 (2000).
3. Population Estimates Program, Population Division, U.S. Census Bureau (Washington, March 9, 2000); http://www.census.gov

4. JA Boughman, PM Conneally, and WE Nance, Population genetic studies of retinitis pigmentosa, *Am. J. Hum. Genet.* **32**:223-235 (1980).
5. CL Bunker, EL Berson, WC Bromley, RP Hayes, and TH Roderick, Prevalence of retinitis pigmentosa in Maine, *Am. J. Ophthalmol.* **97**:357-365 (1984).
6. JR Heckenlively, *Retinitis Pigmentosa* (Lippincott Company, Philadelphia, 1988).
7. M Haim, NV Holm, and T Rosenberg, A population survey of retinitis pigmentosa and allied disorders in Denmark. Completeness of registration and quality of data, *Acta. Ophthalmol.* (Copenh) **70**:165-177 (1992).
8. JA Boughman, M Vernon, and KA Shaver, Usher Syndrome: definition and estimate of prevalence from two high-risk populations, *J. Chron. Dis.* **36**:595-603 (1983).

ABOUT THE AUTHORS

Robert E. Anderson, M.D., Ph.D., is Professor and Chair of Cell Biology, Dean A. McGee Professor of Ophthalmology, and Adjunct Professor of Biochemistry & Molecular Biology and Geriatric Medicine at The University of Oklahoma Health Sciences Center in Oklahoma City, Oklahoma. He is also Director of Research at the Dean A. McGee Eye Institute. He received his Ph.D. in Biochemistry (1968) from Texas A&M University and his M.D. from Baylor College of Medicine in 1975. In 1968, he was a postdoctoral fellow at Oak Ridge Associated Universities. At Baylor, he was appointed Assistant Professor in 1969, Associate Professor in 1976, and Professor in 1981. He joined the faculty of the University of Oklahoma in January of 1995. Dr. Anderson has published extensively in the areas of lipid metabolism in the retina and biochemistry of retinal degenerations. He has edited ten books, nine on retinal degenerations and one on the biochemistry of the eye. Dr. Anderson has received the Sam and Bertha Brochstein Award for Outstanding Achievement in Retina Research from the Retina Research Foundation (1980), and the Dolly Green Award (1982) and two Senior Scientific Investigator Awards (1990 and 1997) from Research to Prevent Blindness, Inc. He received an Award for Outstanding Contributions to Vision Research from the Alcon Research Institute (1985), and the Marjorie Margolin Prize (1994). He has served on the editorial boards of *Investigative Ophthalmology and Visual Science, Journal of Neuroscience Research, Neurochemistry International*, and *Current Eye Research* and is currently on the editorial board of *Experimental Eye Research*. Dr. Anderson has received grants from the National Institutes of Health, The Retina Research Foundation, the Foundation Fighting Blindness, and Research to Prevent Blindness, Inc. He has been an active participant in the program committees of the Association for Research in Vision and Ophthalmology (ARVO) and was a trustee representing the Biochemistry and Molecular Biology section. He has served on the Vision Research Program Committee and Board of Scientific Counselors of the National Eye Institute and the Board of the Basic and Clinical Science Series of The American Academy of Ophthalmology. Dr. Anderson is a past Councilor and Treasurer for the International Society for Eye Research.

Matthew M. LaVail, Ph.D., is Professor of Anatomy and Ophthalmology at the University of California, San Francisco School of Medicine. He received his Ph.D. degree in Anatomy (1969) from the University of Texas Medical Branch in Galveston and was subsequently a postdoctoral fellow at Harvard Medical School. Dr. LaVail was appointed Assistant Professor of Neurology-Neuropathology at Harvard Medical School in 1973. In 1976, he moved to UCSF, where he was appointed Associate

Professor of Anatomy. He was appointed to his current position in 1982, and in 1988, he also became director of the Retinitis Pigmentosa Research Center at UCSF, later named the Kearn Family Center for the Study of Retinal Degeneration. Dr. LaVail has published extensively in the research areas of photoreceptor-retinal pigment epithelial cell interactions, retinal development, circadian events in the retina, genetics of pigmentation and ocular abnormalities, inherited retinal degenerations, light-induced retinal degeneration, and pharmaceutical and gene therapy for retinal degenerative diseases. He has identified several naturally occurring murine models of human retinal degenerations and has developed transgenic mouse and rat models of others. He is the author of more than 120 research publications and has edited 9 books on inherited and environmentally induced retinal degenerations. Dr. LaVail has received the Fight for Sight Citation (1976); the Sundial Award from the Retina Foundation (1976); the Friedenwald Award from the Association for Research in Vision and Ophthalmology (ARVO, 1981); two Senior Scientific Investigators Awards from Research to Prevent Blindness (1988 and 1998); a MERIT Award from the National Eye Institute (1989); an Award for Outstanding Contributions to Vision Research from the Alcon Research Institute (1990); the Award of Merit from the Retina Research Foundation (1990); the first John A. Moran Prize for Vision Research from the University of Utah (1997); and the first Trustee Award from The Foundation Fighting Blindness (1998). He has served on the editorial board of *Investigative Ophthalmology and Visual Science* and is currently on the editorial board of *Experimental Eye Research*. Dr. LaVail has been an active participant in the program committee of ARVO and has served as a Trustee (Retinal Cell Biology Section) of ARVO. He has been a member of the program committee and a Vice President of the International Society for Eye research. He has also served on the Scientific Advisory Board of the Foundation Fighting Blindness since 1973.

Joe G. Hollyfield, Ph.D., is Director of Ophthalmic Research in the Cole Eye Institute at The Cleveland Clinic Foundation, Cleveland, Ohio, and Professor of Cell Biology, Neurobiology and Anatomy, Ohio State University, Columbus, Ohio. He received his Ph.D. from the University of Texas at Austin and was a postdoctoral fellow at the Hubrecht Laboratory in Utrecht, The Netherlands. He has held faculty positions at Columbia University College of Physicians and Surgeons in New York City and at the Cullen Eye Institute, Baylor College of Medicine in Houston, TX. He was Director of the Retinitis Pigmentosa Research Center in The Cullen Eye Institute from 1978 until his move to The Cleveland Clinic Foundation in 1995, and is currently Director of the Foundation Fighting Blindness Research Center at The Cleveland Clinic Foundation. Dr. Hollyfield has published over 150 papers in the area of cell and developmental biology of the retina and retinal pigment epithelium in both normal and retinal degenerative tissue. He has edited nine books, seven on retinal degenerations and one on the structure of the eye. Dr. Hollyfield has received the Marjorie W. Margolin Prize (1981, 1994), the Sam and Bertha Brochstein Award (1985) and the Award of Merit in Retina Research (1998) from the Retina Research Foundation; the Olga Keith Wiess Distinguished Scholars' Award (1981), two Senior Scientific Investigator Awards (1988, 1994) from Research to Prevent Blindness, Inc.; an award for Outstanding Contributions to Vision Research from the Alcon Research Institute (1987); the Distinguished Alumnus Award (1991) from Hendrix College, Conway, Arkansas; and the Endre A. Balazs Prize (1994) from the International Society for Eye Research (ISER). He is currently Editor-in-Chief of the journal, *Experimental Eye Research* published by Academic Press. Dr. Hollyfield has been active in the Association for Research in Vision and Ophthalmology (ARVO) serving on the Program Committee, as a Trustee and as President. He is also a past President and former Secretary of the International Society of Eye Research. He currently serves on the Scientific Advisory Boards of The Foundation Fighting Blindness, Research to Prevent Blindness, The Helen Keller Eye Research Foundation, The South Africa Retinitis Pigmentosa Foundation, and is Co-Chairman of the International Retinitis Pigmentosa Foundation Medical and Scientific Advisory Board.

INDEX